D0906469

Multiple Cropping Systems

Multiple Cropping Systems

Charles A. Francis
University of Nebraska, Lincoln

Macmillan Publishing Company
NEW YORK

Collier Macmillan Publishers
LONDON

Macmillan Publishing Company
866 Third Avenue, New York, NY 10022

Collier Macmillan Canada, Inc.

Printed in the United States of America

printing number year
1 2 3 4 5 6 7 8 9 10 6 7 8 9 0 1 2 3 4 5

Library of Congress Cataloging-in-Publication Data
Main entry under title:

Multiple cropping systems.

Includes index.
1. Multiple cropping systems. I. Francis, C. A.
S603.7.M85 1986 631.5'8 85-19891
ISBN 0-02-948610-6

CONTENTS

FOREWORD

No other human challenge is more critical than that of providing food for an ever-growing human population. And at no time in history has this challenge been more obvious than it is today. Each year some 80 million more people are added to the nearly 5 billion who must be fed. In spite of the unprecedented increases in world food production of the past two decades which scientists have helped stimulate, per capita food production has increased insignificantly. Equally unprecedented growth in human populations has essentially nullified the food production increases. In fact, some areas such as those in Africa, south of the Sahara, have actually lost ground. Per capita food production there has declined during the past 15 years.

Traditionally, increased food production has come from putting more land under cultivation. But in larger areas of the world, and especially in Asia, essentially all the land that can be economically cultivated is already in use. In the future, most of the extra food the world needs must come from higher production from land already being farmed. A major share of this increase will likely come from improved crop cultivars, which will provide higher yields per unit area. But vast areas in the tropics and subtropics also offer a second means of increased food production—increasing the number of crops produced per year on a given field. Such *multiple cropping* offers potential not only to increase food production but to reduce soil erosion by keeping vegetation on the land a higher proportion of the year.

Multiple cropping is not new. In fact, it is the world's oldest cropping system. Long before the modern systems of monoculture came into existence food was being produced in mixed cultures where several different species were harvested from a given land area each year. But in recent years, we have produced a variety of modern tools that greatly increase the potential of multiple cropping. High-yielding, short-seasoned dwarf varieties of food crops, developed originally for monoculture systems, offer great potential for double or triple cropping. Two or even three crops a year are possible where only one was grown before. Similar potential is provided by expansion of land under irrigation, by the availability of fertilizers and pesticides (especially herbicides), and by newly developed minimum tillage practices which can significantly reduce the time between crops.

The potential for increased food production through multiple cropping has led to a marked expansion in research on cropping systems, particularly in the tropics. Scientists from the developing countries have been joined by counterparts from the industrialized nations in these research efforts. They have achieved some noted successes in using multiple cropping to increase food production. At the same time, they have noted the complexity of factors influencing this productivity. This has led to the formation of interdisciplinary research and development teams. Likewise, it has led to a greater realization of the role of farmers in integrating improved technologies into multiple cropping systems.

The chapters of this text illustrate the complexity of multiple cropping systems and the degree to which different scientific disciplines are involved in improving them. They also illustrate two major critical elements that largely determine the suitability of different cropping systems.

First, there is the need for superior component technologies suited specifically to the multiple cropping system in question. Experience has shown that combining weak components does not make for a strong multiple cropping system. For example, the crop varieties included in the system should be high yielding, reasonably resistant to major pests and diseases, and have agronomic characteristics that fit the agroclimatic conditions in the area. Requirements for the multiple cropping system dictate the type of improved varieties to be created, the fertilizer and other chemical regimes to be used, and the soil and crop management to be followed.

The second critical element is the integration of these superior component technologies into equally superior multiple cropping systems and, in turn, superior farming systems. This requires not only interdisciplinary interaction among scientists, but also the participation of farmers in the process of developing the improved system. No agricultural activity requires the integration of so many factors as does those farming systems that include multiple cropping. Some scientists have had to "learn from the farmer" first before they could make headway in developing improved systems. Much of this learning is done by social scientists and economists who in turn can help the biologists focus on farmers' constraints.

The complexity of multiple cropping systems and the range of scientists working on these systems has led to some confusion in concepts, terminology, and understanding of what is required to develop and utilize improved systems. This is due in part to the fact that the publications of the scientists working on a given component technology appear only in disciplinary journals not often read or referred to by scientists in other disciplines. For this reason, a book is badly needed that brings together under one cover a review of what is known of the biological, social, and economic aspects of multiple cropping.

This book should be of great value to scientists and educators around the world as they try to develop improved multiple cropping systems. Reviews of current research results are enlightening, as are discussions of what research is needed in the future. In the long run, the authors of this book will help the most important actors in the drama—the farmers in the developing countries on whose shoulders rests the responsibility for producing the food. A debt of gratitude is owed by many to the scientists who contributed to this volume.

Nyle C. Brady
Cornell University and
U.S. Agency for International Development

PREFACE

Multiple cropping in its many forms provides a substantial proportion of the world's food. Disappearing in some areas due to mechanization of planting and harvest, but increasing in other areas where land is precious and intensive cultivation is the rule, this group of complex cropping systems is just beginning to excite the interest of the scientific community. Farmers, on the other hand, have depended on these systems for centuries. The contributions of multiple cropping systems today are obvious, yet their importance in the future is open to speculation. This book explores both today's systems and the potentials of complex, intensive cropping systems for the future.

Multiple Cropping Systems brings together many of the research results that are not easily available to the reader who is distant from such sources as annual reports and nonpublished data from national programs and international centers. More than a comprehensive, state-of-the-art compilation of historical data, the book critically reviews the existing literature and other information on multiple cropping. Authors of the several chapters were selected because of their professional experience in this field, and the perspective they bring to an evaluation of their own and other work in each discipline. The book works through these topics and presents an assessment of future directions for research and development in each area.

With this degree of detail, the editor and authors hope that the book provides a stimulus for researchers and extension specialists as well directions for potential future research and applications. This is a complex field, and there is need to establish interdisciplinary teams to help sort out some of the complexities. It is equally important to share information with people in various programs, in order to make the most possible progress with limited resources. The ultimate beneficiaries are the small farmers who use these systems.

Acknowledgments

This book is the product of a team effort. From conception of the idea, through the prospectus stage, to communicating with authors, to the final editing and production of the book—this has involved many people on six continents. The reviewers of the prospectus gave helpful suggestions and encouragement. Credit for the technical authenticity of the book goes to the twenty authors. There was an important and unique contribution by each of many colleagues who reviewed and commented on specific chapters. Institutional support from the University of Nebraska, Lincoln, and the Rodale Research Center, Emmaus, Pennsylvania, was invaluable to the project. The editor gives special thanks to Betsy Teselle, Jean Hodges, and JoAnn Collins for enthusiastic secretarial support. Sarah Greene of Macmillan Publishing Company provided technical guidance and stimulus from the conception of the book through to its publication. Finally, the support and encouragement of the editor's family is greatly appreciated.

Charles A. Francis
Lincoln, Nebraska

CONTRIBUTORS

Miguel A. Altieri	Division of Biological Control, University of California, Berkeley, CA 94720
Thomas C. Barker	Wheat Program, CIMMYT (Centro Internacional de Mejoramiento de Maiz y Trigo), Apartado 6-641, Mexico D.F., 0660, Mexico
Stillman Bradfield	Department of Anthropology, Kalamazoo College, Kalamazoo, MI 49007
Jeremy H. C. Davis	Bean Program, CIAT (Centro Internacional de Agricultura Tropical), Apartado Aereo 6713, Cali, Colombia
Charles A. Francis	Department of Agronomy, University of Nebraska, Lincoln, NE 68583
Steven R. Gliessman	Agroecology Program, University of California, Santa Cruz, CA 95064
Robert D. Hart	WIIAD, CARDI FSR/D Project, P.O. Box 971, Castries, Sta. Lucia, West Indies

Matt Liebman	Department of Botany, University of California, Berkeley, CA 94720
John K. Lynam	Cassava Program, CIAT (Centro Internacional de Agricultura Tropical), Apartado Aereo 6713, Cali, Colombia
Stephen C. Mason	Department of Agronomy, University of Nebraska, Lincoln, NE 68583
Roger Mead	Department of Applied Statistics, University of Reading, Whiteknights, Reading, U.K.
Raul A. Moreno	Cassava Program, CIAT (Centro Internacional de Agricultura Tropical), Apartado Aereo 6713, Cali, Colombia
Anne M. Parkhurst	Biometrics Unit, University of Nebraska, Lincoln, NE 68583
Donald L. Plucknett	Consultative Group for International Agricultural Research, World Bank, 1818 H St., NW, Washington, DC 20433
M. R. Rao	ICRISAT (International Center for Research in Semi-Arid Tropics), Patancheru P.O., Hyderabad, AP, 502 324, India
John H. Sanders	Department of Agricultural Economics, Purdue University, West Lafayette, IN 47907
Margaret E. Smith	Maize Program, CIMMYT (Centro Internacional de Mejoramiento de Maiz y Trigo), Apartado 6-641, Mexico D.F., 0660, Mexico
Nigel J. H. Smith	Department of Geography, University of Florida, Gainesville, FL 32611
Brian R. Trenbath	12 New Road, Reading, Berks, RGI 5JD, U.K.
Jonathan N. Woolley	Cassava Program, CIAT (Centro Internacional de Agricultura Tropical), Apartado Aereo 6713, Cali, Colombia

Multiple Cropping Systems

Chapter 1

Introduction: Distribution and Importance of Multiple Cropping

Charles A. Francis

Multiple cropping systems were the first types of organized agriculture. Human beings evolved, with their crops and animal species, from primarily a hunting and gathering society to a sedentary culture. Early multiple cropping systems were characterized by wide diversity of plant species, an integration of plants with animals and people, and a life-style that revolved around the need for the production of food. Thus, multiple cropping has its roots in the history of civilization as we know it today.

Multiple cropping has evolved to fit a nearly infinite number of geographic and climatic niches, so that each farm has some variant of the system which fits the unique microconditions on that farm and the objectives of the farm family (Fig. 1.1). There is no doubt that these complex and intensive systems have filled an important role in the development of agriculture and society. Their continuing importance underscores their value to many farmers in the developing world today. There is a resurgence of interest in crop rotations, intercropping, overseeding legumes into cereals, and double cropping. These methods to intensify production and provide erosion control are growing in importance in the temperate regions of the world.

But what is the future of multiple cropping systems? Some farmers continue to plant crops together for a variety of reasons. In the academic community,

Figure 1.1 Diverse, intercropped home garden with *Colocasia,* sugar cane, papaya, bananas, and other species, near Cañas, Guanacaste, Costa Rica. (Photo by Dr. Stephen Gliessman.)

multiple crops have a strong attraction to the ecologist who likens their structure to the natural ecosystem in its diversity and biological stability (Fig. 1.2). The agronomist and physiologist are intrigued by the potentials of extending resource use through more of the year and the greater total exploitation of the natural environment for food production. Economists are interested in stability of production and income. They see the benefits of diversity of these complex systems which provide buffering for family income through the year. The family diet which can be provided through planting a range of species and integrating their production with animals is of interest to the nutritionist with concern for family welfare. Social scientists are active in their search to understand the threads that link people and their goals with the potential of their environment to help them meet these goals.

There has been a surge of interest in the research community to explore these complex systems in detail, and to see what can be learned from the farmer in our quest to offer improvements based on science and technology. This interest by the research community is illustrated in Table 1.1, where publications on multiple cropping from a literature search in ICRISAT (courtesy of Dr. M. R. Rao) are summarized by 5-year periods. The growing importance to research scientists is clear from the table.

In contrast, other scientists and development people argue that the continuing and even accelerating trend toward specialization, toward monoculture, and toward larger farms, which has been happening for decades, will continue.

Figure 1.2 Complex, multistoried home garden agroecosystem, near Puerto Viejo, Alajuela, Costa Rica. (Photo by Dr. Stephen Gliessman.)

Table 1.1 Accelerating Research on Intercropping
As Illustrated by Numbers of Publications on 14 Crops by 5-year Period Up to 1980 in ICRISAT Resource Lists

		Number of publications							
Crop	First publication	Pre-1950	1951–1955	1956–1960	1961–1965	1966–1970	1971–1975	1976–1980	Total
Cassava	1917	5	1	4	1	4	25	44	84
Chickpea	1914	2	3	2	21	10	5	23	66
Common bean	1956	0	0	3	0	1	14	41	59
Cotton	1940	8	8	19	18	14	28	38	133
Cowpea	1949	1	1	2	5	9	38	123	179
Groundnut	1937	9	4	9	17	14	25	100	178
Mung bean	1954	0	2	0	5	12	30	98	147
Pearl millet	1943	3	1	0	5	11	23	71	114
Pigeon pea	1943	6	4	2	1	4	30	91	138
Rice	1942	2	1	5	4	5	34	35	86
Sorghum	1941	7	4	15	20	21	52	145	264
Soybean	1937	6	2	7	31	20	53	108	227
Sugarcane	1933	2	7	6	18	16	53	40	142
Wheat	1929	9	4	11	50	22	30	43	169
Total		60	42	85	196	163	440	1000	1986

Source: Dr. M. R. Rao, 1984 (unpublished).

Multiple cropping and small farms will slowly disappear, they maintain, as we invest more toward controlling the environment and reaching maximum levels of production from single crops. These specialists conclude that research and development resources should be focused on fine-tuning the high-technology systems that have brought about the "green revolution" in some favored areas.

There is a tendency to polarize around the issue of both multiple cropping and the small farmer. To those who predict greater specialization and solutions through high technology, multiple cropping systems are seen as a vestige of earlier times and a remnant of outdated and economically antiquated systems which will gradually be replaced by mechanized monoculture. Those who study these complex systems often tend to romanticize the life of the small farmer, and some who support the research and dialogue on improving multiple cropping may follow this quest with a missionary zeal. As a result, the scientific method is sometimes abandoned in favor of a nonobjective social statement, which is based not on fact but on belief and conviction about the social wrongs that have been done over the centuries to the small farmer and to poor people in general.

Through a multiauthor approach to the topic of multiple cropping, we have approached these questions from a broad multidisciplinary and multicultural point of view. The contributors represent a wide range of cultural, language, discipline, and geographic experience. They work in universities, international research centers, national research programs, and private industry. We hope that this breadth of experience brings a rich mixture of objective science and intellect with practical field examples to evaluate the current situation of multiple cropping. More than a review of literature, the several chapters attempt to analyze the current state of the art and project where research should go in the future.

DEFINITIONS USED IN CROPPING SYSTEMS

Some of the confusion about multiple cropping is generated by the specialists who research and write about complex systems. To date there is not complete agreement on the terms that should be used to describe specific cropping patterns or mixtures. The first organized attempt to provide guidelines based on the opinions of a wide range of specialists occurred in the Symposium on Multiple Cropping in 1975 at the annual meeting of the American Society of Agronomy (ASA) in Knoxville, Tennessee. The results of the concensus at that meeting were summarized by P. A. Sanchez and presented by Andrews and Kassam (1976), and these terms are repeated here for convenience and clarity (Tables 1.2 and 1.3). This is followed by additional terms that have appeared in the literature and their apparent equivalents from the ASA symposium (Table 1.4). A useful glossary of terms for farming systems, including multiple cropping, is found in Harwood (1979).

There continues to be unnecessary ambiguity in terms and their use, and a frequent invention of new terms rather than acceptance of those already in the literature. The authors argue that this creativity leads to greater precision of description and cultural richness in use of the language. While this may be true,

Table 1.2 Definitions of Principal Multiple Cropping Patterns

MULTIPLE CROPPING The intensification of cropping in time and space dimensions. Growing two or more crops on the same field in a year.

SEQUENTIAL CROPPING Growing two or more crops in sequence on the same field per year. The succeeding crop is planted after the preceding crop has been harvested. Crop intensification is only in the time dimension. There is no intercrop competition. Farmers manage only one crop at a time in the same field.
Double cropping Growing two crops a year in sequence.
Triple cropping Growing three crops a year in sequence.
Quadruple cropping Growing four crops a year in sequence.
Ratoon cropping The cultivation of crop regrowth after harvest, although not necessarily for grain.

INTERCROPPING Growing two or more crops simultaneously on the same field. Crop intensification is in both the time and space dimensions. There is intercrop competition during all or part of crop growth. Farmers manage more than one crop at a time in the same field.
Mixed intercropping Growing two or more crops simultaneously with no distinct row arrangement.
Row intercropping Growing two or more crops simultaneously where one or more crops are planted in rows.
Strip intercropping Growing two or more crops simultaneously in different strips wide enough to permit independent cultivation but narrow enough for the crops to interact agronomically.
Relay intercropping Growing two or more crops simultaneously during part of the life cycle of each. A second crop is planted after the first crop has reached its reproductive stage of growth but before it is ready for harvest.

Source: Andrews and Kassam, 1976, with modifications from P. A. Sanchez, North Carolina State University.

it confuses an already complex situation even more. With the translation of specific terms into Spanish, French, and other languages, and the translation of papers from other languages into English, the confusion grows. In this book we attempt to conform to the terms listed in Tables 1.2 and 1.3. Multiple cropping researchers are urged to use terms that are precise for describing a given cropping pattern or system, but also to use as much as possible those terms that already have some acceptance by other researchers in the field. There is too much confusion due to terminology without adding more, and there are many serious problems that need to be solved in the elaboration and improvement of these systems without the additional burden of communication difficulties.

GEOGRAPHIC IMPORTANCE OF MULTIPLE CROPPING

Patterns of multiple cropping are found in all parts of the world. However, the most diversity and interest is found in the tropics and especially in regions where small farmers operate intensively on a limited land area. The symposium proceedings from the ASA meetings in 1975 (Papendick et al., 1976) presented

Table 1.3 Related Terminology Used in Multiple Cropping Systems

Cropping index The number of crops grown per annum on a given area of land × 100.
Cropping pattern The yearly sequence and spatial arrangement of crops or of crops and fallow on a given area.
Cropping system The cropping patterns used on a farm and their interaction with farm resources, other farm enterprises, and available technology which determine their makeup.
Income equivalent ratio (IER) The ratio of the area needed under sole cropping to produce the same gross income as one hectare of intercropping at the same management level. IER is the conversion of LER into economic terms.
Land equivalent ratio (LER) The ratio of the area needed under sole cropping to one of intercropping at the same management level to give an equal amount of yield. LER is the sum of the fractions of the yields of the intercrops relative to their sole crop yields.
Mixed farming Cropping systems which involve the raising of crops, animals, and/or trees.
Monoculture The repetitive growing of the same sole crop on the same land.
Rotation The repetitive cultivation of an ordered succession of crops (or crops and fallow) on the same land. One cycle often takes several years to complete.
Sole cropping One crop variety grown alone in pure stand at normal density. Synonymous with solid planting; opposite of intercropping.

Source: Andrews and Kassam, 1976, with modifications from P. A. Sanchez, North Carolina State University.

summaries of multiple cropping in tropical Asia, America, Africa, eastern and western United States, and the Middle East. The symposium proceedings should be consulted for more detail on each region, since several chapters give examples of specific crop patterns from the historical and current points of view. The following section presents an overview of geographic importance of cropping systems, in order to provide an introduction to multiple cropping for each major region.

Important Cropping Systems in Asia

The range of systems in which multiple cropping is important has been well documented for Asia (Beets, 1982; Gomez and Gomez, 1983; Harwood, 1979; Harwood and Price, 1976). These are catagorized by Gomez and Gomez (1983) as multiple cropping (1) with lowland rice (*Oryza sativa*), (2) with annual upland crops, (3) with perennial upland crops, and (4) on hilly land. This list can be expanded to include the intensive vegetable cultivation around the homestead area, and the shifting cultivation which is a subset of the "hilly land" category (Harwood and Price, 1976). While rice dominates the wetter areas of south Asia and most of east Asia, the cropping systems of the drier areas of India and Pakistan are dominated by wheat (*Triticum aestivum*), maize (*Zea mays*), sorghum (*Sorghum bicolor*), and millet (*Setaria* spp. and *Pennisetum* spp.) (Harwood and Price, 1976).

Early double cropping in China was described by Beets (1982) from a report of Perkens (1969):

Table 1.4 Additional Terms from the Literature that Are Used to Describe Various Forms of Multiple Cropping
With References to Appearance in Cited Sources

Agrisilviculture The growing of trees for timber but with cultivated crops grown beneath (Harwood, 1979).
Area-time equivalency ratio The ratio of number of hectare-days required in monoculture to the number of hectare-day used in the intercrop to produce identical quantities of each of the components (Harwood, 1979).
Calorie index A single productivity index which incorporates caloric equivalents produced by all crops in a system (Gomez and Gomez, 1983).
Competition effect The competition of intercropped species for light, nutrients, water, CO_2, and other growth factors (Gomez and Gomez, 1983).
Complementary effect The effects of one component on another which enhance growth and productivity, as compared to competition above (Gomez and Gomez, 1983).
Component crops *(or components)* Individual crop species which are a part of the multiple crop system.
Component technology The procedure for growing each component crop (Gomez and Gomez, 1983).
Farm enterprise An individual crop or animal production function within a farming system which is the smallest unit for which resource use and cost-return analysis is normally carried out (Harwood, 1979).
Farming systems research (FSR) The study of whole farm systems, which include all the enterprises on the farm, their biological, economic, and cultural components, and usually implies some involvement of the farmer in the research process.
Interculture Arable crops grown below perennial crops (Harwood, 1979).
Interplanting All types of seeding or planting a crop into a growing stand. It is used especially for annual crops grown under stands of perennial crops (Harwood, 1979).
Maximum cropping The attainment of the highest possible production per unit area per unit time without regard to cost or net return (Harwood, 1979).
Monetary index A single productivity index which incorporates monetary values of all crops in a system (Gomez and Gomez, 1983).
Overyielding The production of component crops in an intercrop which is higher than the sum of appropriate monoculture crops; this is indicated by an LER greater than unity.
Residual effect The effect of the previous crop in a sequential cropping pattern on the productivity of the current crop (Gomez and Gomez, 1983).
Simultaneous polyculture The simultaneous growth of two or more useful plants on the same plot. This includes mixed cropping, intercropping, interculture, interplanting, and relay planting (Harwood, 1979).
Spatial arrangement The physical or spatial organization of component crops in a multiple cropping system.

> The development of an early rice variety in the year 1012 triggered a revolution in growing practices and made cultivation of a second crop possible (in the south). By the Ming period (1368–1644) cold tolerant varieties were developed which could be planted in mid-summer after spring crops or early rice. . . . Further north, in Hunan, it was not until the seventeenth and eighteenth centuries that efforts were made to promote second crops.

In more recent years, the development of short-cycle rice varieties for the tropics have provided this same potential. In the Philippines, the intensive rice-

rice-legume crop sequences in projects in Luzon and Iloilo have brought the potentials of short-cycle crops to many farmers (Gomez and Gomez, 1983).

Lowland rice dominates much of east Asia. Its popularity is attributed to its ability to grow in standing water, its status as an important staple food (with per capita consumption of more than 100 kg per year in most countries), and the fact that it is easily marketed and stored from one harvest to the next (Gomez and Gomez, 1983). Where there is enough water, rice may follow rice, and in some cases three crops per year are possible. More frequently, two crops of rice are followed by a short-season "catch crop" (that is, a crop that uses residual moisture and fertility), such as mung bean (*Vigna radiata*) or sorghum, or if water is limiting, a single crop of rice is followed by the upland crop.

Annual upland crops represent more than 60 percent of the cropped area of Asia, including at least 20 annual food species plus cotton (*Gossypium* spp.) and tobacco (*Nicotiana* sp.) (Gomez and Gomez, 1983), many of which are multiple cropped. They are a part of systems that are characterized by species diversity, production of a low but stable yield which satisfies family food needs, low cash input, and traditional varieties of most crops.

Perennial upland crops are also important components of intercrop systems in much of Asia, including banana and abaca (*Musa* spp.), cacao (*Theobroma cacao*), pineapple (*Ananas comosus*), rubber (*Hevea brasiliensis*), papaya (*Carica papaya*), mango (*Mangifera* spp.), coffee (*Cafea* spp.), and coconut (*Cocos nucifera*). Life span of a system may be from a few years to several decades, and these crops are found on both small and large farms. They may be intercropped with annuals, or more than one perennial may be found together such as coconut with pineapple, banana, cacao, or coffee. Both perennials and annuals are grown on hilly lands, both in continuous culture or in a bush-fallow system.

The areas around houses are often unique. It is not unusual, especially in these homestead areas, to find a mix of many species. Harwood and Price (1976) describe the small, low-income farms of Indonesia and Nepal:

> Those farms are characterized by a diversity of economic plant species, which number 50 to 60 in the more advanced systems. The plant components of the system may include: 5 or 6 tall growing tree species, 5 or 6 medium-height tree species, 5 or 6 bush and shrub species, 4 or 5 root crops, and up to 30 shade-tolerant, short-statured or vine-type annuals.

A variant on the upland culture is shifting cultivation. Several reviews are cited by Harwood and Price (1976). There are many types of systems, and different lengths of the fallow cycle depending on rainfall patterns and amounts, fertility of the soil, and pressure to produce food for a growing population. In general, the systems are very high in efficiency of yield per unit of manual labor input. A difficult but important challenge is to find technology that is appropriate to these shifting cultivation systems, and the alternatives that will allow farmers to shorten or eliminate the fallow cycle while maintaining fertility and some characteristics of the traditional farming systems that have worked well for them. The complexity of fertility and population interactions was explored by Barker

(1984) in trials of sweet potato (*Ipomoea batata*) grown by the Ikalahan in the Philippines. Thus, the systems have evolved in each climatic regime to meet people's needs. Research at the International Rice Research Institute (IRRI, Philippines), International Crops Research Institute for the Semi-Arid Tropics, (ICRISAT, India), and Asian Vegetable Research and Development Center (AVRDC, Taiwan) complements that in several national programs.

Principal Cropping Systems in Africa

Multiple cropping systems are a major component of African agriculture. Elegant summaries of the cropping systems in tropical Africa were presented by Okigbo and Greenland (1976) and by Steiner (1982). They present geographic, rainfall, and vegetation maps of Africa that should be consulted for a more detailed overview of background information and references that are relevant to farming on the continent. Primarily on the basis of rainfall and altitude, Okigbo and Greenland (1976) illustrate zones that are millet-dominant, sorghum-dominant, root crops-dominant, plantain-dominant, maize-dominant, rice-dominant, and livestock-dominant. Some regions are transitional and are characterized by two of these major crop groups. Farming activities are further classified by Okigbo and Greenland (1976) into "traditional and transitional systems" and "modern farming systems," with most multiple cropping systems falling into the former category.

Although shifting cultivation was once prevalent in Africa, as in Asia and Latin America, population pressure has caused this system to evolve into one with ever shorter periods of fallow to where permanent cultivation now predominates (Nye and Greenland, 1960). Especially important in many areas are the compound farms, or kitchen or homestead gardens, where high fertility is due to accumulation and incorporation of crop and animal wastes. These areas are near dwellings, and through them are paths leading to more extensive field cultivation systems and grazing lands. Okigbo and Greenland (1976) point out that,

> The largest number of crop species in mixtures is found in compound farms since the crops there are not only grown for food but also for oils and fats, condiments and spices, masticants and stimulants, drugs, fiber, structural materials, animal feed, demarcation of boundaries, firewood, ornamentals, shade, privacy and protection of the homestead, religious and social functions, and various other uses.

It is in these gardens that intercropping systems are most complex.

Major staple crops may be grown in permanent or semipermanent fields which may be in bush fallow some of the time. The length of the cycle depends on natural soil fertility, distance from the dwelling, additional sources of fertility, and population pressure (Okigbo and Greenland, 1976). Importance of intercropping is illustrated by the following proportions of crops grown associated with others in Nigeria: cowpeas (99 percent), groundnut (95 percent), melon (93 percent), millet (90 percent), cocoyam (86 percent), cotton (80 percent), and maize (76 percent). Patterns of crop culture are similar in Uganda, as reported by Jameson (1970): maize (84 percent), beans (81 percent), pigeon peas (76 percent), coffee (63 percent), cowpeas (62 percent), and groundnuts (56 percent). Diversity is a common characteristic of these cropping systems, with more than one variety of each crop appearing on the farm. According to Miracle (1967), one group in Zaire grew 80 varieties of 30 different species, including 27 varieties of banana and plantain and 22 varieties of yams and other root crops.

Specific cropping systems vary widely among countries and among diverse climatic regions within countries. For example, Nigeria has tropical forest zones in the south and east, characterized by yams, cassava, maize, cocoyams, bananas and plantains, and many vegetable and minor crops (Okigbo and Greenland, 1976; Steiner, 1982). There may be as many as 12 or more crops grown together on mounds and in the lower areas around the mounds (see Fig. 10.2). The middle belt of Nigeria is a transition between the root crops of the south and the dryland cereals of the north. Major crops are sesame, tomatoes, potatoes, and other vegetables, sorghum and millets, yams, cassava, cocoyams, cotton, soybeans, and groundnuts, plus cowpeas which are indigenous to Nigeria. The northern zone is a semiarid region where sorghum and millet, cowpeas, maize, sweet potatoes, and groundnuts are common. Intercropping is widely practiced. The case of Nigeria represents the complexity of one West African country's climatic and crop diversity.

An example of the diversity of cropping systems in East Africa can be illustrated for Tanzania. Again, the country is widely diverse in rainfall patterns and growing seasons (Keswani and Ndunguru, 1982; Monyo et al., 1976; Okigbo and Greenland, 1976). Bananas are commonly planted with coffee in the highland areas near Mt. Kilimanjaro, and beans, cowpeas, or groundnuts are commonly planted with maize in many parts of the country. Millets and sorghum are planted with groundnut, Bambara groundnut, pigeon pea, cowpea, beans, and maize. There is a prevalence of intercropping, although relay intercropping of a range of crops also is common. Cropping intensity near the dwelling is important to family nutrition. Cotton is an important commercial crop, although it is most often intercropped with food crops, and green manure crops may be used in some areas in rotation with cash crops. The primary cereal crop in a specific area depends on amount and distribution of rainfall. In the higher elevations with more rainfall, maize predominates. This overlaps with sorghum as rainfall becomes more limiting, and in the driest zones pearl millet is the preferred crop. Cereals are most often found intercropped with one or more grain legumes.

More attention has been placed on research for intercropping systems in Africa during the past decade. The symposia in Morogoro (Keswani and Ndunguru, 1982; Monyo et al., 1976) and in ICRISAT (1981) outline some of the recent efforts to improve these systems for farmers in Africa. With some research in plant breeding and crop protection, the emphasis in these symposia has focused on agronomy and on farming systems. There has been a concentration on component research, but the new emphasis on farming systems research and a move away from merely descriptive activities signal an important advance toward looking at whole farm systems in this region. Research is active at the International Institute for Tropical Agriculture (IITA, Nigeria) and in several national programs.

Cropping Systems in Tropical America

Intercropping and sequential cropping systems have been practiced in Central America since at least the time of the Mayas, and in South America by the Incas, who grew maize and beans (*Phaseolus vulgaris*) together (Pinchinat et al., 1976). These patterns persist today in many parts of the zone (Fig. 1.3).

The principal agroclimatic zones in tropical America as defined by elevation are the lowlands (sea level to 1000 m), temperate lands (1000 to 2000 m), and the highlands (above 2000 m), according to Hardy (1970). The principal crops and cropping systems in these zones are described by Pinchinat et al. (1976). The specific crops and the cropping patterns in these zones are determined by temperature and elevation and by the amount and distribution of rainfall, as well as the objectives of the farmers in each area. The authors summarize reports on the prevalence of small farms in the region, with half or more of the farmers falling in this category in each country.

Primary food crops found in multiple cropping patterns include several indigenous species: maize, bean, cassava (*Manihot esculenta*), squash (*Cucurbita* spp.), potato (*Solanum* spp.), amaranthus (*Amaranthus* spp.), and quinoa (*Chenopodium quinoa*). Introduced crops that have gained importance in some

Figure 1.3 Multicropped terrace system with peppers, radishes, coriander, lettuce, intercropped maize/beans/squash, and bananas, Finca Loma Linda, Cañas Gordas, Coto Brus, Costa Rica. (Photo by Dr. Stephen Gliessman.)

areas include sorghum, cowpea (*Vigna unguiculata*), pigeon pea (*Cajanus cajan*), and peanut (*Arachis hypogaea*). Several perennial species often identified as export or industrial crops are also important to the small-farm sector and are found in multiple cropping systems: coffee, banana, cacao, and rubber. In addition to these principal crops, there are many vegetable species and minor crops that are produced as a part of the multiple species systems and used for food or condiments.

The cropping systems often are dominanted by one principal food species, such as maize in much of Central America and the highlands of the Andean zone (Castillo, 1974), cassava in the lowlands of Colombia, Venezuela, and Brazil (Soria, 1975), potato in the highlands of the Andean zone, or sorghum in the Northeast of Brazil (Rao, 1984) or El Salvador (Alas, 1974). These principal crops in the pattern may provide a main source of food and income for the farm family. Thus, the objectives are to preserve principal crop yields while not reducing them with an associated crop. They may even be planted as monocultures on small farms, especially if the objective is maximizing production and income from the crop. Pinchinat et al. (1976) state that "diversity and number of species intercropped by the small farmer on his plot tend to increase from the lowlands to the highlands." They cite a number of the aforementioned species and their combinations to support this thesis.

Principal locations where research has been conducted in tropical America include various locations in Mexico, Centro Agronómico Tropical de Investigación y Enseñanza (CATIE, Costa Rica), CIAT and Instituto Colombiano Agropecuario (ICA, Colombia), and several sites in Brazil. Data from many of these experiments are cited throughout the chapters that follow.

Multiple Cropping in Temperate Zones

Some of the systems that are currently used in temperate and subtropical regions were described in the ASA symposium for the United States (Lewis and Phillips, 1976; Gomm et al., 1976) and for the Middle East (Nasr, 1976). These are primarily double cropping systems, with some relay planting under experimental conditions. No-till planting is growing in importance for double-cropped systems. Another common multispecies pattern is the grass/legume mixture often found in pastures and hay cropped land. The small grain nurse crop used to help establish a long-term forage legume is an example of an intercrop or relay system. A common combination is oats with alfalfa or clover. An old technique that is gaining in popularity to improve soil fertility and provide cover through the winter months is the overseeding of a forage or green manure legume into a growing crop of cereal or grain legume, or planting the overseeded species after the harvest of the grain crop. This is becoming more important as the cost of chemical fertilizer increases and concern for preventing erosion becomes a more important part of the decision-making process for farming systems design.

Double cropping in the southeastern United States has increased markedly in the past decade, now surpassing 2 million ha in the region; the majority of this area is in winter wheat and summer soybeans. This system presents few challenges that cannot be met with existing technology and farming methods, although the need for rapid harvest and planting each succeeding crop makes timing a critical consideration. There have been selections of both winter grain cereal and summer legume species to better fit this new pattern. Some research has shown success in relay planting soybeans into growing wheat in the spring, for example, in an attempt to lengthen the potential growing period of both crops and alleviate the constraint of fast land preparation after harvest of the wheat. Although this has shown some success (Lassiter, 1973), the lack of adoption of the practice reflects the fact that yields appear to be reduced below those that are possible with double cropping in which there is no overlap in crop cycles. There has been less success planting maize or other summer cereals into winter cereals, or as a double crop after the winter grains.

Although there are historical reports of double cropping in the Middle East (Nasr, 1976), the greatest growth in this practice appears to have been within the last 100 years. Dalrymple (1971) gives a cropping intensity index for Egypt that has increased from 100.4 to 159.1 between 1879 and 1947. These numbers represent the percentage of land use, and an index of 150 shows that an average of 1.5 crops per year are grown on all the acres in the country. With the increase in irrigation in the past several decades, it is likely that this index has increased even more in Egypt and may currently be near 200 (Nasr, 1976). Winter cereals and cool season vegetables and grain legumes predominate during the colder seasons. Cotton, rice, corn, sorghum, and warm-season vegetables predominate in the hotter seasons in Egypt. Similar double cropping systems are reported for other countries in the region (Nasr, 1976).

Mixed stands of grasses and legumes are an important type of pasture in much of the temperate world. These are described by Gomm et al. (1976), and have been used as examples by Trenbath in Chap. 4. Since there is ample literature on these mixtures, and the primary emphasis of this book is on food and fiber crops, no further review of pasture mixtures is presented.

The overseeding of legume species into cereals or other grain crops for nitrogen production and soil conservation is growing in importance. This system has been tested in maize and soybean with positive results, providing groundcover through the winter months in the temperate zone and up to 100 kg/ha of total nitrogen or more when the legume is incorporated before the next summer crop (Hofstetter, 1984; Brusko, 1985). More details on the potentials of this system are given in Chap. 8. Thus, the growth of multiple cropping or multiple species systems in temperate countries is linked to the need for reducing nitrogen cost and soil erosion and the potential for increasing land use intensity as with double cropping in Egypt and the southeastern United States.

BIOLOGY AND AGRONOMY OF MULTIPLE CROPPING SYSTEMS

Most research on multiple cropping systems, with the increasing volume documented in Table 1.1, has taken a reductionist approach that concentrates on one or a small number of factors in well-controlled agronomic experiments. This reflects the training of agricultural scientists in specific disciplines as well as our administrative organizations that fund and implement research in well-defined fields that concentrate on one aspect of a system at a time. This narrow focus appears to be changing. The research and development activities that are emerging from the farming systems approach are more appropriate for the study of multiple cropping systems. Thus, more of the current research is beginning to look at whole cropping systems, and when a single component is the focus of a particular study, the results are at least interpreted as they relate to the entire cropping and farming system. The biology and agronomy of cropping systems are explored in this context in most of the chapters that follow.

Attention has been given in the research community to biological interactions in cropping systems. Hart (Chap. 3) and Gliessman (Chap. 5) describe the ecological basis for multiple cropping systems, and point out the similarities and differences between these controlled systems and the natural ecosystem. Plant interactions and the complexity of resource use by components in a system are further described by Trenbath (Chap. 4). A series of examples of cereals, starchy roots, and legumes as components of complex systems is presented by Rao (Chap. 6) and by Davis, Woolley, and Moreno (Chap. 7). The complexity of fertility interactions in mixtures of cereals and legumes is explored by Barker and Francis (Chap. 8).

The biological complexity is most apparent when the interactions of crops, insects, pathogens, and weeds are examined. These complex areas are reviewed by Altieri and Liebman (Chap. 9). Rotations of dissimilar crops have been suggested as one method for controlling some crop pests, and this potential is available to farmers across the spectrum of farm size and resource base. Some of the potentials of disease and insect tolerance gained through plant breeding can complement the effects of cropping systems in these complex systems to reduce losses to crop pests. These potentials are described, along with other details about how to adapt crops to complex systems, by Smith and Francis (Chap. 10). These first chapters cover the basic biological context within which multiple cropping systems function, and explore how scientists have begun to understand the details of interactions among crops, pests, climate, and the rest of the environment that affects their success.

The time dimension is important in organizing research and setting priorities in multiple cropping. When seeking answers to constraints or problems that currently limit production, the researcher needs to look not only at systems as they are today, but must predict what systems and components will be important in the future when the results of the research will become available. This time frame may be as short as 2 years if we are concerned about a fertilizer recommendation or a combination of species or plant types that will fit together well in a given environment. However, if the objective is to breed for a component crop that will use light efficiently as a lower story component, it will take 10 cycles (perhaps 10 years) after making a cross to reach the point where the breeder is ready to recommend a new variety for use by farmers. Thus, the current economic and resource situation of the farmer is crucial to the design of cropping systems and the modifications that will be suggested for them; but also we must predict what this resource base will be within 10 years so that our results will not be obsolete before they are ready.

Extrapolation of results is another important biological and agronomic consideration in research and extension of new multiple cropping systems. One dimension is the ability to take results from past research on monoculture and apply them to multiple cropping systems. There is little question about the shape of response curves for fertilizer levels or densities of crops, and this type of principle will no doubt apply to multiple crops. Although the dimensions of a curve may change, the general response of yield to increasing plant densities up to a plateau or optimum level will pertain. The challenge is to decide which of the results are applicable and which components have to be studied anew in these complex systems. Several factors are discussed in the chapters that follow. Another dimension is geographic: there are almost as many potential systems for mixing crops as there are farms and farmers, and it is impossible to research specific results for each system. Thus, one challenge is to determine how far results can move from one region or climatic zone to another, and also how different systems must be before there is need for different research and different results and recommendations. These are among the challenges that face biologists working with multiple cropping systems.

ECONOMIC AND SOCIAL IMPORTANCE OF MULTIPLE CROPPING

The research and extension of results in multiple cropping systems is complicated by the range of factors that is inherent in the systems and by the farmers who use them. The usual concerns about biological yields and economic returns may be less important than such issues as food supply for the family, producing food with minimal investment of capital, minimizing risk, and spreading income and food supply throughout the year (Francis, 1981). This subjective framework is difficult for the researcher and "change agent" to deal with, especially since most of them have been trained in an environment that recognizes yield and net return as the principal yardsticks by which to measure success.

Yet multiple cropping research must take place within this new set of guidelines. Lynam, Sanders, and Mason explore some of the complexities of the economics of multiple cropping in Chap. 11. They address risk as one of the most important considerations that is not well understood, and that should become a more significant part of economic analysis of potential new cropping systems for farmers with limited resources. Bradfield, in Chap. 12, evaluates the complex social factors that also influence the decisions of farmers, as well as the difficulties of forming interdisciplinary teams to address these challenges in national programs and international centers.

Finally, a number of research tools are needed to fully explore the potentials of multiple cropping systems, and the complexities that multiple species introduce into farming. Plant breeding potentials to improve that component of the system are described by Smith and Francis (Chap. 10). The intricacies of statistical design and analysis of results are explored by Parkhurst and Francis (Chap. 13) and by Mead (Chap. 14). There is a need to reduce the complexity to a level that can be understood and researched, without losing the focus on the entire cropping and farming system. There are different points of view on how to do this, from the large and complex factorial design to the simple one- or two-factor test. The importance of linking this activity on experiment stations to supplemental research on the farm is introduced, and the potentials of farming systems research are examined as one solution to this difficult problem.

These economic, social, and statistical considerations all lead to the conclusion that there are no simple answers to complex questions. The solutions to challenges in multiple cropping systems need to fit the social and economic environment of the farmer, and any recommendations that mean a change in the system must solve constraints that are perceived by the farmer and family. The rationality of these decisions is not something that most researchers or extension specialists are prepared to deal with. Yet their careful study and incorporation into a research plan is critical to success of a program.

OVERVIEW OF MULTIPLE CROPPING

Interest in multiple cropping among scientists has increased markedly, as shown by the striking increase in number of published research papers (Table 1.1). This has been accompanied by an interest in publishing comprehensive reviews and books on the subject. The first comprehensive symposium was organized by the ASA in 1975 (Papendick et al., 1976). This has been a ready reference for the past 10 years. The detailed reviews by Willey (1979a, 1979b) remain among the best available that summarize the literature to that date. The symposia in Morogoro (Keswani and Ndunguru, 1982; Monyo et al., 1976) were focused on these systems in semiarid zones, as was the symposium in ICRISAT (1981). Such meetings bring together specialists working in different phases of multiple cropping, and are useful summaries of active research work. Three recent books by Beets (1982), Steiner (1982), and Gomez and Gomez (1983) brought together in an organized manner the most relevant research done to date in Asia and Africa on multiple cropping. Two additional books are in preparation: a comprehensive review by Willey (personal communication) and a detailed treatment of statistical methods for multiple cropping by Federer (personal communication). Judged by the past work of these scientists, the books should be excellent and will provide a valuable complement to this volume.

REFERENCES

Alas L., M. 1974. Breve descripción del sistema de producción del pequeño productor en El Salvador, in: *Conferencia sobre Sistemas de Producción Agrícola para el Trópico,* Centro Agronómico Tropical de Investigación y Enseñanza, Informe Final, CATIE, Turrialba, Costa Rica.

Andrews, D. J., and Kassam, A. H. 1976. The importance of multiple cropping in increasing world food supplies, in: *Multiple Cropping,* (R. I. Papendick, P. A. Sanchez, and G. B. Triplett, eds.), Amer. Soc. Agron. Spec. Publ. 27, pp. 1–10.

Barker, T. C. 1984. *Shifting Cultivation among the Ikalahans,* UPLB-PESAM Working Series No. 1, Program on Environmental Science and Management, University of Philippines, Los Baños, Laguna, Philippines.

Beets, W. C. 1982. *Multiple Cropping and Tropical Farming Systems,* Westview Press, Boulder, Colorado.

Brusko, M. 1985. This year's nitrogen—save up to $80/A, in: *The New Farm,* Regenerative Agric. Assoc., Emmaus, Pennsylvania, February, pp. 18–21.

Castillo, M. 1974. Algunos sistemas de producción agrícola en Guatemala, in: *Conferencia sobre Sistemas de Producción Agrícola para el Trópico,* Centro Agronomico Tropical de Investigación y Enseñanza, Informe Final, CATIE, Turrialba, Costa Rica.

Dalrymple, D. G. 1971. *Survey of multiple cropping in less developed nations,* Foreign Economic Development Service, U.S. Dept. Agric., Washington, D.C.

Francis, C. A. 1981. Rationality of farming systems practiced by small farmers, *Proc. Farming Syst. Res. Symp.,* Kansas State University, Manhattan, Kansas.

Gomez, A. A., and Gomez, K. A. 1983. *Multiple Cropping in the Humid Tropics of Asia,* IDRC, Ottawa, Ontario, Canada, IDRC-176e.

Gomm, F. B., Sneva, F. A., and Lorenz, R. J. 1976. Multiple cropping in the Western United States, in: *Multiple Cropping,* (R. I. Papendick, P. A. Sanchez, and G. B. Triplett, eds.), Amer. Soc. Agron. Spec. Publ. 27, pp. 103–115.

Hardy, F. 1970. *Suelos Tropicales, Pedologia Tropical con Énfasis en América,* Herrero Hermanos Sucs., S. A., Mexico, 334 pp.

Harwood, R. R. 1979. *Small Farm Development,* Westview Press, Boulder, Colorado.

Harwood, R. R., and Price, E. C. 1976. Multiple cropping in tropical Asia, in: *Multiple Cropping,* (R. I. Papendick, P. A. Sanchez, and G. B. Triplett, eds.), Amer. Soc. Agron. Spec. Publ. 27, pp. 11–40.

Hofstetter, R. 1984. *Overseeding research results, 1982–1984,* Agronomy Department, Rodale Research Center, RRC/AG-84/29, Kutztown, Pennsylvania.

ICRISAT (International Crops Research Institute for the Semi-Arid Tropics). 1981. *Proc. International Workshop on Intercropping,* Hyderabad, India, 10–13 January 1979, ICRISAT, Patancheru P.O., A.P 502 324, India.

Jameson, J. D. 1970. *Agriculture in Uganda,* Uganda Government Ministry of Agriculture and Forestry, Oxford University Press, London, U.K.

Keswani, C. L., and Ndunguru, B. J. 1982. *Intercropping: Proc. Second Symposium on Intercropping in Semi-Arid Zones,* Morogoro, Tanzania, 4–7 August, 1980, IDRC-186e, Ottawa, Ontario, Canada, 168 pp.

Lassiter, F. 1973. Plant beans into standing grain, *No-Till Farmer,* June, pp. 1, 19.

Lewis, W. M., and Phillips, J. A. 1976. Double cropping in the Eastern United States, in: *Multiple Cropping,* (R. I. Papendick, P. A. Sanchez, and G. B. Triplett, eds.), Amer. Soc. Agron. Spec. Publ. 27, pp. 41–50.

Miracle, M. P. 1967. *Agriculture in the Congo Basin,* University of Wisconsin Press, Madison, Wisconsin.

Monyo, J. H., Ker, A. D. R., and Campbell, M. eds. 1976. *Intercropping in Semi-Arid Areas,* Symposium at Faculty of Agriculture, Forestry, and Veterinary Science, University of Dar es Salaam, Morogoro, Tanzania, IDRC-076e, Canada.

Nasr, H. G. 1976. Multiple cropping in some countries of the Middle East, in: *Multiple Cropping,* (R. I. Papendick, P. A. Sanchez, and G. B. Triplett, eds.), Amer. Soc. Agron. Spec. Publ. 27, pp. 117–127.

Nye, P. H., and Greenland, D. J. 1969. *The soil under shifting cultivation,* Commonwealth Agric. Bur. Tech. Commun. 51, Harpenden, U.K.

Okigbo, B. N., and Greenland, D. J. 1976. Intercropping systems in tropical Africa, in *Multiple Cropping,* (R. I. Papendick, P. A. Sanchez, and G. B. Triplett, eds.), Amer. Soc. Agron. Spec. Publ. 27, pp. 63–101.

Papendick, R. I., Sanchez, P. A., and Triplett, G. B. eds. 1976. *Multiple Cropping,* Amer. Soc. Agron. Spec. Publ. 27.

Perkens, D. H. 1969. *Agricultural Development in China,* Aldine Publishing Co., Hawthorne, New York.

Pinchinat, A. M., Soria, J., and Bazan, R. 1976. Multiple cropping in Tropical America, in *Multiple Cropping,* (R. I. Papendick, P. A. Sanchez, and G. B. Triplett, eds.), Amer. Soc. Agron. Spec. Publ. 27, pp. 51–61.

Rao, M. R. 1984. Prospects for sorghum and pearl millet in the cropping systems of Northeast Brazil. *Proc. Farming Systems Workshop,* CIMMYT/ICRISAT/INTSORMIL, Mexico, September 19.

Soria, J. 1975. Sistemas de producción bajo varias condiciones ecológicas en América Latina, con énfasis en el mejoramiento de la agricultura tradicional de pequeños productores, in: *Consulta de Expertos sobre Investigación Agrícola en América Latina,* FAO, Panama.

Steiner, K. G. 1982. *Intercropping in Tropical Smallholder Agriculture: with Special Reference to West Africa,* German Agency for Technical Cooperation (GTZ), D-6236 Eschborn, Germany.

Willey, R. W. 1979a. Intercropping—its importance and research needs. Part 1. Competition and yield advantages. *Field Crop Abstr.* 32:1–10.

———. 1979b. Intercropping—its importance and research needs. Part 2. Agronomy and research approaches. *Field Crop Abstr.* 32:73–85.

Chapter 2

Historical Perspectives on Multiple Cropping

Donald L. Plucknett
Nigel J. H. Smith

We will never know what the first cultivated fields looked like. The origins of our crop plants, their diffusion paths, and ancient agricultural systems have been explored in numerous publications. But agricultural scientists, anthropologists, archeologists, and geographers rarely discuss the composition of cultivated areas at the dawn of agriculture. This reluctance to speculate on the appearance of prehistoric fields is understandable, given the paucity of hard evidence.

For our purposes, multiple cropping is defined as the simultaneous growing of two or more crops or varieties of a single crop in a field. As we reach farther back in time, the agricultural picture becomes hazier, so that much of what we will have to say is speculative. This chapter begins by confronting the scientific problems involved in trying to decipher early cropping patterns, and then examines the origins of agriculture and multiple cropping. A brief section is devoted to uninvited field guests: weeds. Spontaneous colonizers of cultivated areas have proved to be both a nuisance and a blessing. Weeds constitute a form of multiple cropping when they are tolerated and provide benefits. Multiple cropping is then explored in a historical context in tropical and temperate regions. Finally, reasons for the decline of multiple cropping in this century are outlined and future trends are highlighted.

ANALYTIC PROBLEMS

Pitfalls in confirming the presence of plants in a given region at different times abound, and establishing the associations of plants in the past is even more difficult. Pollen samples taken from soil profiles or the bottoms of lakes or ponds, for example, can reveal the dominant families and sometimes the species of plants in a region at various times. Pollen analysis can pinpoint the appearance of new plants in a region, such as introduced crops, or large-scale vegetation changes provoked by climatic shifts or clearing. This technique can uncover the presence of crops in an area, but it cannot reveal whether the cultigens were grown in the same field. Furthermore, some important food crops, such as pulses and many varieties of cassava (*Manihot esculenta*), produce little if any pollen (Turner and Miksicek, 1984).

Excavation of ancient habitation sites sometimes provides intriguing insights into the antiquity of crop plants and helps document the spread of cultigens. In dry areas, desiccated plant remains as well as straw, sticks, and occasional seeds employed in building wattle-and-daub houses and storage bins indicate the presence of plants in an area. In humid areas, on the other hand, plants soon perish except for carbonized seed; many crops that could have been important in an area may thus leave no trace. If the people occupying an archeological site made pottery, accidental or deliberate impressions of plants on ceramics before they were baked can provide hints as to the presence of crops. Still, pottery making came on the cultural scene long after people had been manipulating the plant world to their advantage. The presence of stone grinders at archeological sites does not necessarily indicate cultivation; wild and domesticated plants are ground to make flour or meal.

Even in archeological sites containing botanical material and where transects have been laid out through cultural layers, results can be misleading. Plant roots are notorious for mixing the soil, particularly if the site is or was humid, and ants shift seeds throughout the soil profile. With new carbon-14 dating techniques, such as tandem accelerator mass spectrometry, organic material as small as seed can now be dated directly, thus overcoming the problem of stratigraphic mixing (Betancourt et al., 1984). But even if this new dating method becomes widely used, the composition of fields will still be a matter of conjecture.

AGRICULTURAL ORIGINS AND MULTIPLE CROPPING

The tending of plants by relatively small human groups occurred long before the beginning of agriculture, usually assigned as some 10,000 years ago. The transition from hunting and gathering to settled farming was a gradual process, extending far back into the Paleolithic, and occurred at different rates in widely scattered regions (Table 2.1). Humankind did not suddenly wake up one day in the Neolithic to a checkerboard pattern of open fields. People had been tending plants for a considerable time before someone tossed seed onto a prepared bed or pushed cuttings into moist soil.

Table 2.1 Crop Domestication Stages from Gathering to Commercial Farming

Stage	Characteristic
Gathering	Wild plants in native stands.
Protection of preferred plants	Wild plants in native stands; volunteer plants around camps and along trails.
Gardening	Transplanted seedlings, roots, cuttings of wild plants, planting of seed crops.
Subsistence farming	Trees, shrubs, herbs, and grasses, usually grown in polycultural assemblages under shifting agriculture conditions.
Subsistence and cash cropping	Polyculture common in tropics, less so in temperate areas; cash crops often grown in separate fields.
Commercial farming	With tropical tree crops, polyculture still common, but trend is toward monocropping.

Camps, rather than cleared fields, served as sites for incipient agriculture. Discarded plant and animal material, as well as human excreta, around the perimeter of camps enriched the soil and provided a favorable environment for the spontaneous sprouting of seeds, tubers, and stems (C. O. Sauer, 1947; J. D. Sauer, 1967; Baker, 1970a; Hawkes, 1970; Plucknett, 1976). Although hunting and gathering groups were small, they were not necessarily always on the move. Some camps may have been occupied for several years, and even when abandoned to take advantage of more plentiful resources elsewhere, many would have been reoccupied. Soil enrichment was thus pronounced.

Edges of trails radiating from camps would also provide open, if less fertile, ground for the germination of wild seeds and the rooting of certain rhizomes and tubers spilled from containers. Women probably did much of the collecting of wild seeds, fruits, and nuts, as they do today in many parts of the world, and they surely paid particular attention to these strips or islands of colonizing plants. It probably did not take long before spontaneous kitchen midden gardens were harvested and trips were undoubtedly made to abandoned campsites to gather seed, nuts, fruits, and tubers. Given the random nature of sprouting, such gardens contained a variety of useful plants.

Although it seems unlikely that spontaneous gardens on campsite garbage mounds provided a major portion of human nutritional requirements, they acted as catalysts in the early evolution of agriculture. Opportunistic plants growing in the nutrient-rich earth are likely to have been generally larger and thus more noteworthy than plants of comparable age growing away from campsites. Furthermore, kitchen midden plants were close at hand and thus especially convenient. Women, in particular, are likely to have cared for such desirable plants by pulling out unwanted weeds; they undoubtedly noticed that plants responded well to such care. Prolonged association with semidomesticated plants eventually nurtured the idea of planting.

Spontaneous campsite gardens were not only biologically diverse, but they provided several useful products in addition to food. Plants employed in interacting with the supernatural realm, as well as species used in treating the sick and for making twine, baskets, and dyes, were at least as important as food-producing plants.

The less control people have over nature, the more likely their culture will be steeped in mysticism. The earliest cultigens may well have been consciousness-altering plants, such as tobacco, used to communicate with spirits, to further hunting success and to decipher the causes of illnesses. Plants relatively rare in nature, rather than the species providing the bulk of the food, were especially likely candidates for early planting.

The composition of campsite gardens during the incipient stage of agriculture would vary from region to region depending on culture and opportunities for food, dyes, and drugs provided by the surrounding vegetation. In the humid tropics, tuber-bearing plants, herbs, and trees providing fruits, nuts, and dyes were probably heavily represented in spontaneous kitchen midden gardens. In the Zagros-Taurus mountains of the Middle East, for example, these campsite resource islands consisted of wild emmer (*Triticum dicoccoides*), wild einkorn (*Triticum boeoticum*), wild barley (*Hordeum spontaneum*), and pulses such as peas (*Pisum* spp.) and lentils (*Lens culinaris*) (Wright, 1976). In the Middle East, numerous cereals important in gathering were spilled on kitchen middens and sprouted or dispersed naturally from surrounding areas. At Tell Mureybat in the headwater region of the Euphrates in Syria, for example, archeologists have found a 10,000-year-old site containing seeds of 18 species of collected plants, including the ancestors of wheat and barley (M. Harris, 1978).

Such mixed garden plots, with varying degrees of care, were widespread throughout the world during the Paleolithic, possibly stretching as far back as hundreds of thousands of years. Starting some 10,000 years ago, or possibly earlier, various cultures in many different areas began to plant. Why it took so long for humankind to get around to cultivation is not known, but opportunity costs probably had a lot to do with it. The idea of planting surely arose long before the Neolithic, but humans only began cultivating plants when the returns exceeded those that could be achieved by gathering. Local overpopulation resulting in depletion of food resources in an area or environmental change may have triggered the move to cultivation. Whatever the forces behind the shift to farming, early fields, just like spontaneous campsite gardens, were probably polycultural and relatively small.

The fertile crescent of the Tigris and Euphrates basins is often pinpointed as the earliest agricultural site, mainly on the basis of a better archeological record and more intensive research. The first farms, however, were likely to have been in the tropics. Carl Sauer (1969) has argued that root crop agriculture predates seed farming because tuberous cultigens are easier to care for and harvesting can be staggered. Cuttings or corms could be easily slipped into the ground after harvesting the ancestors of taro (*Colocasia esculenta*), cocoyam (*Xanthosoma sagittifolium*), sweet potato (*Ipomoea batatas*) and other tuber-bearing plants.

Cereals and pulses are unlikely to have been the first domesticated crops. Wild grasses, for example, have a brittle rachis that shatters when seeds are mature, thereby scattering them. Wild legumes have pods that crack open (dehisce) when mature, thus spilling their contents. Another complication with domesticating cereals and pulses is that seeds are dormant in many cases; this evolutionary strategy ensures survival of some progeny in the event of an exceptionally unfavorable growing season (Rindos, 1984). Nonbrittle rachises, indehiscent capsules, and the loss of dormancy were among the earliest features selected for when grasses and legumes were taken into cultivation. Cereals and pulses, in spite of their preeminent importance in the world food picture today, were not the easiest plants to domesticate and are unlikely to have been the first cultigens.

Johannessen (1970) supports Sauer's contention that root crops were the first cultigens by arguing that varietal integrity is generally easier to maintain with crops that are propagated vegetatively. Echoing the idea that tuber cultivation predates seed farming, Gade (1975) suggests that root cropping east of the Andes is the oldest form of agriculture in South America. Archeological evidence for Sauer's assertion that cereal farming is not the earliest type of agriculture has emerged from a spirit cave in northwestern Thailand. The oldest cultural layers in this limestone cavern, which have been dated at 7000 B.C., contain remains of fruits, nuts, pulses, and root crops (Gorman, 1969).

The earliest fields were probably in alluvial areas subject to light or rare floods and in lightly wooded transitional zones between thick forest and grassland. With primitive tools such as stone axes and wooden digging sticks, it would have been easier to cultivate woodland soils after felling the smaller trees and burning the debris than to till the thick turf of open savanna. Dense jungle would be more difficult to clear and burn than forest transitional to savanna. And several of the important root crops, such as cassava and yams (*Dioscorea* spp.), have tubers that survive dry seasons (D. R. Harris, 1972, 1976).

Early fields were modest because human groups were relatively small and because hunting and gathering still provided a considerable portion of food requirements. They were polycultural because that is the way kitchen midden gardens developed and because mixed plots provided variety in the diet. Polycultural gardens were also ecologically more stable, an important advantage as the subsistence base shifted to farming.

Some of the early cultivated areas may have been barely visible as fields. Hints of what early cultivated plots may have looked like in the tropics can be obtained by turning to subsistence activities of the Kayapó Indians of the Brazilian Amazon. The Kayapó practice open field agriculture cleared from the jungle, but they continue to rely heavily on hunting and gathering for food. The Kayapó traditionally conduct extended hunting trips in the jungle and construct temporary overnight shelters. Desirable plants spotted on hunts, such as trees bearing fruit and those used for making pigments, are uprooted while they are still in the seedling stage and taken to the makeshift camps where they are planted. The forest around the camps is not cleared, but light gaps created by tree falls are exploited for transplanting. Such camps are visited repeatedly, and eventually a sizable cluster of planted trees and shrubs accumulates; at least 54 plant species have thus been semidomesticated by the Kayapó (Posey, 1983).

The Kayapó also create resource islands of trees, shrubs, herbs, and root crops in open grassland. These resource islands, which contain at least 100 species transplanted from the forest, are built up gradually from small, mulched patches (Kerr and Posey, 1984; Posey, 1984). To the untrained eye, these transplanted enclaves may appear to be part of the natural landscape. Trails through the forest are also planting sites for the Kayapó. A 13-km survey of a jungle trail leading from a Kayapó village to a garden revealed 185 planted trees comprising at least 15 species, as well as approximately 1500 medicinal plants and 5500 food plants representing numerous species. Agriculture in the Yucatán Peninsula may have started in similar fashion beginning with selective manipulation of certain forest trees, shrubs, and vines (D. Harris, 1978a).

UNINVITED GUESTS: WEEDS

When people transplant under forest cover, weeds are not much of a problem, but as soon as vegetation is cleared, unwanted sun-loving plants invade open spaces. Weeds were an early nuisance, and fields are virtually never free of them even today with the use of herbicides. Although weeds surely reduced crop yields at the onset of farming, as they still do, they were sometimes tolerated and even harvested. Early fields were polycultural by design and default; plots were purposely planted to several cultigens and volunteer plants soon joined them.

Weeds in the history of agriculture have been a mixed blessing. Although the persistant invaders have reduced yields, they have occasionally contributed useful genes to crops and have sometimes been domesticated themselves. Weed-infested fields have served as direct recruiting grounds for new crops. Several important cultigens started as field weeds. Rye (*Secale cereale*) and oats (*Avena* spp.) developed from weeds in barley and wheat fields in the Middle East and in northern and western Europe. In Central America, two cultivated amaranths (*Amaranthus hypochondriacus* and *A. cruentus*) and chenopodium (*Chenopodium nuttaliae*) probably originated as weeds in cultivated plots (Wolf, 1959). Carl Sauer (1969) has suggested that seed crops—amaranths, beans (*Phaseolus* spp.), squashes (*Cucurbita* spp.), and maize (*Zea mays*)—which dominate the subsistence base for much of Central America, began as weeds in fields planted to cassava and sweet potato. Root crop cultivation penetrated Central America from the south and probably predates seed cropping. In West Africa, the cultivation of yams along the forest fringe probably arose before seed cropping; some cereal cultigens may have started as weeds in yam fields (Shaw, 1976).

Weeds destined to become crops pass through three main stages. First, they are destroyed. Then as they proliferate, some are tolerated because they are too troublesome to pull, or because they provide useful products such as food or straw. The next step, planting, occurs when the weed has accompanied humans long enough to find one or more uses and is considered to be worth cultivating. The new crop may be planted in separate fields, as in the case of rye and oats, but these are in turn invaded by weeds.

Multicropped fields invaded by weeds have served as primitive "laboratories" in the development of crop plants. In the case of cultivated potatoes (*Solanum* spp.), for example, weedy *Solanum* diploids have interbred with cultigens to form new domesticates. The ploidy level is not always affected. For example, a diploid potato (*S. ajanhuiri*) cultivated between 3800 and 4100 m elevation by the Aymara of southern Peru and northern Bolivia arose as a result of natural hybridization between a cultivated potato (*S. stenotomum*) and a wild species (*S. megistacrolobum*) (Huamán et al., 1980; Hawkes, 1981). The latter species is responsible for the frost resistance of *S. ajanhuiri*.

The common potato (*Solanum tuberosum*) is tetraploid and resulted from a cross between a diploid cultigen (*S. stenotomum*) and a weedy diploid (*S. sparsipilum*). The latter species imparted resistance to root-knot nematodes and the common potato adapts to a wider range of conditions than its parents. Resistance to more virulent pathotypes of potato cyst nematode (*Heterodera rostochiensis*) has also reached the common potato through crossing. Resistance to this destructive pest spread from a wild tetraploid (*S. oplocense*) to weedy *S. sucrense,* and then to *S. tuberosum* (Hawkes, 1977). Weedy potatoes, in and around fields, have thus contributed to the development of new potato cultigens and have upgraded resistance to pests and severe climates.

Weeds have been a constant source of new genes for resistance to pests, diseases, and adverse weather in other crops as well. Bread wheat (*Triticum aestivum*), for example, is a cross between emmer wheat (*T. dicoccum*) and *Aegilops squarrosa* (syn. *Triticum tauschii*); the latter, a wild grass, has conferred cold tolerance to bread wheat (Anderson, 1952; Feldman, 1976; Hawkes, 1980). Barley (*Hordeum vulgare*) has benefitted from a continual introgression of genes from weedy relatives growing within and along the borders of fields (Baker, 1970b). In the Middle East, lentils (*Lens culinaris*) have crossed with wild relatives as they spread (Cubero, 1984a). Wild relatives of lentils probably accompanied the crop and have provided a storehouse of fresh genes. In East Africa, finger millet (*Eleusine coracana*) hybridizes with a weedy relative (*E. indica*) in fields (Hussaini et al., 1977).

Wild relatives sometimes mimic crops so closely that farmers leave them in the field. For example, teosinte (*Zea mexicana*), the grass from which maize originated, grows in and around maize fields in Mexico and Guatemala. Some populations of teosinte can only be safely separated from maize when they mature. Farmers consequently only weed teosinte between furrows, leaving others in rows with maize (Wilkes, 1972). Although a weed, teosinte benefits traditional maize growers. Teosinte is harvested along with maize and is fed to livestock; seeds that escape grinding teeth are reintroduced to fields in manure.

The weed has introgressed with maize for thousands of years, thereby contributing vigor to the crop. Farmers in the Nobogame Valley in Mexico's Sierra Madre Occidental recognize that teosinte benefits maize and sow seeds of both plants together (Wilkes, 1977).

Accidental crosses with wild and domesticated potatoes, wheats, barley, lentils, finger millet, and maize were spotted by farmers, and some of them were saved for propagation. With the widespread use of herbicides in modern agriculture, one wonders what evolutionary opportunities are being foreclosed. On the other hand, scientific breeders have now largely substituted for natural hybridization and herbicides have clearly boosted crop yields.

MULTIPLE CROPPING IN THE TROPICS

Agriculture and multiple cropping probably not only started in the tropics, but polycultural systems are generally more diverse in equatorial regions. Rich polycultural assemblages have long been the tradition in rainforest regions, such as in the Amazon and Orinoco drainage basins (Lathrap, 1970; D. R. Harris, 1971; Eden, 1974). As a general rule, root crop cultivation is more biotically diverse than seed-based farming and more people in the tropics depend on tuber crops for a livelihood than in the temperate zone. The diversity of multicropping patterns declines poleward or as altitude increases within the tropics, and as rainfall amounts decrease (D. R. Harris, 1976).

With human migrations, and particularly with the advent of European colonization in the sixteenth and seventeenth centuries, multicropping assemblages have been enriched, although this process is now slowing with the increased attention paid to market farming. In the humid, lowland tropics, root crops intercropped with other cultigens have been the mainstay of people's diets for millennia. In the Americas, cassava, sweet potato, cocoyam, and to a lesser extent cush-cush yam (*Dioscorea trifida*) were among the first food plants domesticated and soon formed the main staples in polycultural plots (Fig. 2.1). These root crops were interplanted with numerous other cultigens that provided fruit, dyes, and drugs. Later, maize spread into parts of the lowland Neotropics to enrich polycultural fields, but the cereal is more susceptible to diseases and pests than the hardier tuber crops. During the colonial period, new crops, such as bananas and sugarcane, were incorporated in multicropped fields.

The composition of polycultural plots varies widely according to cultural and environmental differences, and only a few examples will be explored here. A survey of 11 fields, belonging to the Andoke and Witoto of the Colombian Amazon and ranging from 0.5 to 1 ha, revealed from 5 to 18 intercropped species, with cassava the main staple (Eden, 1980). The Arara, who live in the Xingu valley of Amazonia, illustrate the complexity of polycultural fields in the lowland Neotropics and the acquisition of at least one Old World cultigen. A 1-ha field of this Carib group visited near km 80 of the Altamira-Itaituba stretch of the Transamazon Highway in 1972 revealed various crop layers (Fig. 2.2). Groundcover was provided by squash, sweet potato, and two varieties of pine-

Figure 2.1 Cassava and cacao (*Theobroma cacao*) intercropped along Brazil's Trans-amazon Highway 80 km west of Altamira, 1974.

apple, while cassava and a ginger (*Renealmia occidentalis*) formed the middle story. Three varieties of banana [a red-skinned type, a plantain, and a landrace (a primitive variety) with small, sweet fruits] and papaya (*Carica papaya*) rose above the rest of the crops. The bananas were clumped, whereas the papaya appeared to have been sown randomly. Sometimes the arrangement of crops in a field is elaborate. The Kreen-Akrore, who also live in the Brazilian Amazon, plant their multicropped fields in geometric patterns, undoubtedly in response to cosmological beliefs (Cowell, 1974).

In Central America, root crops are still important, but the maize/beans/squash trio emerged as the dominant farming pattern in prehistoric times (Reichel-Dolmatoff, 1973). The ecological and dietary advantages of this crop partnership, which probably arose in Mexico, explain its wide adoption in highland and drier areas: maize provides a trellis for beans that enrich the soil with nitrogen, while squash provides groundcover, thereby reducing soil erosion, compaction, and weed growth.

Planted backyard gardens are often even more botanically diverse than open fields (Fig. 2.3). As in the case of spontaneous campsite gardens in the Paleolithic, they contain a mixture of food, medicinal, dye, and fiber plants. Furthermore, ornamental plants, a more recent luxury, are often common in such planted dooryard gardens. In Central America, for example, it is not uncommon to find two dozen cultivated plants in backyard gardens covering only 0.1 ha (Wilken, 1976). Also, for each species in a garden, several varieties may be grown. In Martinique, for example, a dooryard garden may contain as many as

Figure 2.2 Polycultural field of the Arara containing papaya (*Carica papaya*), banana, cassava, and other cultigens, near km 80 of the Altamira-Itaituba stretch of Brazil's Transamazon Highway, 1972.

Figure 2.3 Dooryard garden on Amazon floodplain containing cassava, vegetables, banana, mango, and other crops. Careiro Island, near Manaus, Brazil, 1972.

six cocoyam cultivars (Kimber, 1978). Some of the varieties are grown for their starchy rhizomes, whereas others are cultivated primarily as leafy vegetables. Such a high degree of selection indicates that the crop has been around a long time.

In West Africa, several root crops were domesticated in the transition zone from forest to savanna and became the subsistence base for the numerous tribes occupying the region. Yellow guinea yam (*Dioscorea cayenensis*), white guinea yam (*D. rotundata*), kafir potato (*Plectranthus esculentus*), yampea (*Sphenostylis stenocarpa*), and piasa (*Solenostemon rotundifolius*) were probably the main components of polycultural plots in humid areas during prehistoric times. African oil palm (*Elaeis quineensis*), a native of West African jungles, is traditionally spared when clearing and is a common component of polycultural fields in the region. Another indigenous species, African rice (*Oryza glaberrima*), may have been planted with other crops, but in general multiple crop farming in the humid zone of West Africa appears never to have been as biotically diverse as comparable climatic regions in the New World.

Rice (*O. sativa* and *O. glaberrima*) was grown in West Africa before the arrival of Europeans, but the main enrichment of cropping patterns in this region occurred during the colonial period with the introduction of maize and cassava. Maize and cassava now dominate food production in West Africa and in many other parts of the continent. In upland areas of Guinea, Sierra Leone, and Liberia, for example, cassava and several varieties of African and Asian rice are often intercropped (D. R. Harris, 1976).

In drier parts of Africa, several seed crops were domesticated, such as sorghum (*Sorghum bicolor*), several millets, and tef (*Eragrostis tef*), and simple multiple cropping patterns have arisen based on cereals. In Ethiopia, for example, weedy oats are tolerated in barley and emmer fields; mixed seeds are sown and the plants are harvested together (Harlan et al., 1976). But as is typical in areas where environmental stresses are severe, multicropping is much less pronounced than in hot, humid regions. Fewer plants can withstand drought than moist conditions, and agroecosystem diversity parallels the decline in plant species diversity of natural vegetation as one proceeds from the equator to drier or cooler areas.

In tropical Asia, vegeculture and cultivation of trees also probably predates cereal farming. In humid areas, taro, giant taro (*Alocasia macrorrhiza*), giant swamp taro (*Cyrtosperma chamissonis*), and the greater yam (*Dioscorea alata*) were early domesticates and were often planted with other crops. Breadfruit (*Artocarpus altilis*), jackfruit (*Artocarpus heterophyllus*), and coconut (*Cocos nucifera*) have also been components of multiple cropping associations in Asia and the Pacific for a long time. In Southeast Mindoro Island in the Philippines, for example, the Hanunoo plant up to 40 species of trees, shrubs, herbs, tubers, and cereals in a single field, a complex pattern that must have taken a long time to develop (Conklin, 1954). Tree crops, like tuberous cultigens, require less attention than cereals and were thus easier to manage by early farmers. They also provide a variety of products, such as fiber and dyes, in addition to food.

Coconut and pasture combinations are common in coastal regions of the Pacific and Indian oceans (Plucknett, 1984). In Sri Lanka, for example, cows are tethered to the palms at night and enrich plantation soils with dung.

After rice (*Oryza sativa*) was domesticated, probably in southern China, Asian polycultural systems began losing ground to the adaptable cereal. As in the case of Africa, multiple cropping is generally not as diverse as in the Americas. The reduced incidence of multiple cropping in Asia is partly due to widespread cultivation of paddy rice (D. R. Harris, 1978b). Although fields in tropical Asia are not usually as rich in cultigens as in traditional farming areas of the lowland Neotropics, paddy rice fields often contain several landraces.

TEMPERATE ZONE MULTIPLE CROPPING

Seed-based farming has always been more important in temperate areas than in the tropics, so polyculture is less diverse in the cooler regions. Temperate zone farmers have generally attempted to grow crops such as wheat, barley, rye, and oats in separate fields. Diversity in time, through rotational cropping, rather than diversity in space as in polycultural patterns, has been the predominant farming pattern in temperature regions (Chang, 1983). Alternating crops annually or every few years has helped temperate zone farmers overcome problems caused by deterioration of soil fertility and structure and the buildup of pests and diseases. Furthermore, more crops have been domesticated in the tropics and subtropics than in the colder climates, and mirroring natural ecosystems, temperate farms are usually biotically less diverse than tropical ones.

Although monocultural cropping is prevalent in temperate regions, especially in recent years, polycultural assemblages are, or have been, common in certain areas. The maize/beans/squash pattern, discussed in its Central American hearth, spread as far north as the St. Lawrence River long before the arrival of Europeans in North America (C. O. Sauer, 1969). And as recently as 1923, 60 percent of the soybean (*Glycine max*) crop in Ohio and Illinois was intercropped with maize (Kass, 1978). Mechanized harvesting has now largely done away with that practice.

In the Middle East, multiple cropping with several cultigens or landraces in the same field has been, and still is, relatively common. By biblical times, *Lolium temulentum* was an established weed in wheat fields; seed of the grassy volunteer was so difficult to separate from wheat that farmers sowed and harvested them together (De Wet and Harlan, 1975). Today in Syria, farmers often sow several varieties of wheat in the same field (Fig. 2.4); weeds are also harvested in such fields for livestock consumption (ICARDA, 1983). Perennial crops are frequently grown together; in the vicinity of Aleppo, for example, combinations of grape/almond, olive/grape, and olive/fig are common. Such associations were probably developed well before the time of Christ.

In the Mediterranean region, primitive varieties of vetch (*Vicia sativa*) are frequently grown with wild relatives (Cubero, 1984b). In Turkey, chickpea (*Cicer arietinum*) fields are often invaded by a wild relative (*C. bijugum*), but

Figure 2.4 Several varieties of bread wheat sown in the same field, near Aleppo, Syria, 1984.

farmers do not attempt to weed the latter out and both species are harvested together (Ladizinsky, 1979). Wherever farmers save their own seed for planting the following cycle, unintentional mixing of extraneous varieties or wild species sometimes occurs leading to polycultural farming by default. Occasionally, farmers will deliberately plant several varieties in the same field as a hedge against crop failure.

POLYCULTURAL PROGNOSIS

Throughout most of agricultural history, multiple cropping has been generally gaining in complexity. But in this century, three main forces have arrested this trend: changes in land management, demands of the food processing industry, and commercialization of farming. Polycultural fields are labor intensive, both when planting and at harvest time. When farmers turn to implements and machinery to streamline their operations, cropping patterns are simplified. Livestock-drawn ploughs or tractors employed to prepare seed beds and to destroy weeds are difficult to manage when more than one crop is grown in a field. Combine harvesters are designed to operate with specific crops and commercial farmers have adjusted their planting patterns accordingly. Furthermore, modern watering systems, such as drip irrigation from pipes, are often tailored to the requirements of specific crops and are difficult and much costlier to implement for more than one crop per field.

The food processing industry has also influenced farming patterns. Transportation and factory processing are facilitated when crops are standardized. In the United States, for example, tomatoes have been bred for thick skins to withstand travel by truck to canning factories and to supermarkets. The potato chip industry prefers potatoes of uniform shape, color, cooking quality, taste, and texture. Specialization and standardization of crops is easier in monoculture than polyculture.

Market farming also tends to promote uniformity because crop treatments are easier and wholesalers find it more convenient to deal with a few standard varieties than with a bewildering array of landraces with different colors, shapes, and cooking qualities. Even farmers who cultivate polycultural plots for subsistence needs will often plant crops destined for market in separate fields (Wilken, 1969, 1977; Ruddle, 1974; Smith, 1978).

The trend to monocultural cropping is universal. It is most advanced in the temperate world, but also occurs in the better lands of the Third World, areas endowed with better soils, a good water supply, and proximity to markets. Isolated areas, such as remote jungle regions, and many marginal zones suffering from frequent drought, are still strongholds for multiple cropping.

Although polycultural farming has been on the retreat in this century, numerous cases, discussed in subsequent chapters in this book, illustrate that a countercurrent has developed that is reintroducing biotic diversity on some farms. Cereal crop breeders, for example, are developing multiline varieties that share the same agronomic properties but have different resistance genes. In this way,

Figure 2.5 Japanese-Brazilians cultivating cupuacu (*Theobroma grandiflorum*), a relative of cacao, under the shade of nitrogen-fixing *Erythrina* trees, Tome-Acu, Brazil, 1977.

Figure 2.6 Alley cropping with maize and leucaena (*Leucaena leucocephala*) at the International Institute of Tropical Agriculture, Ibadan, Nigeria, 1983. Leucaena supplies nitrogen to the soil and can be used for livestock fodder and fuel wood.

farmers can plant crops that are compatible with machinery and the food processing industry. At the same time, they introduce more genetic variability in their fields and thereby reduce pest and disease damage. As the cost of chemical control of insects, diseases, and weeds climbs, farmers are seeking economically feasible means to keep these problems under control. Multiple cropping strategies can help them achieve this goal.

Agroforestry is also receiving renewed attention by agricultural scientists and development planners and this rekindled interest will lead to the further evolution of multiple cropping systems (Figs. 2.5 and 2.6). Spurred by fuel wood shortages, desertification, rampant soil erosion, and a need to protect watersheds, scientists are investigating ways to incorporate perennial crops into food producing systems (Eckholm, 1979; Smith, 1981, 1985; Brown et al., 1984). In this manner, farmers increase their income earning opportunities and help stabilize productivity, since not all eggs are in the same basket as in monocropping. Multicropping, the oldest cultivation pattern, will thus continue to supply a diverse, and ever changing, assortment of products for societies throughout the world.

REFERENCES

Anderson, E. 1952. *Plants, Man and Life,* University of California Press, Berkeley, California.

Baker, H. G. 1970a. *Plants and Civilization,* Wadsworth Publishing Company, Belmont, California.

———. 1970b. Taxonomy and the biological species concept in cultivated plants, in: *Genetic Resources in Plants: Their Exploration and Conservation,* (O. H. Frankel and E. Bennett, eds.), Blackwell Scientific Publications, Oxford, U.K., pp. 49–68.

Betancourt, J. L., Long, A., Donahue, D. J., Jull, A. J. T., and Zabel, T. H. 1984. Pre-Columbian age for North American *Corispermum* L. (Chenopodiaceae) confirmed by accelerator radiocarbon dating, *Nature London* 311:653–655.

Brown, L., Chandler, W., Postel, S., Starke, L., and Wolfe, E. 1984. *State of the World: A Worldwatch Institute Report on Progress Toward a Sustainable Society,* W. W. Norton, New York.

Chang, T. T. 1983. The origins and early cultures of the cereal grains and food legumes, in: *The Origins of Chinese Civilization,* (D. N. Keightley, ed.), University of California Press, Berkeley, California, pp. 65–94.

Conklin, H. C. 1954. An ethnoecological approach to shifting agriculture, *Trans. N.Y. Acad. Sci.* 17(2):133–142.

Cowell, A. 1974. *The Tribe that Hides from Man,* Stein and Day, New York.

Cubero, J. I. 1984a. Taxonomy, distribution and evolution of the lentil and its wild relatives, in: *Genetic Resources and their Exploitation: Chickpeas, Faba Beans and Lentils,* (J. R. Witcombe and W. Erskine, eds.), Martinus Nijhoff/Dr W. Junk, The Hague, pp. 187–203.

———. 1984b. Utilization of wild relatives and primitive forms of food legumes, in: *Genetic Resources and Their Exploitation: Chickpeas, Faba Beans and Lentils,* (J. R. Witcombe and W. Erskine, eds.), Martinus Nijhoff/Dr W. Junk, The Hague, pp. 73–84.

De Wet, J. M. J., and Harlan, J. R. 1975. Weeds and domesticates: Evolution in the man-made habitat, *Econ. Bot.* 29:99–107.

Eckholm, E. 1979. *Planting for the Future: Forestry for Human Needs,* Worldwatch Institute, Paper 26, Washington, D.C.

Eden, M. J. 1974. Ecological aspects of development among Piaroa and Guahibo Indians of the upper Orinoco basin, *Antropológica* 39:25–56.

———. 1980. A traditional agro-system in the Amazon region of Colombia, *Trop. Ecol. Develop.* 1:509–514.

Feldman, M. 1976. Wheats, in: *Evolution of Crop Plants,* (N. W. Simmonds, ed.), Longman, London, U.K., pp. 120–128.

Gade, D. W. 1975. *Plants, Man and Land in the Vilcanota Valley of Peru,* Dr W. Junk, The Hague.

Gorman, C. F. 1969. Hoabinhian: A pebble complex with early plant associations in southeast Asia, *Science* 163:671–673.

Harlan, J. R., De Wet, J. M. J., and Stemler, A. 1976. Plant domestication and indigenous African agriculture, in: *Origins of African Plant Domestication,* (J. R. Harlan, J. M. J. De Wet, and A. B. L. Stemler, eds.), Mouton, The Hague, pp. 107–153.

Harris, D. R. 1971. The ecology of swidden cultivation in the upper Orinoco rain forest, Venezuela, *Geogr. Rev.* 61(4):475–495.

———. 1972. The origins of agriculture in the tropics, *Amer. Sci.* 60:180–193.

———. 1976. Traditional systems of plant food production and the origins of agriculture in West Africa, in: *Origins of African Plant Domestication,* (J. R. Harlan, J. M. J. De Wet, and A. B. L. Stemler, eds.), Mouton, The Hague, pp. 311–356.

———. 1978a. The agricultural foundations of lowland Maya civilization: A critique, in: *Pre-Hispanic Maya Agriculture,* (P. D. Harrison and B. L. Turner, eds.), University of New Mexico Press, Albuquerque, New Mexico, pp. 301–322.

———. 1978b. The environmental impact of traditional and modern agricultural systems, in: *Conservation and Agriculture,* (J. G. Hawkes, ed.), Duckworth, London, U.K., pp. 61–69.

Harris, M. 1978. *Cannibals and Kings: The Origins of Cultures,* Vintage Books, New York.

Hawkes, J. G. 1970. The origins of agriculture, *Econ. Bot.* 24(2):131–133.

———. 1977. The importance of wild germ plasm in plant breeding, *Euphytica* 26:615–621.

———. 1980. The taxonomy of cultivated plants and its importance in plant breeding research, in: *Perspectives in World Agriculture,* Commonwealth Agricultural Bureaux, Farnham Royal, Slough, pp. 49–66.

———. 1981. Biosystematic studies of cultivated plants as an aid to breeding research and plant breeding, *Kulturpflanze* 29:327–335.

Huamán, Z., Hawkes, J. G., and Rowe, R. P. 1980. *Solanum ajanhuiri:* An important diploid potato cultivated in the Andean altiplano. *Econ. Bot.* 34(4):335–343.

Hussaini, S. H., Goodman, M. M., and Timothy, D. H. 1977. Multivariate analysis and the geographical distribution of the world collection of finger millet, *Crop Sci.* 17:257–263.

ICARDA (International Center for Agricultural Research in the Dry Areas). 1983. *ICARDA Annual Report 1983,* Aleppo, Syria.

Johannessen, C. L. 1970. The dispersal of *Musa* in Central America: The domestication process in action, *Ann. Assoc. Amer. Geogr.* 60(4):689–699.

Kass, D. C. L. 1978. *Polycultural Cropping Systems: Review and Analysis,* Cornell International Agriculture Bull. 32, Cornell University, Ithaca, New York.

Kerr, W. E., and Posey, D. A. 1984. Informações adicionais sobre a agricultura dos Kayapó, *Interciencia* 9(6):392–400.

Kimber, C. 1978. A folk context for plant domestication: Or the dooryard garden revisited, *Anthropol. J Can.* 16(4):1–11.

Ladizinsky, G. 1979. Seed dispersal in relation to the domestication of Middle East legumes, *Econ. Bot.* 33(3):284–289.

Lathrap, D. W. 1970. *The Upper Amazon,* Thames and Hudson, London, U.K.

Plucknett, D. L. 1976. Edible aroids, in: *Evolution of Crop Plants,* (N. W. Simmonds, ed.), Longman, London, U.K., pp. 10–12. 1984.

———. Tropical tree crops in crop/livestock systems, paper presented at the Annual Meeting of the American Association for the Advancement of Science, New York, N.Y., 24–29 May.

Posey, D. 1983. Indigenous knowledge and development: An ideological bridge to the future, *Ciênc. Cult. Sao Paulo* 35(7):877–894.

———. 1984. A preliminary report on diversified management of tropical forest by the Kayapo Indians of the Brazilian Amazon, *Advan. Econ. Bot.* 1:112–126.

Reichel-Dolmatoff, G. 1973. The agricultural basis of the sub-Andean chiefdoms of Colombia, in: *Peoples and Cultures of Native South America,* (D. R. Gross, ed.), Doubleday/The Natural History Press, Garden City, New York, pp. 28–36.

Rindos, D. 1984. *The Origins of Agriculture: An Evolutionary Perspective,* Academic Press, Orlando, Florida, p. 88.

Ruddle, K. 1974. *The Yupka Cultivation System: A Study of Shifting Cultivation in Colombia and Venezuela,* Ibero-Americana 52, University of California Press, Berkeley, California.

Sauer, C. O. 1947. Early relations of man to plants, *Geogr. Rev.* 37(1):1–25.

———. 1969. *Agricultural Origins and Dispersals: The Domestication of Animals and Foodstuffs,* M.I.T. Press, Cambridge, Massachusetts, p. 64.

Sauer, J. D. 1967. The grain amaranths and their relatives: A revised taxonomic and geographic survey, *Ann. Mo. Bot. Gard.* 54(2):103–137.

Shaw, T. 1976. Early crops in Africa, in: *Origins of African Plant Domestication,* (J. R. Harlan, J. M. J. De Wet, A. B. L. Stemler, eds.), Mouton, The Hague, pp. 107–153.

Smith, N. J. H. 1978. Agricultural productivity along Brazil's Transamazon Highway, *Agro-ecosystems* 4:415–432.

———. 1981. *Wood: An Ancient Fuel with a New Future,* Worldwatch Institute, Paper 42, Washington, D.C. 1985.

———. 1985. Trees and food for a hungry world, *Food Policy* 10(1):82.

Turner, B. L., and Miksicek, C. H. 1984. Economic plant species associated with prehistoric agriculture in the Maya lowlands, *Econ. Bot.* 38(2):179–193.

Wilken, G. C. 1969. Drained-field agriculture: An intensive farming system in Tlaxcala, Mexico, *Geogr. Rev.* 59:215–241.

———. 1976. Management of productive space in traditional farming, paper presented at the Symposium on La Dynamique des Systèmes Culturaux Traditionnels en Amérique Tropicale, XLII Congrès International des Americanistes, Paris, 2–9 September.

———. 1977. Integrating forest and small-scale farm systems in Middle America, *Agro-ecosystems* 3:291–302.

Wilkes, H. G. 1972. Maize and its wild relatives, *Science* 177:1071–1077.

———. 1977. Hybridization of maize and teosinte in Mexico and Guatemala and the improvement of maize, *Econ. Bot.* 31(3):254–293.

Wolf, E. 1959. *Sons of the Shaking Earth,* University of Chicago Press, Chicago, Illinois.

Wright, H. E. 1976. The environmental setting for plant domestication in the Near East, *Science* 194:385–389.

Chapter 3

Ecological Framework for Multiple Cropping Research

Robert D. Hart

It is often forgotten that agronomy is an applied subdiscipline of ecology. Although agronomists and ecologists both claim such nineteenth century scientists as Liebig with his "law of the minimum" as founding fathers, in more recent times agronomy and ecology have taken separate paths. Agronomy is often taught without explicitly discussing its ecological foundation, while agriculture is usually presented in ecology texts as a human activity that is responsible for the destruction of "natural" ecosystems. Liebig's "law" is discussed by ecologists as a general systems principle applicable to steady-state ecosystems; it is used by agronomists as a guideline for fertilizer recommendations. The connection between general systems theory and formulations of fertilizer recommendations is not immediately clear to many agronomists and ecologists.

Interest in multiple cropping systems has brought ecologists and agronomists closer together. It is not surprising that basic ecological principles relating to population interactions (such as competition and allelopathy) should be useful as an organizing framework for agronomic research on multiple cropping systems; what is surprising is how little of the basic ecological methods and techniques (e.g., system analysis, computer models) have actually been used by agronomists doing multiple cropping research. An important characteristic of multiple cropping research is that it is farmer-targeted. For this reason agronomists doing multiple cropping research have often had more contact with social scientists who share their interest in people than they have had with ecologists who share their interest in understanding how complex biological systems function.

In this chapter I attempt to summarize some of the key concepts of ecology that are directly related to agronomic research on multiple cropping systems. Ecologists are interested in a wide range of natural phenomena from physiological processes occurring at the cellular level to processes at the level of organism, population, community, and ecosystem. When this hierarchy of systems (organism-population-community-ecosystem) is extended to include the social systems that are important in agriculture, it can be used as a conceptual framework for agricultural research and development, and consequently as a framework for multiple cropping research.

This chapter is organized as follows. First, a brief description of the hierarchical system conceptual framework is presented. This is followed by a review of some of the basic ecological principles or areas of ecological research that are directly relevant to multiple cropping research. The hierarchical systems that are a shared interest of ecology and agronomy (populations, communities, and ecosystems) are used to organize this review. The review is followed by a discussion of the potential use of the conceptual framework as a tool to organize multiple cropping systems research.

HIERARCHICAL AGRICULTURAL SYSTEMS

Development-oriented agricultural scientists continually face the problem of how to integrate a diverse array of research activities that span a continuum from the analysis of soil microbes to the study of regional markets. Multiple cropping, cropping systems, and farming systems research require the integration of physical, biological, and social science research. A prerequisite for successful multidisciplinary team research is a shared conceptual framework.

Figure 3.1 depicts a hierarchy of agricultural systems that can function as a conceptual framework for development-oriented agricultural research (see Hart, 1982, for a description of the framework, and Hart and Pinchinat, 1982, for guidelines for its application). Agricultural research that is development oriented, and furthermore, targeted at a specific geographic region and a specific type of community and farm, must allocate resources to the analysis of these macroagricultural systems since they are the environment in which the crop and livestock agricultural ecosystems function. The predominant agricultural ecosystems are the larger systems (suprasystems) in which any new crop systems (with single or multiple crop components) must function.

As suggested in Fig. 3.1, the region, community, and farm systems are social systems, and the methods applied in the analysis of these systems as part of development-oriented research should be based on the application of social science (economics, sociology, anthropology) concepts. But farm systems are a combination of ecological and socioeconomic subsystems and their analysis

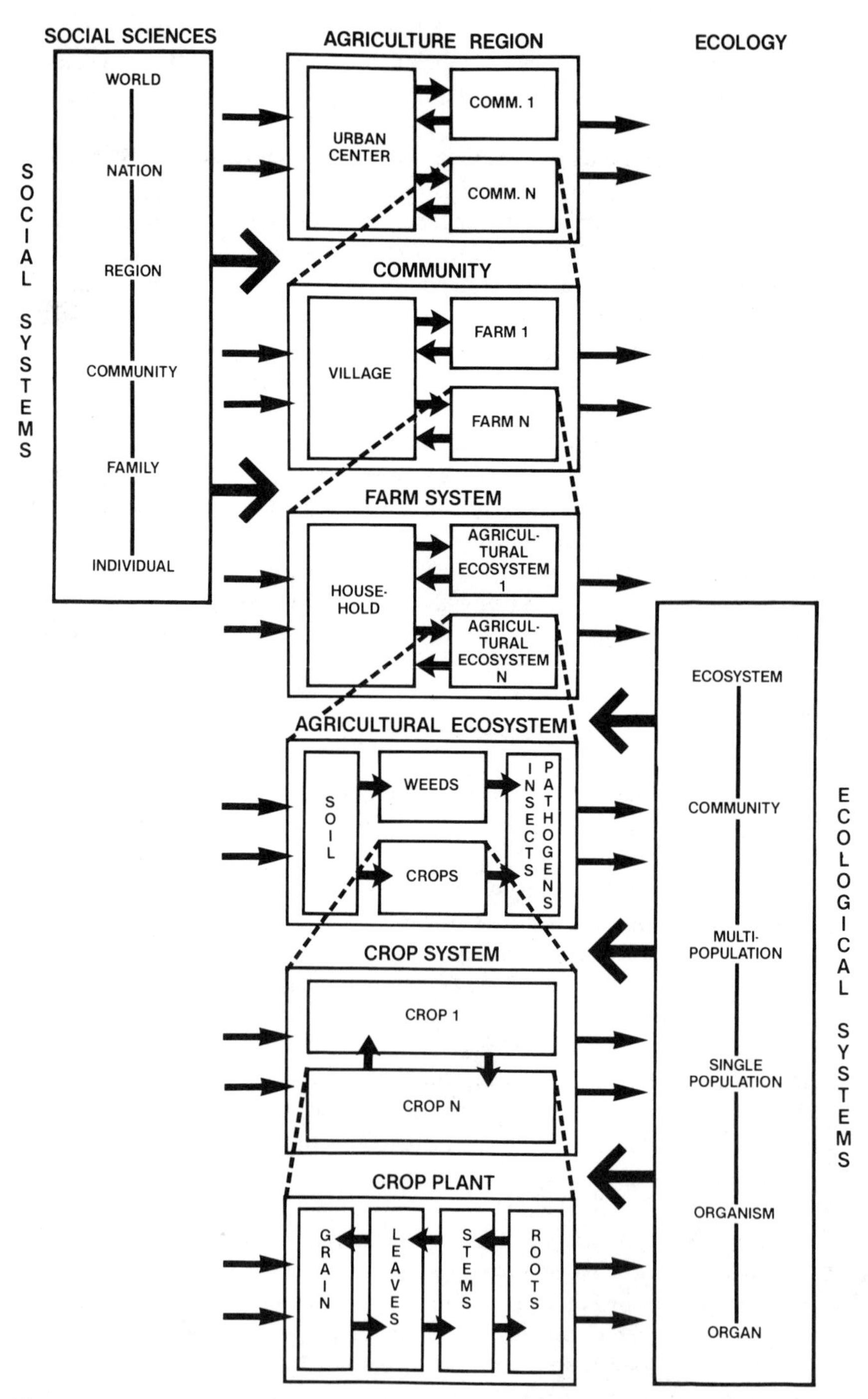

Figure 3.1 Conceptual framework for agricultural research and development that combines concepts from ecology and the social sciences to form a hierarchy of agricultural systems.

requires both social science and ecological concepts. The methods applied in the analysis of agricultural ecosystems and their crop and livestock subsystems must be based on the application of ecological concepts. For purposes of this discussion multiple cropping research is defined as a set of activities directed at designing or improving crop systems with more than one crop population. The agriculture ecosystem—crop system—crop plant hierarchy, depicted as the bottom three systems in Fig. 3.1, is the ecological framework for multiple cropping research.

Ecologists subdivide the physical and biological domain of ecology into a hierarchy of systems. An "ecosystem" is defined as a biological community and the physical environment with which it interacts; a "community" is defined as a set of interacting populations; and a population is defined as a set of interacting taxonomically related organisms. Ecologists group themselves as either population or community ecologists and often disagree as strongly as commodity oriented agronomists do with cropping system oriented agronomists.

Population, community, and ecosystem ecological concepts and basic principles are clearly relevant to multiple cropping research. The subsequent sections of this chapter summarize the general approach taken by ecologists in the study of populations, communities, and ecosystems. Each section briefly presents the approaches taken by agronomists in the study of agricultural populations, communities, and ecosystems, and then two examples are selected to illustrate these concepts and show where the development of ecological principles and research approaches are directly relevant to multiple cropping research.

POPULATION-LEVEL ECOLOGICAL PRINCIPLES

The plant population has always been the central subject of agronomic research. Crop breeders try to develop crop populations with "better" genetic potential, and production agronomists do research on how to "manage" a given population (when to plant, what density to plant) in order to get the most out of a given genetic potential. In general, crop ecologists emphasize the study of the effect of individual environmental factors (e.g., low night temperature) on the yield of one crop population (e.g., citrus). Recently, however, crop ecologists have made considerable progress in the development of computer simulation models that explain the behavior of one plant as it is affected by multiple interacting environmental factors. One-crop population models that include population-level variables (such as distance between plants) are also being developed.

Population-level ecological principles can be conveniently separated into principles relating to single-population dynamics and principles relating to multiple-population dynamics. A central interest of ecologists studying one population is the study of population growth (also of considerable interest to agronomists). Population cycles, fluctuations, and dispersal, while important interests of ecologists, are of less interest to agronomists (particularly agronomists working with annual crops) since seed selection and seed dispersal are managed by farmers in agricultural ecosystems.

Ecological research on the interaction between two populations emphasizes both the relationships between populations at the same trophic level (for example, competition between two plant populations) and relationships between populations at different trophic levels (for example, predator/prey or host/parasite relationships). Agricultural entomologists and plant pathologists have gone through a period when emphasis was placed on finding chemicals to kill predators and parasites and the relationship between their respective preys and hosts was less studied. However, the current interest in integrated pest management has brought ecologists and agricultural scientists closer together. The study of competition between two plant populations is, of course, one of the primary interests of agronomists doing research on multiple cropping systems. To illustrate the link between basic population-level ecological research and applied multiple cropping research, basic ecological principles related to population growth and competition between two populations are summarized below.

Population Growth

Population growth is conceptualized by ecologists as the net result of natality N, immigration I, mortality M, and emigration E. The intrinsic rate of natural increase r in a population is equal to $(N + I) - (M + E)$. A population P_1 changes over time t as described by the following equation:

$$\frac{dP_1}{dt} = rP_1$$

This simple equation implies that environmental factors do not affect or limit growth. If an environmental limit k is assumed and the change in P_1 approaches zero as this limit is reached, then the growth of a population is described by the equation:

$$\frac{dP_1}{dt} = rP_1\frac{k - P_1}{k}$$

This equation, and variations of it, are referred to by some authors as a logistic growth curve (H. T. Odum, 1983). The equation was first proposed by Verhultst in 1834 and later rediscovered by Pearl and Reed in 1920 (E. P. Odum, 1971). The logistic equation is only the starting point for much of the ecological research on single population dynamics. May (1976), in his review of models of single populations, emphasized that the equation should "not be taken seriously" and proceeded to demonstrate that time-delay factors that affect both r and k must be introduced before the models can be of applied value.

One-population growth models have been used to a limited extent by agricultural scientists. Both ecologists and agricultural entomologists commonly characterize insect populations as type r or type k depending on whether the population growth is regulated by intrinsic biological factors or by external resource availability, respectively. Crop physiologists have also developed crop growth models for single plants, but the emphasis has been placed on internal physiological processes (photosynthesis, respiration) and on anatomic characteristics (leaf area, stomata opening).

Competition between Populations

The different types of interactions between and among populations have been classified by many authors. E. P. Odum (1971) identified nine types of interactions including neutralism (no interaction) and commensalism (one population benefits; the other is not affected). He identified two types of competition, direct inhibition and indirect inhibition through competition for the same resource.

Beets (1982), in his book, *Multiple Cropping and Tropical Farming Systems,* reviewed both the ecological and agricultural literature on plant competition and identified four types of interrelationships. Various authors assign different names to the different relationships, but the criteria for classification is again the negative, neutral, or positive effects of the interrelation on the species involved.

It is impossible in this brief review to do justice to the large body of ecological research on competition between two populations. J. Maynard Smith's book *Models in Ecology* (1974) and Robert M. May's book *Theoretical Ecology* (1976) are only two examples of reviews of the ecological literature on this subject. But, as in the case of single population models, many of the ecological principles relating to two or more populations also evolved from the pioneering work of ecologists who did not have the benefit of computer technology.

Much of the ecological research on competition between populations began with the work of Lotka and Volterra done in the 1920s. The equations are similar to the logistic equation for one population that was described above except that factors related to the competing populations have been introduced. The equations they propose are described below using the forms and terminology used by E. P. Odum (1971):

$$\frac{dP_1}{dt} = r_1 P_1 \frac{k_1 - P_1 - aP_2}{k_1}$$

$$\frac{dP_2}{dt} = r_2 P_2 \frac{k_2 - P_2 - bP_1}{k_2}$$

where P_1 and P_2 are two populations, a and b are the inhibitory effects of the competing population, and k_1 and k_2 are environmentally defined limits. A great deal of ecological research has been done on the stability of Lotka-Volterra models. If intraspecific competition is stronger than interspecific competition and if the resource requirements of the two populations are not identical, there is an equilibrium point where both species can coexist. The concept of species niche and the principle of competitive exclusion (two species cannot occupy the same niche) were developed as ecologists postulated different variations of these equations.

Hart (1974), Trenbath (1976), Beets (1982), and many others have discussed the application of basic two-population ecological concepts to multiple cropping research. Agronomic indices that show higher total yield when two crops are planted together than when planted separately, such as "relative yield totals" and "land equivalent ratios," can be explained in terms of different niche requirements and lower interspecific than intraspecific competition.

Agronomic research on two-crop systems primarily emphasizes the study of mixtures of forage crops and of intercropped annual crops. As might be expected, the dynamics of forage crop mixtures is much more easily explained by ecological models since many of these systems are similar to natural nonagricultural systems in that they are often self-propagated and persist long enough so that populations can oscillate around a steady-state condition. Intercropped annuals are qualitatively different from natural vegetation in that farmers regulate the timing, densities, and distances between plants. Also, the architecture of many annual crops can be extremely variable. All of these factors must be added to two-population ecological models before they can begin to explain the dynamics of an annual crop multiple cropping system.

Figure 3.2 summarizes the concepts shared by ecologists and agricultural scientists doing population-level research. In the figure, the same ecological concept is depicted as being applicable to both natural and agricultural populations.

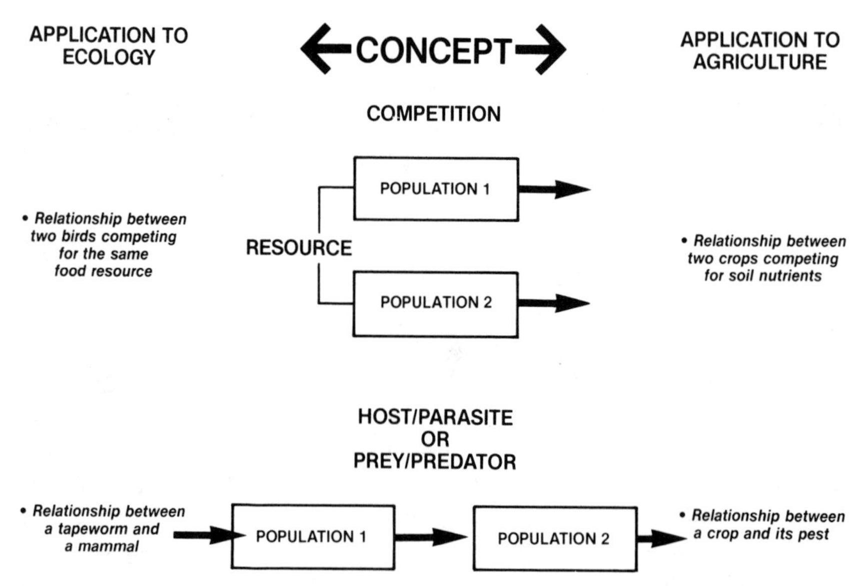

Figure 3.2 Examples of the application of two basic population level concepts to agricultural and ecological phenomena.

COMMUNITY-LEVEL ECOLOGICAL PRINCIPLES

Agronomists have not traditionally attempted to analyze the biological community formed by interacting crop, weed, insect, and microorganism populations. However, as agronomic research has been directed toward multispecies (two or more populations) cropping systems, and as integrated pest management schemes have evolved that explicitly consider the manipulation of the structure of a plant community as a means of regulating pests, it has become increasingly obvious that there is a need for more community-level agricultural research.

A biological community is a set of interacting plant and animal populations. As a system, a community has both structural and functional characteristics and, to do community analysis, a priority of ecologists has been to develop indices to quantitatively describe community structure and community function.

Ecological analysis of biological communities has the objective of formally relating community structure to community function, but general ecological principles are only just beginning to emerge. Community-level models can be developed either by linking together population models (as was done for the two-population models) or by combining populations with similar structure (such as grasses, shrubs, or trees) or similar function (e.g., autotrophs or heterotrophs). Structural indices can then be related to functional indices.

A community-level process that has been the subject of considerable ecological research is natural succession. Secondary succession is the orderly process of change in community structure and function that occurs when a climax or steady-state community is suddenly removed from a site and the community must progress through a series of stages until it returns to a similar state.

The relationship between community diversity and community productivity, and the understanding of the forces that regulate natural succession are two areas of ecological research with implications for agronomists conducting multiple cropping research. These two types of ecological research are discussed below to illustrate how community-level ecological principles could be applicable to multiple cropping research.

Diversity/Productivity Relationships

One of the most overstated simplifications of the popular ecology press is that diversity is directly related to productivity. The relationship between diversity and productivity is far from simple, and the value of diversity as a rationale for adding more crop species to a cropping system cannot be easily quantified. The first issue is how to measure diversity and productivity. As stated above, this has been a key area of community-level ecological research.

Ecologists distinguish between community "richness" (number of species per unit area) and the "evenness" or distribution of these populations. There have been a number of indices proposed that combine measurements of richness and evenness (see E. P. Odum, 1971). There are strong differences of opinion over the value of the different indices. Perhaps the most commonly used measurement of diversity is the Shannon index:

$$\overline{H} = -\Sigma \frac{N_i}{N} \log \frac{N_i}{N} \qquad \text{or} \qquad -\Sigma\, P_i \log P_i$$

where N_i is the importance value (e.g., number of individuals, biomass) of each species, N is the total of the importance values, and P_i is the importance probability for each species (N_i/N). One reason for the interest in this index is its relationship to information theory, although May (1976) describes the link as "by an ectoplasmic thread." In simplistic terms the objective of the index is to compare real population "connectedness" (relationships between populations) with potential connectedness. Margalef (1968) sees diversity as a measurement of linkages among populations and relates higher diversity with longer food chains, more parasitism, and symbiosis.

Community-level productivity is usually measured by ecologists in terms of primary (plants only) and secondary (animals and microorganisms) gross and net productivity. Gross primary productivity is the total rate of photosynthesis, while net community productivity is the rate of storage of plant tissue. What is subtracted to go from "gross" to "net" is the plant organic matter that is respired by the plants and that which is consumed by animals and other organisms along the food chain. Thus the amount of plant biomass that can be harvested from a given area is a measurement of net community production. Since productivity is measured in units per area per unit time, the measurement of standing biomass must be done for at least two points in time. In most cases such harvest methods have been used to measure only aboveground production. Mitchell (1984), in his comparison of the ecological productivity of agricultural and natural communities, used aboveground net production only, since very few estimates of the total net production from agricultural communities have included belowground productivity.

The relationship between diversity and productivity has been analyzed by ecologists from two perspectives. One approach has been to compare the diversity and productivity of natural communities in different environments; the other approach has been to do computer simulation of community-level models. There is little doubt that both diversity and productivity are, in general, higher in the humid tropics than in the wet/dry tropics, and that temperate-climate communities usually have lower diversity and lower productivity than tropical-climate communities. But, on closer inspection, these generalizations become almost meaningless.

In both natural and agricultural plant communities the highest production is often obtained from low-diversity communities. Mitchell (1984) points out that these high productivity areas (e.g., estuaries, ocean upwellings, irrigated rice, and intensive sugarcane production) are all characterized by high nutrient subsidy. It would be a gross simplification to conclude that multiple cropping systems are appropriate only on poor soils, but many cropping systems research projects have found that to improve a farmer's high diversity multiple cropping system, it is necessary to increase inputs (e.g., fertilizer) and decrease the diversity of the cropping system (e.g., plant in monoculture strips, rather than seemingly random intercropping). It is important not to confuse the relationship between diversity, nutrient input, and productivity with the relationship between diversity and productivity.

An important question for agronomists doing research on multiple cropping systems is how to separate the indirect effect of community diversity on predator/prey and host/parasite relationships from the more direct effect of complementary resource use. Two populations of maize and beans arranged in different spatial patterns have different diversity indices. The different spatial arrangements of crops "communicate" different amounts of information to their would-be predators and parasites. It may be possible to relate the information content of these patterns to yield loss due to pests.

Natural Succession

When a climax community is removed from a site and the site is allowed to return to its original state, the community develops through an orderly series of stages with distinct changes in the community structure and function. E. P. Odum (1971) has summarized the changes that occur during this process. One of the changes that could be particularly relevant to research on multiple cropping systems is that as the community matures, diversity increases while net community production decreases.

During the early successional stage, nutrient availability is usually high and a low-diversity plant community composed of fast-growing "sun-loving" plant populations takes root. This first stage is relatively inefficient in terms of nutrient conservation and it has a low level of population interaction. As succession progresses, colonizing shade-tolerant populations begin to occupy different niches, "connectedness" among populations increases, and the overall organization of the community increases. While these structural changes occur, gross productivity is increasing, but net community productivity decreases as increasing amounts of photosynthate are used to maintain existing structure. From the point of view of a herbivore, the yield per unit area per unit time has decreased.

The possibility of designing successional multiple cropping systems has been discussed by Hart (1974, 1980a, 1980b) and others. In general, agricultural communities must be maintained at early successional stages to have high net productivity. In fact, agriculture is based on the exploitation of early successional plants. It is obvious that for a given area of tropical forest, fewer people can be supported by a hunting and gathering system than can be supported by the introduction of a slash-and-burn cropping system. It is equally obvious that even more people can be supported if a nutrient subsidy is introduced and early-successional crops are planted sequentially. The primary arguments for the use of successional multiple cropping systems are based on the desirability of extending the cropping period by replacing fallow vegetation with similar economic plants while continuing to maintain the natural resource base, not on the argument that a successional cropping system can be more productive than a nutrient-subsidized sequential cropping system. Thus, short-term high productivity that can be gained by sequentially planting the same crops, resulting in a deleterious effect on the resource base, is given up in exchange for long-term or lower productivity that does not reduce the natural resource base.

ECOSYSTEM-LEVEL ECOLOGICAL PRINCIPLES

An ecosystem has both biological and physical components. The biological components (populations) can be grouped together to form a community (a subsystem of an ecosystem) that interacts with the physical environment. In addition to the interactions among organisms that occur at the population level, and the interactions among populations that occur at the community level, there are also interactions between the biological components and physical components such as soil nutrients, soil water, and solar radiation.

Ecosystem-level research has been done on systems as small as an aquarium and as large as a watershed. However, the emphasis of ecological research has been on units such as a pond, a meadow, or a forest. These units are similar in structure and function to a field of wheat, or a field of intercropped maize and beans, or a multistory household garden that might be found on a farm. In the discussion that follows, any reference to agricultural ecosystems will be directed at these types of farm subsystems. There has been very little ecosystem-level research with agricultural ecosystems.

Ecosystem-level processes, because of the very size of ecosystems, are difficult to study. Since their components include phenomena as different as soil microbes and large mammals, the study of ecosystems requires a multidisciplinary team. All of the diverse types of data must be integrated so that the function of the ecosystem as a unit can be analyzed. The two ecosystem-level processes that have been most studied in an attempt to explain ecosystem structure and function have been energy flows and nutrient cycles. General ecological principles relating to these processes and their implications for multiple cropping research are discussed below.

Energy Flow

The flow of energy through an ecosystem is a process that begins with the capture of diffuse solar radiation (photosynthesis) to build highly structured organic molecules which, when oxidized (respiration), release metabolizable energy (and heat) that can be used to build other organic molecules that can also be broken down to release energy (and heat) until all of the energy embodied in the organic structure has been converted into heat. Sources of energy other than solar radiation, such as wind and lightning, provide minor energy inputs, but can be very important to specific ecosystem subprocesses. The flows of energy through the different trophic levels of ecosystems have been quantified for many different ecosystems (see H. T. Odum, 1983).

The flow of energy through an ecosystem is not a passive process. The behavior of ecosystems is in many ways similar to that of organisms. They have evolved different structures and subprocesses to maximize their long-term survival. Energy is channeled into different compartments and stored to be called on later. Organisms on the top of food chains are inefficient to maintain and often play a key regulatory role in the maintenance of ecosystem homeostatis.

The application of energy flow concepts to agricultural ecosystems has been done primarily to compare the efficiency of different agricultural ecosystems. Usually, traditional are compared to modern agricultural systems (Pimentel, 1984). These studies usually do not include energy inputs from solar radiation; therefore, the studies usually show more energy output than input. Human energy, fossil fuel inputs, and the output or organic matter are converted into energy units and the total energy inputs and outputs are compared. Most of the energy analyses of agricultural ecosystems do not systematically trace the flow of energy through the physical and ecological components of an agricultural ecosystem.

In agricultural ecosystems where crop residue is not recycled and organic fertilizer is not an important input, the detritus pathways, which are often a key energy flow in natural ecosystems, are probably of little importance. However, the present interest in minimum tillage and use of organic fertilizer as an integrated part of multiple cropping systems will, undoubtedly, cause agronomists to give more consideration to energy flow analyses, including the flow of energy through the detritus pathway.

Energy flows in both natural and agricultural ecosystems are regulated by information processes. Information processes in agricultural ecosystems are significantly different than those found in natural ecosystems (see Hart, 1984). Many of the regulatory organisms in natural ecosystems are defined as pests in agricultural ecosystems and farmers try to minimize the flow of energy to those populations.

Nutrient Cycles

While energy flow is a one-way process, nutrients can cycle and recycle. The tracking of a nutrient element as it moves from a reservoir pool has produced important insights on how ecosystems function. These biochemical cycles are of two types, those that draw on elements from the atmosphere, and those that draw on elements from the soil (E. P. Odum, 1971).

The nitrogen cycle is an example of a nutrient cycle that draws upon elements from a reservoir in the atmosphere. The largest reservoir of nitrogen is in the atmosphere, but all along the cycle as the nitrogen moves from plants and animals to microorganisms and back to plants, there are pools of nitrogen in ammonia, nitrite, and nitrate forms. The bacteria that move the nitrogen from one pool to another (for example, from nitrite to nitrate) require an energy source, and energy flows and nutrient cycle pathways are continually intertwined.

The phosphorus cycle is an example of a nutrient cycle that draws upon a reservoir that is in the earth's crust. The movement of phosphorus from phosphate rock to dissolved phosphates to be taken up by plants or washed down rivers to the sea to be deposited as phosphate rocks is, of course, extremely slow relative to agricultural timetables.

A key issue in multiple cropping research is the study of competition between crops for soil nutrients. The fact that the availability of these nutrients depends on soil biological processes that are regulated by both the physical environment and the availability of organic inputs is often not given sufficient attention. Ecological research on these processes is of direct relevance to agronomy.

Much of "modern" agriculture is based on the design of agricultural ecosystems that subsidize the nitrogen and phosphorus cycles. Electrical energy is used to produce nitrogen fertilizer that is injected into various pools along the nitrogen cycle. Phosphate rock is mined and applied as fertilizer in order to release phosphates more quickly. As the prices of these inputs increase, research programs have placed more emphasis on efficiency in converting inputs to outputs. One of the most difficult, but potentially most promising areas of multiple cropping research is related to the design of new agricultural ecosystems. A better understanding of ecosystem-level nutrient cycles is surely a good place to start.

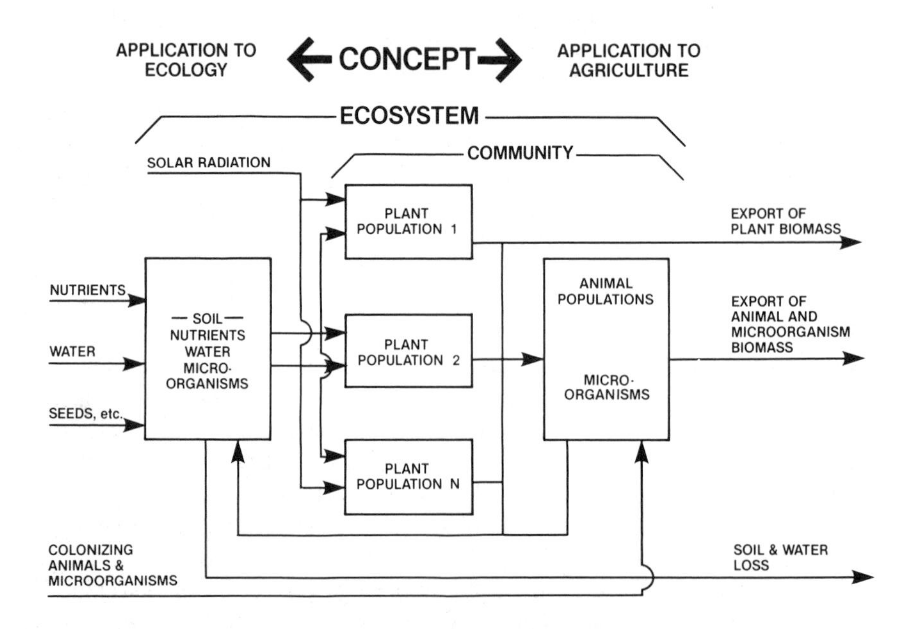

Figure 3.3 Examples of the application of community and ecosystem concepts to agricultural and ecological phenomena.

Figure 3.3 summarizes the concepts shared by ecologists and agricultural scientists doing commodity and ecosystem-level research. As in Fig. 3.2, the diagram in the center of the figure is applicable to both natural and agricultural ecosystems.

USING THE HIERARCHICAL SYSTEMS' CONCEPTUAL FRAMEWORK

Agricultural research during the past decades has primarily been directed at increasing crop production by (1) increasing nutrient availability by adding fertilizer at one point in the ecosystem nutrient cycle, (2) breeding and selecting crop varieties that can take advantage of these changes in the ecosystem nutrient cycles, (3) designing cropping patterns that can take advantage of the new varieties, and (4) identifying chemicals that can control the weeds, insects, and microorganisms which, through self selection, also try to take advantage of the changes that occur.

Multiple cropping research could take an approach that is quite different from the approach described above. It should be possible to systematically apply what is known about ecosystems, communities, and populations to the design of hypothetical agricultural ecosystems that are consistent with farmers' objectives. Research can then be directed toward the development of the technology needed to put together the multiple cropping systems that fit into the agricultural ecosystems that in turn fit into farm systems and larger regional agricultural systems.

While it is possible to conduct multiple cropping research as part of a basic research program, most multiple cropping research is done with development as well as research objectives. If farm, community, and regional development is a research objective, the conceptual framework for multiple cropping research must be extended above the ecosystem level to include farms and macrosocioeconomic community and regional systems. Just as this chapter has described basic ecological concepts at the population, community, and ecosystem levels and suggested that these concepts are applicable to the study of agricultural systems, basic socioeconomic concepts could be described and shown to be applicable to the study of these larger agricultural systems.

It was suggested at the beginning of this chapter that the hierarchy of agricultural systems depicted in Fig. 3.1 could function as a conceptual framework for multiple cropping research. If it is accepted that the ecological concepts developed by ecologists and the socioeconomic concepts developed by social scientists can be integrated, then the next important question is, how can these concepts be used to do multiple cropping research? One possible answer is to apply the concepts separately and continue to do either good population, community, ecosystem, farm, and regional-level research and hope that integration occurs naturally. But this is extremely inefficient since any one level in the hierarchy is simultaneously another system's subsystem or suprasystem and a scientist focusing on one level (e.g., population-level studies) would benefit from the research occurring at the next level up and the next level down. What is needed is a strategy to integrate the research conducted at different levels.

Figure 3.4 is a simplified development-oriented research strategy. It was assumed that most development-oriented multiple cropping systems research programs have the basic objective of combining what farmers have learned through centuries of experience with the knowledge that has been gained through agronomic, ecological, and social science research. Traditionally, agronomic research has focused on one-crop populations and the physiological and genetic processes occurring within one population. Recently, considerable knowledge on the principles of multiple cropping systems has become available. Ecological

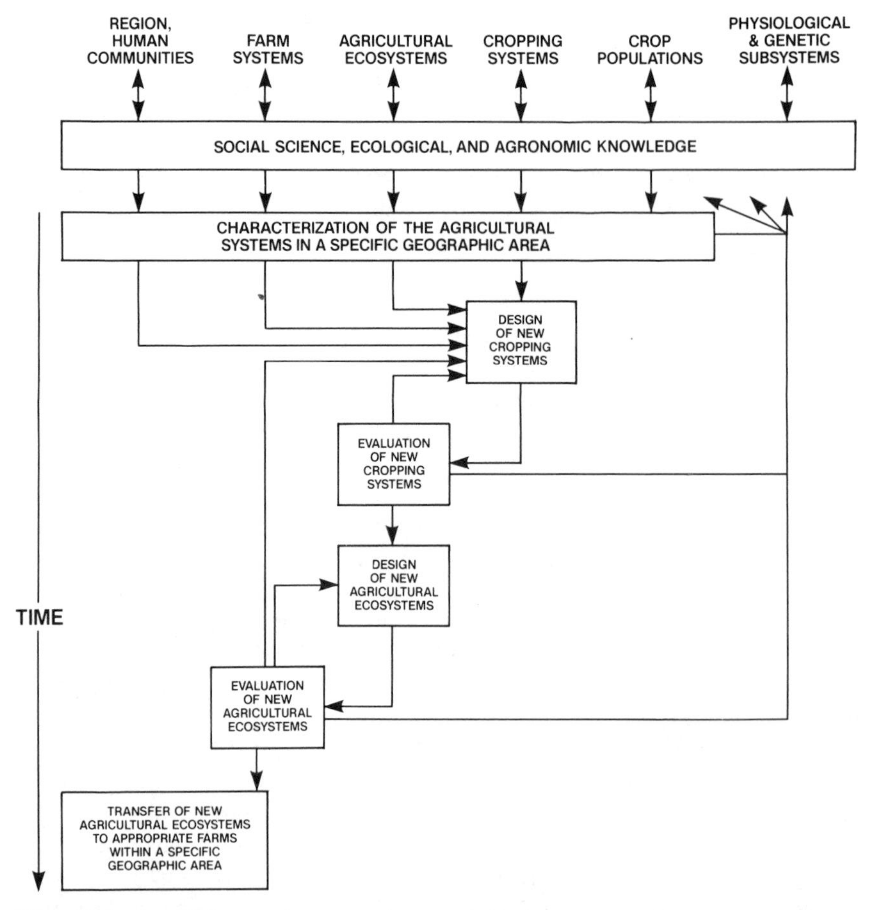

Figure 3.4 A multiple cropping research strategy that integrates social sciences, ecology, and agronomy concepts, and combines this knowledge with farmer experience to design and evaluate new cropping systems and agricultural ecosystems.

research has focused on population, community and ecosystem phenomena. Social sciences have developed a considerable body of knowledge relating to macrosocioeconomic systems.

The strategy described in Fig. 3.4 begins by combining what is known about the farmers' systems with basic agronomic, ecological and social science knowledge to design alternative cropping systems. These cropping systems are evaluated by how well the agricultural ecosystem performs with the new cropping system inserted. Knowledge from this evaluation is used for a redesign (if necessary), feeding back basic information to the "knowledge pool," and designing alternative agricultural ecosystems that can take advantage of the characteristics of the alternative cropping system.

The alternative agricultural ecosystem is evaluated on the basis of the performance of a farm system with the new ecosystem inserted. This evaluation also feeds back information for redesign and adds to the stock of basic knowledge. If the alternative agricultural ecosystem meets the objective of farmers, the alternative technology can then be transferred to other farmers with similar farming systems in the region.

The simplified research strategy described above is, in reality, an extremely complex process requiring an integration of social scientists, ecologists, and agronomists. The biological research requires an integration of ecological and agronomic disciplines. However, the exchange of information between ecology and agronomy should not be a one-way flow in which agronomists simply apply ecological concepts. The design of agricultural ecosystems will understandably raise many questions and provide theoretical insights that will also be applicable to natural ecosystems. One of the most important contributions of multiple cropping research has been to bring ecologists and agronomists closer together. Future research on multiple cropping systems will definitely continue to benefit from this interaction.

REFERENCES

Beets, W. C. 1982. *Multiple Cropping and Tropical Farming Systems,* Westview Press, Boulder, Colorado, 156 pp.

Hart, R. D. 1974. The design and evaluation of a bean, corn, and manioc polyculture cropping system for the humid tropics, Ph.D dissertation, University of Florida, Gainesville, Florida, University Microfilms order no. 75-19, 341, 175 pp.

———. 1980a. *Agroecosistemas,* Centro Agronómico Tropical de Investigación y Enseñanza (CATIE), Turrialba, Costa Rica, 211 pp.

———. 1980b. A natural ecosystem analog approach to the design of successional crop system for tropical forest environments, *Biotropica (12) Trop. Succession,* pp. 73–82.

———. 1982. An ecological systems conceptual framework for agricultural research and development, in: *Readings in Farming Systems Research and Development,* (W. W. Shaner, P. F. Philipp, and W. R. Schmehl, eds.), Westview Press, Boulder, Colorado, pp. 44–58.

———. 1984. Agroecosystem determinants, in: *Agricultural Ecosystems: Unifying Concepts,* (R. Lowrance, B. R. Stinner, and G. J. House, eds.), John Wiley & Sons, New York, pp. 105–119.

Hart, R. D. and Pinchinat, A. M. 1982. Integrative agricultural systems research, in: *Caribbean Seminar on Farming Systems Research Methodology,* IICA, San Jose, Costa Rica, pp. 555–565.

Margalef, R. 1968. *Perspectives in Ecological Theory,* University of Chicago Press, Chicago, Illinois, 111 pp.

May, R. M. 1976. *Theoretical Ecology: Principles and Applications,* W. B. Saunders Company, Philadelphia, Pennsylvania, 317 pp.

Mitchell, R. 1984. The ecological basis for comparative primary production, in: *Agricultural Ecosystems: Unifying Concepts,* (R. Lowrance, B. R. Stinner, and G. J. House, eds.), John Wiley & Sons, New York, pp. 13–53.

Odum, E. P. 1971. *Fundamentals of Ecology,* 3rd ed., W. B. Saunders Company, Philadelphia, Pennsylvania, 574 pp.

Odum, H. T. 1983. *Systems Ecology: An Introduction,* John Wiley & Sons, New York, 644 pp.

Pimentel, D. 1984. Energy flow in agroecosystems, in: *Agricultural Ecosystems: Unifying Concepts.* (R. Lowrance, B. R. Stinner, and G. J. House, eds.), John Wiley & Sons, New York, pp. 121–132.

Smith, J. M. 1974. *Models in Ecology,* Cambridge University Press, Cambridge, U.K., 146 pp.

Trenbath, B. R. 1976. Plant interactions in mixed crop communities, in: *Multiple Cropping,* ASA Special Publ. 27. Amer. Soc. Agron., Madison, Wisconsin, pp. 129–169.

Chapter 4

Resource Use by Intercrops

Brian R. Trenbath

During growth and development, crop plants must intercept and absorb growth factors—light energy, water, and nutrients—and use them in processes that produce biomass. Some part of this biomass is the harvestable yield. Since crop growth depends on the use of light, water, and nutrients, these factors are vital resources for most agricultural activity.

Although growth factors are distributed variously in space and time, plant species are able to intercept and absorb them with parts of their shoot and root systems adapted specifically for this. Provided that certain conditions are fulfilled (e.g., soil water content is favorable for nutrient uptake), the growth factors are available to the plants of a crop and they are absorbed. Provided that further conditions are fulfilled (relating mainly to temperature and the balance of supplies of factors), the absorbed factors are utilized in growth, and potentially harvestable matter accumulates.

One of the principal contributions of crop physiologists to agronomy has been to provide information on how different crop species and varieties differ in their capacity to benefit from a given pattern of resource availability. In ecological terms, the morphological and physiological differences among species result in their ability to occupy different niches. Observed differences include the timing of resource interception (Fig. 4.1a), its location (Fig. 4.1b), rate of absorption (Fig. 4.1c), rate of growth (Fig. 4.1d), and response of yield to variation in the level of resource availability (Fig. 4.1e).

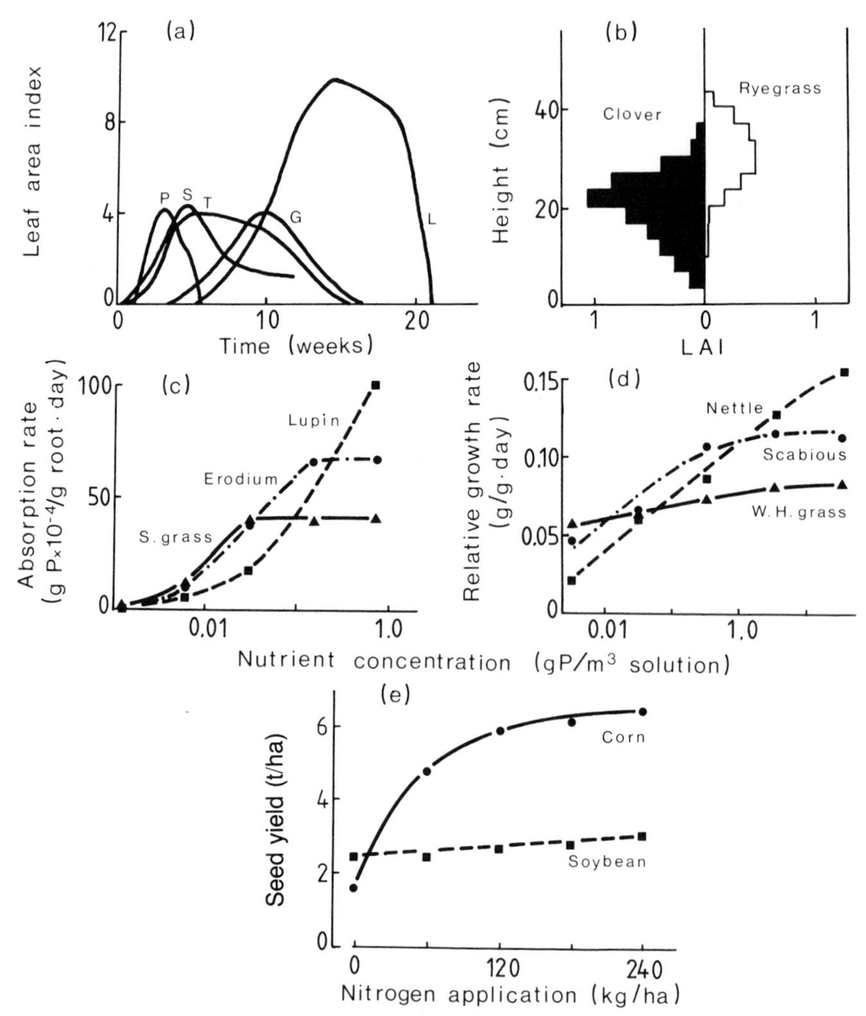

Figure 4.1 Morphological and physiological differences among crop and pasture species that allow them to occupy different ecological niches. (a) Patterns of leaf area index (LAI) duration in five root crops (species of *Solanum, Ipomoea, Xanthosoma,* and *Dioscorea*). Crops are: P, potato; S, sweet potato; T, tamia; G, greater yam; L, lesser yam. (Redrawn from Wilson, 1977.) (b) Canopy heights in two pasture species, subterranean clover and Wimmera ryegrass. (Redrawn from Stern and Donald, 1962, with permission from CSIRO Publications.) (c) Specific absorption rates of phosphorus by roots of three pasture species (fresh-weight basis). S. grass is silver grass; Lupin is West Australian blue lupin. (Data of Loneragan and Asher, 1967.) (d) Relative growth rates (whole-plant, dry-weight basis) of three plant species found in pastures. Scabious is small scabious; W.H. grass is wavy hair grass. (Data of Rorison, 1968.) (e) Responses of a leguminous and a nonleguminous crop to nitrogen fertilizer. (Data of IRRI, 1975.) The abscissae of (c) and (d) are on logarithmic scales.

In view of the differences among crop types, the varied distributions of growth factors in space and time suggest the hypothesis that in many agricultural environments those factors could be more completely absorbed and converted to biomass by a mixed stand of crops than by a pure stand. Indeed, the success of many traditional cropping patterns seems to be based on this principle (Aiyer, 1949; Baldy, 1963).

Recently, more detailed knowledge of crop types and environments has allowed experiments in designing new intercropping systems to be done on a more rational basis. To design experiments that will produce a maximum amount of information on the way resources are used in mixed stands, I first consider the components of resource utilization efficiency (RUE), and then propose a framework in terms of RUE for analyzing productivity of crop mixtures (intercrops) and of associated plots of their plant components grown in pure stand (sole crops). A derived value, the land equivalent ratio (LER), can then be calculated and resolved into four additive numerical components based on absorption and conversion rates of the most limiting resource. I will present an example where these numerical components of LER have been estimated. Finally, to show the directions that resource use research has taken with respect to the resource most intensively studied, namely light, a number of ways are outlined for estimating absorption (or a closely related proxy, interception) and conversion rates of light in intercrops. These results are used to suggest new designs for improved performance of intercrops.

RESOURCE UTILIZATION EFFICIENCY (RUE)

Although the efficiency with which a crop uses any individual resource depends greatly on the level of supply of the others, the utilization efficiency of any single factor can, given appropriate data, be calculated as the product of two other efficiency measures: the efficiencies of capture by crop plants and the efficiency of conversion into yield. In many cases, each of these two measures can be further resolved: capture efficiency is the product of interception and absorption; conversion efficiency is the product of resource conversion into whole-plant biomass and harvest index. Symbols for these relationships can be summarized:

$$\text{RUE} = \text{capture efficiency} \times \text{conversion efficiency} \tag{4.1}$$

with

$$\text{Capture efficiency} = \frac{R_i}{R_o} \cdot \frac{R_a}{R_i} \tag{4.2}$$

and

$$\text{Conversion efficiency} = \frac{B}{R_a} \cdot H \tag{4.3}$$

where R_o, R_i, and R_a are per-land area measures of the quantities of resource potentially available, intercepted, and absorbed in the growing season, respectively; B is whole-plant biomass; and H is the whole-plant harvest index, or proportion of the whole-plant biomass present at harvest that provides economic yield. In this chapter, the focus is on annual crops; the component ratios of RUE are denoted by symbols.

Where one resource is in such short supply that it limits growth and yield, it is the utilization efficiency of this resource that can be most usefully analyzed. Although the analytic treatment proposed below can accommodate the effects of minor shortages of other factors, this chapter mainly considers intercrops with a single, major, limiting factor. It is recognized that in the real world of the small farm this is seldom the case. However, the methodology is useful to begin to understand how complex systems function.

In studies of sole-crop performance, the usefulness of isolating conversion efficiency or B/R_a from RUE rests largely on the observation that, for a single crop type, production of biomass and/or some part of it is often strictly propor-

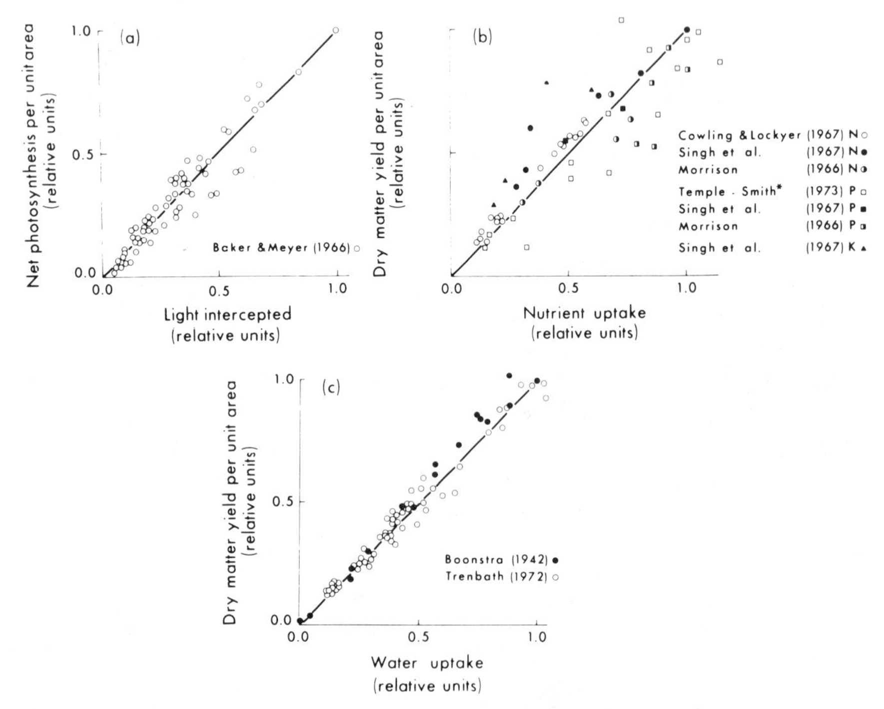

Figure 4.2 Responses of crops to quantities of limiting growth factor absorbed. (a) Crop net photosynthesis and light in cotton. (b) Dry-matter yield and nutrients in various species. (c) Dry-matter yield and water in potatoes (filled symbols) and wheat (open symbols). Each set of data has been scaled so that the observation with the highest y value is plotted as (1,1). In two cases where this point lay well off the linear regression line, the observation with the highest x value was taken instead. In (b), observations suggesting nutrient toxicity effects have been excluded. (*Source*: Trenbath, 1976. Reproduced by permission of the American Society of Agronomy.)

tional to the quantity of a limiting growth factor absorbed. This has been shown clearly for crops where the principal limiting factor is either light (Monteith, 1972; Biscoe and Gallagher, 1977) (Fig. 4.2a) or water (de Wit, 1958) (Fig. 4.2c). Data for the absorption of limiting nutrients (Fig. 4.2b) suggest that for final biomass the same proportionality is true for nitrogen but not for phosphorus and potassium.

On a graph of the type shown in Fig. 4.2, a value on the vertical axis is given by a corresponding value on the horizontal axis multiplied by the slope of the line. Since the slope of the line has units of growth or yield per unit of resource absorbed, it must equal either B/R_a or the conversion efficiency $(B/R_a)H$, depending on what is plotted.

The data in Fig. 4.2 and the literature cited suggest that the efficiency with which absorbed limiting light or water is converted into biomass or yield by a given crop type can be expected to stay essentially constant over a wide range of levels of absorption of the resource. Where the resource is a major nutrient, this conversion efficiency may remain constant, as in the case of limiting nitrogen, or may vary as with phosphorus or potassium.

ANALYSIS OF RUE IN INTERCROPS

Until recently, interpretations of intercrop advantage seldom went beyond an assumption that it was due in some way to more efficient use of resources (e.g., Osiru and Kibira, 1981). Studies are presently underway that measure the absorption of individual resources. These studies will allow the advantages or disadvantages of an intercrop to be traced to differences between intercrops and sole crops in overall capture efficiency and/or conversion efficiency of specified resources. For example, the advantage of a sorghum/pigeon pea intercrop over comparable sole crops at ICRISAT was traced partly to a higher conversion efficiency of light early in growth (Fig. 4.3a); the same intercrop also benefited from capturing more water (Natarajan and Willey, 1980). Assuming that the amount of water potentially available (R_o) is the same in all plots, level of capture (i.e., of absorption), R_a, is a measure of capture efficiency R_a/R_o.

To show how information on RUE variables can be fitted into an overall analysis of how the intercrop uses a limiting resource, the approach to sole crops described in the previous section is applied to intercrops. Assuming that conversion efficiency for each plant component is the same in an intercrop as in a sole crop, measurement of resource absorption and yield for each component in the two kinds of plant stand might well produce a yield/resource absorption graph resembling Fig. 4.3b. Here, the absorptions in the sole crops A_{ii} and A_{jj} are drawn to differ somewhat; the conversion efficiencies, the slopes E_{ii} and E_{jj} of the lines, also differ. The resource absorptions in the intercrop A_{ij} and A_{ji} are drawn to be, respectively, 40 and 60 percent of the species' absorptions in sole crop. Denoting the per-area yields of i and j in sole crops as y_{ii} and y_{jj} and the corresponding yields in intercrop as y_{ij} and y_{ji}, the land equivalent ratio is given by

$$\text{LER} = \frac{y_{ij}}{y_{ii}} + \frac{y_{ji}}{y_{jj}} \tag{4.4}$$

Substituting into this the products of the relevant absorptions and conversion efficiencies (the latter here are assumed to be the same in intercrop and sole crops) gives

$$\begin{aligned} \text{LER} &= \frac{A_{ij}\,E_{ii}}{A_{ii}\,E_{ii}} + \frac{A_{ji}\,E_{jj}}{A_{jj}\,E_{jj}} \\ &= \frac{A_{ij}}{A_{ii}} + \frac{A_{ji}}{A_{jj}} \\ &= 0.4 + 0.6 \\ &= 1 \end{aligned} \tag{4.5}$$

The origin of this result is made clearer if the transformation $a_i = (A_{ij}/A_{ii}) - 1$ is introduced so that

$$\begin{aligned} \frac{A_{ij}}{A_{ii}} + \frac{A_{ji}}{A_{jj}} &= (1 + a_i) + (1 + a_j) \\ &= 1 + (1 + a_i + a_j) \\ &= 1 + (1 - 0.6 - 0.4) \end{aligned}$$

This shows the LER value to be the sum of two terms, unity and a deviation term representing the relationships among the absorption values. The 60 and 40 percent reductions of absorption and of yield below sole crop values cause the deviation term to be zero.

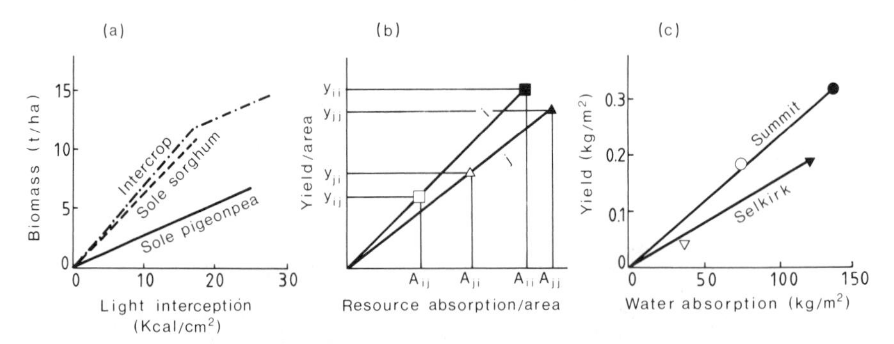

Figure 4.3 Responses in intercrop and sole crop of plants to quantities of limiting growth factor intercepted or absorbed. (a) Sorghum and pigeon pea biomass responding to cumulative interception of light energy. (Redrawn from Natarajan and Willey, *J. Agric. Sci.*, 1980.) (b) Theoretical situation for crop types *i* and *j* in intercrop (open symbols) and sole crops (filled symbols) (see text). (c) Wheat varieties (cultivars Summit and Selkirk) responding to water absorption in intercrop (open symbols) and sole crops (filled symbols). (B. R. Trenbath, unpublished data.)

We next relax the assumption that the conversion efficiencies in the intercrop E_{ij} and E_{ji} are the same as those in the sole crops, E_{ii} and E_{jj}. In this case,

$$\text{LER} = \frac{A_{ij}E_{ij}}{A_{ii}\,E_{ii}} + \frac{A_{ji}\,E_{ji}}{A_{jj}\,E_{jj}} \tag{4.6}$$

Introducing another transformation $e_i = (E_{ij}/E_{ii}) - 1$ gives

$$\begin{aligned} \text{LER} &= (1 + a_i)(1 + e_i) + (1 + a_j)(1 + e_j) \\ &= 1 + (1 + a_i + a_j) + (e_i + e_j) + (a_ie_i + a_je_j) \end{aligned} \tag{4.7}$$

The first two terms of the right-hand side are again unity and deviation from unity due to absorption, while the third term is a deviation term due to differences between conversion efficiencies in sole crop and intercrop, and the fourth term represents deviations due to interactions between absorption relationships and the differences in conversion efficiency. Although the fourth term may seem difficult to interpret, the second and third terms summarize the direct contributions of the main RUE variables to the success of the intercrop. The contribution of each plant component to each term is also apparent, so that a two-way analysis, according to RUE variable and plant component, is also possible.

As an alternative to the almost completely additive model for LER (Eq. 4.7), another approach could be based on the partially multiplicative model of Eq. 4.6. In fact Marshall and Willey (1983) have used this equation, adapted to consider per-plant rather than per-area values, in an analysis of a millet/groundnut mixture. They have emphasized the significance of the terms of the type A_{ij}/A_{ii} and E_{ij}/E_{ii} by naming them respectively resource capture ratio (RCR) and conversion efficiency ratio (CER), and by presenting their analysis as a two-way table. Their data show that the advantage of the intercrop was due to a 10 percent better interception of light by the millet and a 46 percent better biomass production efficiency by the groundnut than in the respective sole crops. The analysis does not reveal how much of the high LER value is due to the direct effects of two individual inflated RCR and CER values and how much is due to their interaction. By setting each in turn to unity, it can be shown that their interaction accounts for most of the deviation of LER from unity. With this in view, there seems to be an advantage in using the additive model of Eq. 4.2 where the interaction term is available to be calculated.

To illustrate the use of the additive model of LER, I present data of a mixed intercrop of two wheat varieties grown together with comparable sole crops on stored soil moisture in a high-radiation environment. Competition for water is likely to have determined the course of the biological interaction in intercrop and so the data of yield (ear weight) and water absorption by the individual varieties are analyzed in terms of water utilization. The quantities of water taken up were estimated from silica contents of whole shoots (Hutton and Norrish, 1974) and calibration curves were obtained from a parallel experiment using pure stands of the same varieties grown in the same soil.

The illustrative data on yield and water absorption (Table 4.1a) allow a yield/absorption graph to be plotted (Fig. 4.3c). The intercrop point of cultivar Selkirk falls well below the line drawn through its sole-crop point and hence the presence of plants of cultivar Summit has markedly lowered its conversion efficiency. As seen in Table 4.1b, the LER of the intercrop is 0.8000 and calculation of its additive components shows that the part of the depression of LER due to the third term ($e_1 + e_2$) is about twice that due to the second term ($1 + a_1 + a_2$). In other words, the low LER can be attributed mainly to reduced conversion efficiency in the intercrop rather than lowered capture efficiency. The values of e_1 and e_2 show that only Selkirk was affected.

The size of the second term ($1 + a_1 + a_2$) is by no means negligible and indicates that without any reduction in conversion efficiency, the LER would still have been as low as 1 − 0.1361 = 0.8639. This is necessarily the same value that is derived if the LER calculation is based on values of absorption (capture) rather than yield.

If biomasses, in addition to resource absorption, are known, then the LER value can be resolved into still more components with terms measuring the direct effect of absorption, of conversion into biomass B/R_a, and of harvest index H. With respect to the two components of conversion efficiency, it appears that B/R_a tends to be the same in intercrop and sole crop whereas H can differ dramatically.

Table 4.1 Water Use in an Intercrop of Two Wheat Varieties

	Intercrop		Sole crop	
Variety	**Selkirk**	**Summit**	**Selkirk**	**Summit**
	(a) Resource-use efficiency variables			
Seed yield (g ears/m^2)	42	181	188	314
Water absorption (kg H_2O/m^2)	39	76	123	139
Conversion Efficiency (g ears/kg H_2O)	1.08	2.38	1.53	2.26

(b) Additive numerical components of LER based on water use[a]

						=	
Unity						=	1.0000
Absorption term ($1 + a_1 + a_2$)	1.0000	−	0.6829	−	0.4532	=	−0.1361
Conversion term ($e_1 + e_2$)		−	0.2952	+	0.0544	=	−0.2408
Interaction ($a_1e_1 + a_2a_2$)			0.2016	−	0.0247	=	0.1769
LER (sum of the 4 terms)						=	0.8000

[a]Subscripts 1 and 2 denote Selkirk and Summit cultivars, respectively.

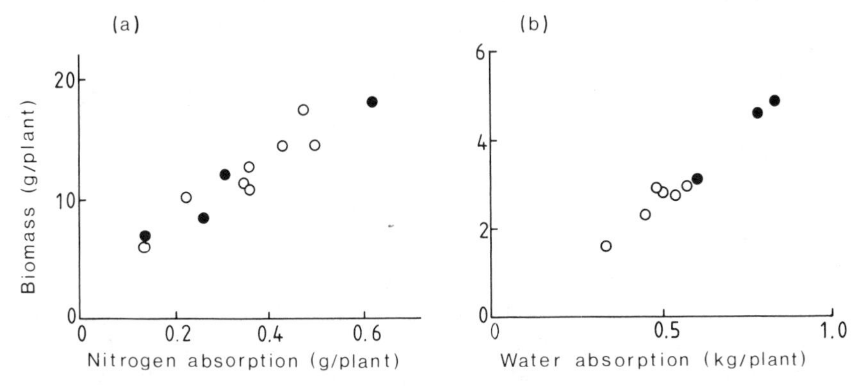

Figure 4.4 Responses to quantities of limiting growth factor absorbed in mixed intercrops (open circles) and sole crops (filled circles) of cereal varieties. (a) Rice, cultivar Taichung Native-1. (Redrawn from Kawano and Tanaka, *J. Fac. Agr. Hokkaido Univ., Sapporo,* 1967.) (b) Wheat, cultivar Selkirk (B. R. Trenbath, unpublished data).

The suggestion of constancy in B/R_a is based on two data sets. The first (Kawano and Tanaka, 1967) relates to nitrogen use by the components of mixed intercrops of rice varieties growing under conditions where competition for nitrogen presumably occurred. Analysis of nitrogen in the biomass of the components allows Fig. 4.4a, a graph equivalent to Fig. 4.2, to be drawn. A regression line would show that intercrop points are indistinguishable from sole crop points. The second set relates to water use by the mixed intercrop of wheat varieties growing on stored soil water. Figure 4.4b shows that, as in Fig. 4.4a, intercrop and sole-crop points are on indistinguishable regression lines which pass close to the origin.

Variation in harvest index has often been found. Although harvest indices of the components were the same in intercrop as in sole crops in a millet/groundnut experiment (Reddy and Willey, 1981), a 71 percent difference was observed in one component of a sorghum/pigeon pea combination (Natarajan and Willey, 1980; Rao and Willey, 1983). Among many marked effects on harvest index in trials with yam, maize, melon, and cowpea, the largest differences from sole crop values were −93 and +175 percent in two 4-component intercrops (data of Okigbo, 1981). Indeed, reproductive effort in plants is so variously and strongly affected by competitive status that where the yield is a reproductive structure, change in harvest index must be considered a likely reason for high or low LER values (Trenbath, 1976).

RUE CONCEPTS IN LIGHT USE BY INTERCROPS

Light use efficiency (LUE) can be defined in terms of the RUE concepts introduced above:

$$\text{LUE} = \frac{I_i}{I_o}\frac{I_a}{I_i}\frac{B}{I_a}H \qquad (4.8)$$

where I_o, I_i and I_a are, respectively, the quantities of photosynthetically active radiation incident, intercepted, and absorbed by the crop stand (units of, say, GJ/m^2/time unit). An instantaneous measure of LUE that can be used before harvestable parts form a significant proportion of the biomass is:

$$\mathrm{LUE} = \frac{I_i}{I_o}\frac{I_a}{I_i}\frac{P_n}{I_a} \tag{4.9}$$

where P_n is the net photosynthesis rate of the stand.

In studies of water or nutrient utilization, it is easier to determine the quantities absorbed than the quantities intercepted. In studies of light the reverse is true. For the purpose of the following section which treats only light use, LUE will therefore be considered to have just two measurable components:

$$\mathrm{LUE} = \frac{I_i}{I_o}\frac{P_n}{I_i} \tag{4.10}$$

As in the work at ICRISAT (Reddy and Willey, 1981; Natarajan and Willey, 1980; Marshall and Willey, 1983) and elsewhere (Newman, 1984b), the quantity of light intercepted by the plants of a crop is estimated as the difference in flux density of photosynthetically active radiation (PAR) to horizontal, suitably filtered radiation meters set above and below the crop canopy. Although the lower sensor receives some light transmitted or reflected downward after interception, the quantity of this is relatively small (Cowan, 1968) and is usually disregarded in calculating I_i.

The examples of methods of estimating light use parameters of individual components in intercrops are considered in two groups: methods for finding I_i and hence I_i/I_o, and methods in which both I_i/I_o and P_n/I_i can be estimated.

Estimation of Light Interception

When, as in intercropped coconut, the canopies of the mixture components are distinct vertically (Nelliat et al., 1974), light interception by individual components can be measured with horizontal PAR meters placed above and below each of the canopies. To overcome sampling problems due to the clumping of leaves within canopies, large numbers of measurements need to be made which are then integrated over space and time (Nair and Balakrishnan, 1976; Charles-Edwards and Thorpe, 1976). Such data can be collected by using mobile meters mounted on tracks (Charles-Edwards and Thorpe, 1976; Marshall and Willey, 1983) or by using a large number of low-cost units (Newman, 1984a). The successful incorporation of inexpensive integrators in the units (S. M. Newman, personal communication) at last removes much of the cost objection to a detailed characterization of light climates within crops. Care must also be taken to solve as far as possible the multitude of detailed technical problems associated with light measurements in vegetation (Anderson, 1982).

It may be feasible to judge the potential advantage of changes in intercrop canopies by taking photographs from low in the stand using a camera with a fisheye lens (Anderson, 1966b). In predominantly sunny climates, it will be useful to superimpose tracks of the sun's apparent movement overhead (Anderson, 1966a). Study of such photographs may suggest how a change of planting pattern or a canopy type or some intervention such as pruning can increase interception or improve the balance of interceptions by the intercrop components. An automated method can be used on fisheye photographs to estimate interception under various sky conditions (Bonhomme and Chartier, 1972).

When, as in young intercrops, the component canopies are distinct in the horizontal plane, interception by individual components can in principle be estimated using PAR meters but the sampling problems are made formidable by the need to consider light inputs from all parts of the sky hemisphere to all parts of the canopies. Again, mobile and automatically integrating sensors are likely to be useful.

An alternative, but rather rough estimate of light-sharing between such canopies can be obtained using a computer program SUNEYE (Trenbath et al., 1977) which reverses the viewpoint from that of a fisheye camera. Here the computer generates views of the schematized plants of the intercrop as seen from a series of points in the sky (Fig. 4.5). Making the crude assumption that light interception by the components is proportional to the area of schematic plant exposed to light from the sky point chosen, light interceptions associated with a series of representative points can be averaged using beam intensities (Anderson, 1966b) from the points as weightings.

If most of the light arrives as direct solar radiation, the sky points for this program can be chosen to lie on the solar track and given the beam intensities of sunshine from those points. An example of this approach has been given elsewhere (Fig. 1 in Trenbath, 1981) where two row-orientations were compared from a single sun position equivalent to a half-hour before noon. In that example, the addition of a short-statured component increased total interception from 23 up to 56 percent in the north-south planting and from 39 up to 60 percent in the east-west planting. Thus, in the two plantings, the sharing of intercepted light between the short and tall components was estimated to be 23:33 and 39:21, respectively. At that time of day, the north-south planting clearly favored the

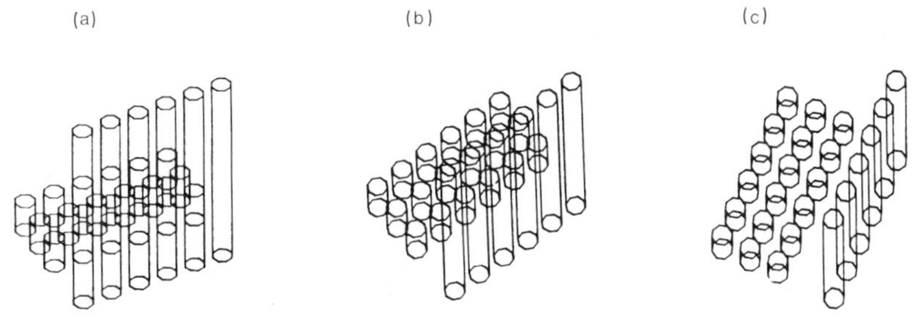

Figure 4.5 Sun-eye views, drawn by computer, of a theoretical corn/soybean intercrop, planted in north–south rows, at 3.5°N on 15 June. (a) $2\frac{1}{2}$, (b) $1\frac{1}{2}$, and (c) $\frac{1}{2}$ hours before noon.

short component. However, time of day affects the situation; this is seen in the three views in Fig. 4.5 which are of this same intercrop. Estimates of sharing should therefore be integrated over times of day and weighted according to the relative frequencies of sunny and cloudy conditions.

Where canopies overlap vertically and/or horizontally, the task of estimating how light interception is shared between intercrop components becomes much more difficult. A method of measuring sunlit leaf area by spraying the foliage with a photosensitive chemical has been used in sole crops of vine (R. E. Smart, personal communication). Where sunny conditions prevail, this ingenious approach could be adapted to intercrops, with suitable integration over time of day. The sharing of indirect radiation from the sky would however be ignored and the true interception by heavily shaded components would be greatly underestimated. The use of substances with a more graduated response to level of illuminance might be explored.

If the canopies are not too tall, a visual point quadrat method devised by Baeumer and de Wit (1968) can be used. For this, an observer uses a special viewer to identify the component that intercepts each of a large number of random "light pencils" entering the intercrop from above. The frequencies of interceptions from each angle can be weighted by the brightness of the corresponding part of the sky; changing solar elevations can also be taken into account. However, in the example that these authors presented, direct sunlight was ignored and all light was supposed to come from an overcast sky with a brightness distribution corresponding to a fixed solar elevation. The good agreement between shoot weights and expectations according to a growth model based on the light interception measurements suggests that in this case the simplified method was satisfactory.

Even where there is some overlap among canopies, it may be possible to treat the intercrop components as if they are confined to nonoverlapping volumes. In their study of light use in a millet/groundnut intercrop, Marshall and Willey (1983) used this approach. Although some rather arbitrary assumptions had to be made to separate the interception by the two components, the results were probably at least as accurate as those of Baeumer and de Wit (1968). The finding that the biomass/interception relationships of the two species in intercrop matched those in sole crop suggests consistency in the method; it is hoped that the interesting sigmoid relationship found in the legume will be checked by an independent method and its significance explained.

Under certain experimental conditions, the light interception of individual components of an intercrop may be measured directly using PAR meters placed on the surface of a sufficiently large sample of the leaves of each component. This extremely laborious approach has been used in sole crops (Kumura, 1965) but not yet, so far as I am aware, in intercrops. Movement of upper leaves by wind introduces enormous variations through time in measured values at points lower in a canopy, and limitation of movement of monitored leaves due to the placing of a sensor on them introduces unmeasured biases. This method can only be used in a controlled environment where, under the conditions of the experiment, leaves are almost motionless.

Where the components' canopies have similar morphological characteristics (leaf inclination, leaf size, degree of clumping, light-scattering properties) and canopies are continuous horizontally, the partitioning of light interception between mixture components can be estimated using vertical "light profiles" and using information on the vertical distributions of leaf area of the components (Fig. 4.1b; Black, 1958). Light profiles commonly have been obtained by measuring PAR on a horizontal sensor at 10 to 15 equally spaced levels just above, just below, and within the mixed canopy. Leaf area distributions have been measured using the stratified clip method in which canopies are cut successively at the same 10 to 15 levels, foliage of components is separated, and leaf areas are determined (Monsi and Saeki, 1953; Monsi, 1968; Donald, 1963). On the assumption that the apparent interception within any layer can be attributed to component canopies in the layer in proportion to their leaf areas, interception is estimated for all layers; these estimates are summed to show the overall proportions of the incident light which are intercepted by the components and by the soil surface. Although interception estimates made in this way in mixed clover swards (Black, 1958, 1960, 1961) are likely to be accurate, use of the same method in mixtures of strongly contrasting canopy types may give less reliable results because the proportionality constants for interception dependent on leaf area will not be the same for the various components. This difficulty must arise, for instance, in mixtures of a cereal and a legume with sun-following leaves where, under sunny conditions, interception by the legume will be underestimated.

A final approach to estimating interception by intercrop components involves the detailed measurement of canopy structure and the use of a complex mathematical model of light interception. Although detailed measurements of leaf inclination and/or orientation have been made in several sole crops (de Wit, 1965; Fukai and Loomis, 1976), no comparable analyses seem to have been made in intercrops. A number of methods are available (reviewed in Trenbath and Angus, 1975), but high labor requirements have discouraged their use. A possible means of avoiding the need to measure the detailed structure in both sole crops and intercrops might be to estimate the canopy structure of the mixture from those of the pure stands. Such an estimation would need to be fully validated by measurements in the intercrop since plant form in a component can differ radically between sole crop and intercrop (Black, 1961).

The mathematical models initially proposed for estimating light interception assumed horizontal uniformity in the canopy (de Wit, 1965; Duncan et al., 1967), but since in real row crops in the field leaves are often clumped both within the rows and within the plants, such models must tend to underestimate light penetration (Monsi, 1968). More recent models take into account horizontal discontinuities in the canopies in row crops (Smart, 1973; Fukai and Loomis, 1976; Palmer, 1977) or in low densities of individual plants (Charles-Edwards and Thornley, 1973; Oikawa and Saeki, 1977). When data are available on the canopy structure of intercrops, there will be no shortage of models for estimating the implications of the data in terms of light interception, or, as the next section demonstrates, in terms of photosynthesis and P_n/I_i.

Estimation of Light Interception and Conversion Efficiency

Among the few studies where one of the above methods of estimating light interception has been used and related to observed growth, those of Black, Donald, and colleagues at the Waite Institute, Adelaide, are outstanding. Competitive success in varietal mixtures of clover was related qualitatively and sometimes quantitatively to the percentage of incident light intercepted by mixture components (Black, 1958, 1960, 1961); growth of a component remained positive only as long as it intercepted more than 1 percent of the incident light. This kind of effect was shown to explain much of the response of grass/clover mixtures to nitrogen fertilizer (Stern and Donald, 1962). The percentage of incident light intercepted was shown to be a critical determinant of growth in a sole crop of wheat (Puckridge and Donald, 1967) and in an intercrop of clover and wheat (Santhirasegaram and Black, 1968); both groups of authors presented data in a form that agrees with Fig. 4.2a.

These practical demonstrations of the relationship between light interception and growth in intercrops have promoted interest in using light interception models coupled with the photosynthesis/irradiance relationships of individual leaves to predict growth rates of components in different kinds of intercrops. The first such combined model proposed for sole crops (Saeki, 1960) can be applied with little modification to estimate instantaneous net photosynthesis rates P_n in mixed stands where the two canopies can be treated as vertically distinct, continuous horizontal layers. As an example of the kind of results obtained we consider a case where a theoretical mixed stand was taken to consist of two vertically distinct canopies each of LAI = 2. Comparable to this mixed stand, theoretical pure stands of the same tall- and short-statured crops were considered, each with a canopy of LAI = 4. To estimate the effect of various types of upper canopy in such an intercrop, two important upper-canopy variables were systematically varied over their observed ranges and the effects on P_n and P_n/I_i were calculated by the model. The two upper-canopy variables were the maximum light-saturated leaf gross photosynthesis rate $P_{g\infty}$ and the canopy extinction coefficient for light (the greater the value of the latter, the more efficiently leaves intercept light). Some further details of the model are given in Trenbath (1981).

The results of this exercise (Fig. 4.6a) show that high total P_n seems to be favored by both increased maximum $P_{g\infty}$ and lowered extinction coefficient. In other words, under the conditions assumed, fastest growth is expected in a stand where the upper canopy has (1) leaves with a very high photosynthetic capacity (e.g., C_4-type) and (2) leaves set near vertically such that they let much of the incident light pass through to a lower canopy with less erect leaves. The results for P_n/I_i show that to maximize conversion efficiency (and also P_n), an intermediate to low interception efficiency is needed in the upper canopy, with a leaf

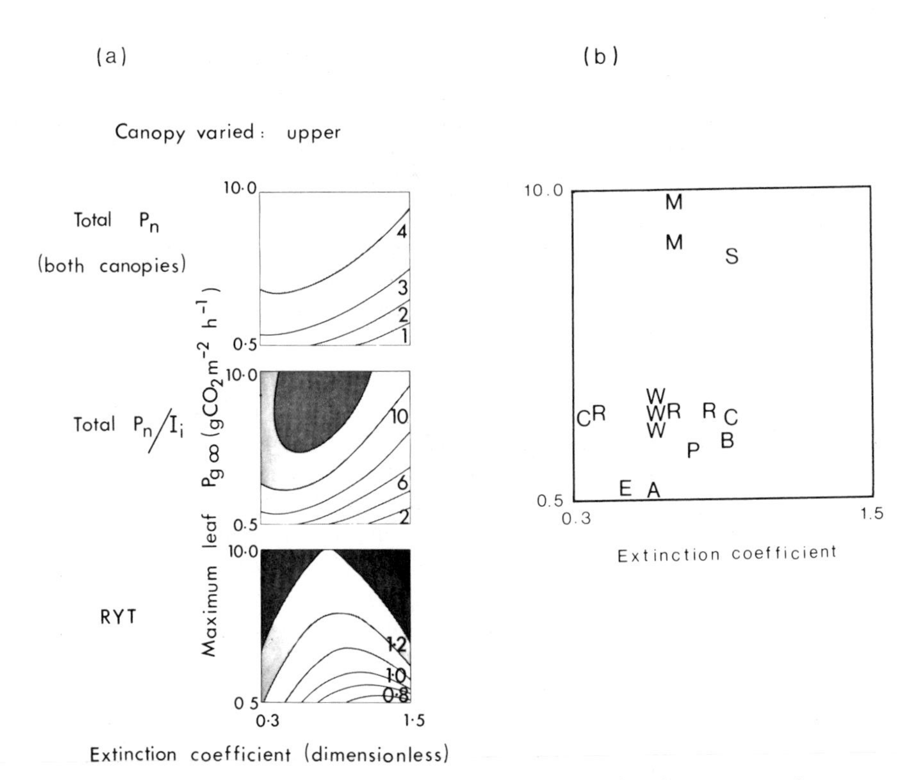

Figure 4.6 (a) Response surfaces for a two-layer intercrop of total canopy net photosynthesis (P_n, g $CO_2/m^2\cdot$ hour), conversion efficiency (P_n/I_i, μg CO_2/J) and RYT (based on P_n) calculated with respect to variation of two parameters of the upper canopy. Values of the two parameters, light extinction coefficient and the maximum leaf light-saturated rate of gross photosynthesis ($P_{g\infty}$), are held at standard values in the lower canopy. Calculations of P_n were made using the model of Saeki (1960). (b) Ordination of crop species according to measured values of extinction coefficient and P_g. The ranges of the parameters used in (b) match those in (a) so that crop species can be identified that have leaf characteristics corresponding to various parts of the squares shown in (a). The species are: M, maize; S, sunflower; W, wheat; R, rice; C, cocksfoot (orchard grass); B, buckwheat; P, sweet potato; E, coffee; A, cocoa.

photosynthetic capacity as high as possible. Values of the relative yield total RYT (de Wit and van den Bergh, 1965) based on P_n show that high values are indeed expected where P_n/I_i has been maximized. The reason for surprisingly high RYT values, where the upper canopy has high maximum $P_{g\infty}$ and extinction coefficient, is apparently the respiration model used which assumes leaf dark respiration to be proportional to $P_{g\infty}$ (Tooming, 1977). In a sole crop (LAI = 4) of the upper-canopy component, much of such a canopy is below compensation point, thus giving the sole crop a very low calculated P_n.

The potential usefulness of Fig. 4.6a becomes more obvious when data of actual crop species are plotted on an equivalent square according to their values of leaf $P_{g\infty}$ and extinction coefficient (Fig. 4.6b). Among the relatively few crops for which estimates of both values could be derived from the literature, the position of corn in the square suggests that it would be best for an upper canopy, with sunflower as second best. As emphasized by Bavappa and Jacob (1982), the choice of crop combinations depends at present more on art than science. However, an elaboration of this preliminary approach could indicate ways in which science might contribute. If some difficulties can be overcome, rapid screening of physiological and morphological criteria may allow some unsuitable genotypes and species to be eliminated before they are entered in costly long-term experiments. While the relation between growth rates and yield is often uncertain, this approach would estimate growth *potential* in simulated combinations.

Where the canopy layers are essentially uniform horizontally, the approach of Black and co-workers or an adapted Saeki (1960) model can be used. To estimate light interception and growth, taking into account horizontal heterogeneity due to planting pattern and species differences, introduces great complexities. Elaborate treatments based on detailed consideration of the distribution of foliage and of light sources have been developed and applied to sole crops (Oikawa, 1977) and intercrops (Trenbath et al., 1977). These two models, originating in Tokyo and Canberra, respectively, rely on the same method of Monte Carlo integration in which interception is calculated for a large number of random light pencils entering the canopy.

The Canberra model also calculates the photosynthetic effect of each light pencil so that by averaging within each plant layer and summing within each plant over layers and hours of the day, a value is derived for each plant of estimated daily gross photosynthesis (EDGP). Applied for sunny conditions to the intercrop of Fig. 4.5, this model showed that the photosynthetic disadvantage of the shorter component was only slightly less in an east-west planting than in a north-south one, that is, the EDGP ratio of short plants to tall plants was only 5 percent less. Indeed, the EDGPs of the whole stands planted in the two orientations were within 1 percent of being the same. The model's output of individual plant photosynthesis values showed the presence of the shadow of the tall plants (Puckridge and Donald, 1967) as it crept across the triple rows of short plants. In the north-south planting, the shadow's disappearance near noon (Fig. 4.5b) resulted in the estimated photosynthesis of the newly sunlit row increasing by 17 percent. Although no detailed calculations of light interception were made, it was noted that the skylight contribution to leaf irradiance in the short plants was slightly less than half that in the top layer of the six-times taller tall plants. Under overcast conditions, the EDGP ratio of short plants to tall plants is thus likely to be much more than the values of about 0.38 expected under sunny conditions. Overcast conditions will favor short components (Trenbath, 1976).

Returning to the situation where canopy layers can be treated as horizontally homogeneous, a development of the Saeki (1960) model of crop P_n combines in the model the leaf photosynthesis/irradiance relationships with a consideration of the detailed geometry of light sources (mainly sun and sky) and of foliage (de Wit, 1965; Duncan et al., 1967). An adaptation of the Duncan model (Trenbath, 1972) to allow the simulation of P_n in a two-component intercrop is able to use leaf-area inclination distributions together with data determining sun position and brightness to calculate light interception and P_n. The output shows interception and P_n by sunlit and shaded leaf area, by leaf inclination class, by layer, and by component. Figure 4.7 illustrates the results of a theoretical study of the dynamics of light interception (I_i) and conversion efficiency P_n/I_i through a sunny day in an intercrop. To simplify presentation, the canopy is two-layered as in Fig. 4.6. As expected, away from noon the quantity of light intercepted by both components falls (Fig. 4.7a). In the upper component, photosynthesis is mostly due to the activity of sunlit leaves, whereas in the lower component the activity of shaded leaves is more important (Fig. 4.7b). Away from noon, P_n/I_i generally rises (Fig. 4.7a). However, extreme shading of the lower canopy in the intercrop causes P_n/I_i to suddenly plunge. This can be traced in the model's output to a large proportion of the leaves falling below their compensation points.

As the dynamics of growth within intercrops are subjected to more detailed

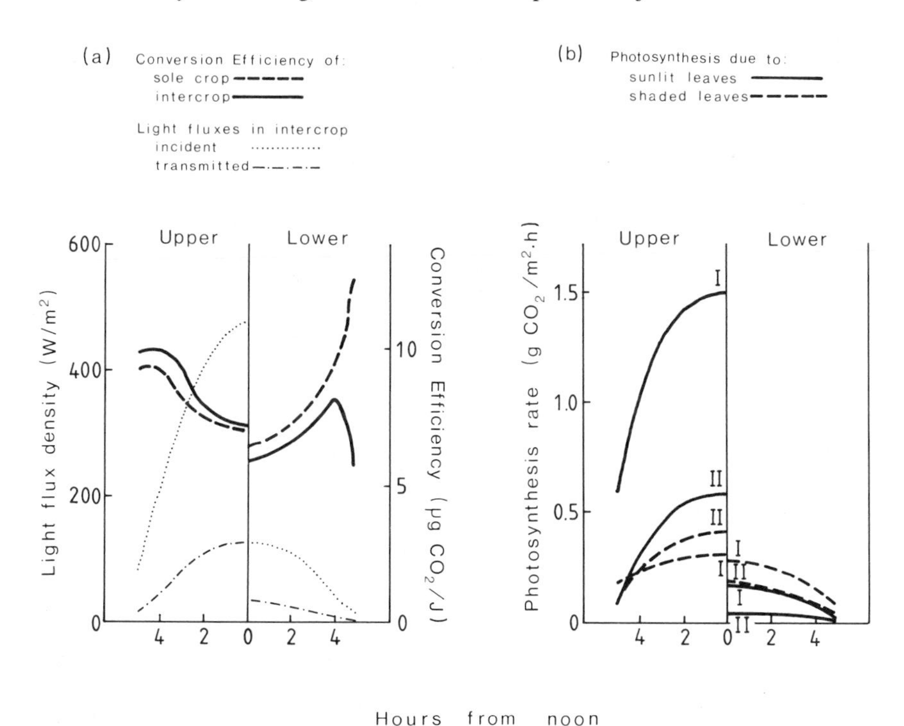

Figure 4.7 Simulation of light interception and net photosynthesis (P_n) by a two-layered intercrop where the upper and lower canopies each contain 2 LAI units. The light regime is for a clear day on latitude 23.5° at the equinox; "hours from noon" therefore correspond to sun elevations of 90°, 75°, 60°, and so on. (a) Light interception and conversion efficiency (P_n/I_i). Light interception by each canopy is given by the difference between the light fluxes entering it (incident) and leaving it (transmitted). Conversion efficiencies are shown for the two intercrop canopies and also for comparable sole crops with 4 LAI units. (b) Net photosynthesis due to sunlit and shaded leaves in the upper half (I) and lower half (II) of each of the two intercrop canopies.

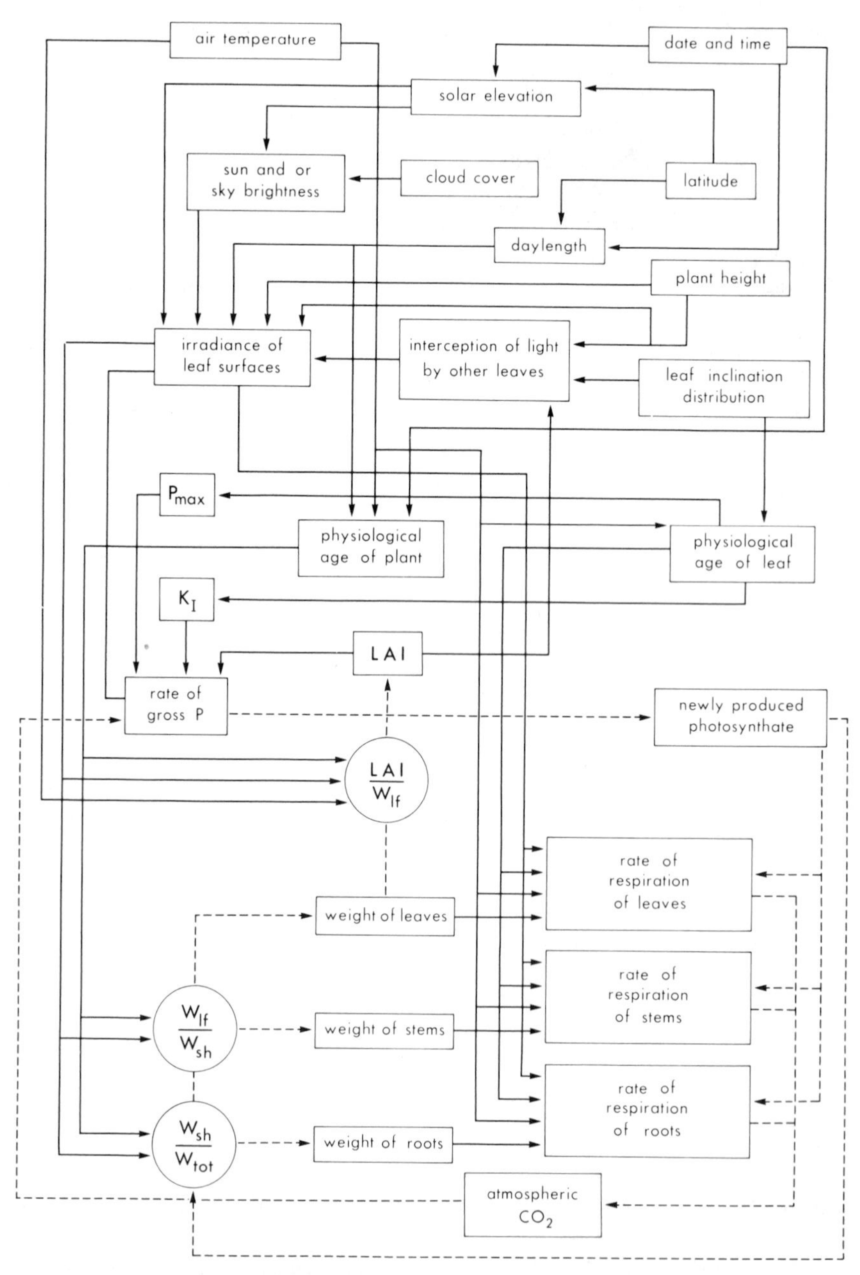

Figure 4.8 Relational diagram of the model GRODEV (Trenbath, 1974) which simulates growth and vegetative development of wheat plants, either in sole or intercrop under conditions where competition is only for light. Flows of carbon are shown as broken lines and flows of control as solid lines. P_{max} and K_I are respectively leaf light-saturated gross photosynthesis rate and the level of leaf irradiance which gives a rate of $P_{max}/2$. Three ratios (circled) involve total leaf weight (W_{lf}), shoot weight (W_{sh}) and total plant weight (W_{tot}). Items in boxes without arrows entering them are provided as inputs for the model, where necessary as functions of time.

study (Marshall and Willey, 1983; Trenbath, 1983), mathematical models of growth processes in the components are likely to become useful. Such a model has been constructed (Trenbath, 1974), based on physiological data for cereals to produce, in effect, a computerized plant growth model in which the theoretical components compete for light throughout the period of vegetative growth. The factors included in the model are diagrammed in Fig. 4.8.

Among applications of this model, it has been shown that the theoretically optimum configuration of an intercrop canopy, as judged by instantaneous P_n (see Fig. 4.6), may be difficult to maintain on a long-term basis (Trenbath, 1981). The erect-leaved taller component tends to suppress the more prostrate-leaved shorter component and may cause it to die before it can set seed. Another application showed a theoretical basis for expecting that similar cereal types, growing in replacement series and competing only for light, will produce biomass according to the de Wit (1960) model with an RYT of unity (Trenbath, 1978). Simulation with dynamic models of this type may allow researchers to identify morphological and physiological characteristics which are critical for successful performance in intercrops. Such characteristics could then provide selection criteria for use with genotypes grown in sole crop.

CONCLUSIONS AND FUTURE DIRECTIONS FOR RESEARCH

The hypothesis that intercrops will show yield advantages has been amply proved by many observations of LER exceeding unity, but the extent to which ecological differences between species contribute to these advantages is often unclear. Mathematical models show how the differences *might* lead to high LERs but in the case of a given field example it is usually difficult to identify the mechanism.

Postulated mechanisms are often expressed in terms of resource use efficiency. Measurements of RUE itself or of its multiplicative components throw light on the mechanisms and can show whether postulations have to be revised. When intercrop productivity is likely to be related to how a single limiting resource is utilized, the approach described in the first part of the chapter can in principle be applied to measurements of the efficiencies of resource capture and of its conversion into biomass or yield. This approach allows, for each species, a calculation of three additive contributions to the intercrop's LER. The three contributions correspond respectively to how much each species captures of the limiting resource (a measure of capture efficiency), to the species' conversion efficiencies, and to the mathematical interactions between these contributions. If necessary, the approach could be extended to multispecies intercrops and to take account of three or more multiplicative components of RUE.

An illustrative analysis of the LER of a binary (varietal) intercrop has been presented using two RUE components, the amounts of water captured and the conversion efficiencies. This first reported analysis of water use by individual plant components in an intercrop was made possible by the recognition that silicon is taken up passively by the species involved (Jones and Handreck, 1967). If such water-use analyses are to be made in a wider range of species, it seems necessary to seek substances that more species absorb passively. The alternative might be to try to develop some easily integrated physical measure of water loss from individual plants.

The analysis of LER in terms of light use efficiency presents difficulties since it is hard to measure the amount of light that is actually captured (absorbed) by a plant component in an intercrop. Strictly, this amount of light needs to be known for the conversion efficiency to be meaningful. If, however, measurements of intercepted light could be satisfactorily transformed into values of absorbed light by subtracting estimates of the amounts of light reflected upward, capture and conversion efficiencies could be calculated which would be comparable with those for water. For this, the use of a light reflection (albedo) model might be explored. It is possible that some of the nonlinearity in the cumulative dry matter/light interception curve of the groundnut in Marshall and Willey's (1983) experiment might be due to ontogenetic changes in leaf posture and hence albedo. For the measurement of amounts of light absorbed, a further possibility to be explored is the use of photosensitive chemicals sprayed onto the plant surfaces.

Analysis of LER in terms of nutrient utilization would appear relatively straightforward. Indeed, since data on nutrient absorption (capture efficiency) and tissue concentration (reciprocal of conversion efficiency) are routinely collected by agronomists, they could immediately provide material for study. For parameters of nutrient use to be comparable with those of light and water, the measure of nutrient absorption by an intercrop component should include nutrient held in its root biomass. Either root nutrient content will have to be estimated or some compromise must be struck on comparability of parameters. LER analyses based on nitrogen in legume/nonlegume intercrops will be particularly illuminating.

Since LER analysis is a new technique, its usefulness in improving intercrop performance has still to be demonstrated. In the case of the mixed intercrop of wheat with an LER of 0.8, the analysis showed that cultivar Selkirk's water conversion efficiency was lowered by 30 percent in intercrop and that the aggregate absorption of water by the intercrop was 7 percent less than that of the sole crop with lower absorption (Table 4.1). Further experimentation has suggested that, in the intercrop, Selkirk had suffered in early competition for nutrients and that its root system had been cut off from access to stored water by a layer of soil dried by the earlier-growing roots of the other component (Trenbath, 1972). Assuming that this explanation is correct, a guideline for choosing intercrop components can be offered: when intercrops are to be grown on stored soil resources, components should be chosen so that earlier uptake of water and nutrients in the upper layers of the profile by one component should not be so efficient as to suppress the growth of later-developing root systems of other components that are potentially more effective exploiters of the lower layers.

The results of Marshall and Willey's (1983) analysis of the LER of a millet/groundnut intercrop show that in the intercrop both per-plant light interception and a component of groundnut's conversion efficiency, biomass production efficiency, were increased relative to sole crop. These data pose a problem however as to how they can be used to improve intercrop performance. One suggestion might be that since these two features are numerically responsible for the high LER, they should both be enhanced in some way. A contrary suggestion might be that since the law of diminishing returns commonly applies to improvements of any sort, it would be more effective to attempt instead to enhance the *other* features that contribute to the high LER.

The basic results of Marshall and Willey can be interpreted qualitatively as being due (1) to individual millet plants spreading more widely in intercrop than in sole crop and so intercepting more light, and (2) to groundnut being shaded more in intercrop, being therefore further down in its photosynthesis/light response curve, and hence converting light more efficiently. The first suggestion above would lead to experimentation with an even more spreading millet type and the second suggestion would lead to use of a more compact, erect type. This confused situation suggests that even in this well-studied intercrop, theoretical understanding is not enough for results from a single treatment to be usefully exploited. The agronomist is in the position of a control engineer seeking an optimum on a graph of unknown form of which only one point is given. To solve this difficulty, it seems essential that the valuable experimental and theoretical analysis of the millet/groundnut intercrop should be continued using experiments carried out with several levels of the same factor. In this regard, an LER analysis for water use would add greatly to the understanding of the origin of the spectacular rise in LER as sorghum/groundnut plots are irrigated with less and less water (Willey et al., 1983).

The most useful ideas to emerge from the detailed consideration of light utilization are suggested from the results of the most complete models: where an intercrop is two-layered, the highest LERs based on net photosynthesis rate will probably come from combinations of a taller component having high photosynthetic capacity and steeply inclined leaves with a shorter component having lower photosynthetic capacity and more prostrate leaves. Combinations with the highest LER are not necessarily the most productive (Fig. 4.6). Compass orientation of crop rows is probably unimportant for light use by intercrops. Overcast conditions will favor shorter components. As incident light level and P_n fall, conversion efficiency (P_n/I_i) rises at first, but falls sharply at very low light levels (Fig. 4.7). An intercrop canopy structure that is near optimal with respect to growth rate at some given time will probably remain optimal only if the taller component is continually pruned. A promising technique to aid in selecting beneficial combinations of crop types (species or cultivars) was described that uses morphological and physiological attributes, relevant to light utilization, measurable in sole crop. Before it can be used, it will need further development and validation (see discussion in Trenbath, 1981).

As a general comment, when scientists of various disciplines, extension workers, and farmers are considering aims and means in relation to multiple cropping and its farm context, process-oriented models of resource dynamics seem likely to greatly aid communication and understanding. If the technical facilities are available, ideas and hypotheses that emerge during discussion can be subjected to immediate theoretical test. Similarly, but of greater interest to the scientists, proposed experiments can be set up in advance for simulation by computer model. When run with a range of assumed weather conditions, the model can produce sets of possible theoretical results. Experimental designs can then be modified to allow for previously unforeseen contingencies.

The advantages from working with mathematical models, coupled with the deepening of insight obtained in the process of constructing and testing them, suggest that integrated models are now needed to account for both water and nutrient dynamics along with those of light. Providing that they can be acceptably simple and accurate, I believe that they will help greatly to test and sharpen hypotheses about multiple cropping, and also facilitate the fruitful interaction of experiment and theory that will allow improvement in the efficiency with which limiting environmental resources are used to provide food and income.

ACKNOWLEDGMENTS

I thank my wife for typing the manuscript, and the editor for making many useful suggestions.

REFERENCES

Aiyer, A. K. Y. N. 1949. Mixed cropping in India, *Indian J. Agric. Sci.* 19:439–543.

Anderson, M. C. 1966a. Some problems of simple characterization of the light climate in plant communities, in: *Light as an Ecological Factor,* (R. Bainbridge, G. C. Evans, and O. Rackman, eds.), Blackwell Scientific Publications, Oxford, U.K., pp. 77–90.

———. 1966b. Stand structure and light penetration. II. A theoretical analysis, *J. Appl. Ecol.* 3:41–54.

———. 1982. *Reflections on the shade cast by trees,* Treephysindia 82, International Workshop on Special Problems in Physiological Investigations of Tree Crops, Kottayam, India, 22 pp.

Baeumer, K., and de Wit, C. T. 1968. Competitive interference of plant species in monocultures and mixed stands, *Neth. J. Agric. Sci.* 16:103–122.

Baldy, C. 1963. Cultures associées et productivité de l'eau, *Ann. Agron.* 14:489–534.

Bavappa, K. V. A., and Jacob, V. J. 1982. High intensity multi-species cropping, *World Crops* 34:47–50.

Biscoe, P. V., and Gallagher, J. N. 1977. Weather, dry matter production and yield, in: *Environmental Effects on Crop Physiology,* (J. J. Landsberg and C. V. Cutting, eds.), Academic Press, New York, pp. 75–100.

Black, J. N. 1958. Competition between plants of different initial seed sizes in swards of subterranean clover (*Trifolium subterraneum* L.) with particular reference to leaf area and the light microclimate, *Aust. J. Agric. Res.* 9:299–318.

———. 1960. The significance of petiole length, leaf area, and light interception in competition between strains of subterranean clover (*Trifolium subterraneum* L.) grown in swards, *Aust. J. Agric. Res.* 11:277–291.

———. 1961. Competition between two varieties of subterranean clover (*Trifolium subterraneum* L.) as related to the proportions of seed sown, *Aust. J. Agric. Res.* 12:810–820.

Bonhomme, R., and Chartier, P. 1972. The interpretation and automatic measurement of hemispherical photographs to obtain sunlit foliage area and gap frequency, *Israel J. Agric. Res.* 22:53–61.

Charles-Edwards, D. A., and Thornley, J. H. M. 1973. Light interception by an isolated plant. A simple model, *Ann. Bot.* 37:919–928.

Charles-Edwards, D. A., and Thorpe, M. R. 1976. Interception of diffuse and direct-beam radiation by a hedgerow apple orchard, *Ann. Bot.* 40:603–613.

Cowan, I. R. 1968. The interception and absorption of radiation in plant stands, *J. Appl. Ecol.* 5:367–379.

de Wit, C. T. 1958. Transpiration and crop yields, *Versl. Landbouwk. Onderz.* 64(6):1–88.

———. 1960. On competition, *Versl. Landbouwk. Onderz.* 66(8):1–82.

———. 1965. Photosynthesis of leaf canopies, *Versl. Landbouwk. Onderz.* 663:1–57.

de Wit, C. T., and van den Bergh, J. P. 1965. Competition between herbage plants, *Neth. J. Agric. Sci.* 13:212–221.

Donald, C. M. 1963. Competition among crop and pasture plants, *Adv. Agron.* 15:1–118.

Duncan, W. G., Loomis, R. S., Williams, W. A., and Hanau, R. 1967. A model for simulating photosynthesis in plant communities, *Hilgardia* 38:181–205.

Fukai, S., and Loomis, R. S. 1976. Leaf display and light environments in row-planted cotton communities, *Agric. Meteorol.* 17:353–379.

Hutton, J. T., and Norrish, K. 1974. Silicon content of wheat husks in relation to water transpired, *Aust. J. Agric. Res.* 25:203–212.

IRRI (International Rice Research Institute). 1975. *Annual Report for 1974,* Los Banos, Philippines.

Jones, L. H. P., and Handreck, K. A. 1967. Silica in soils, plants, and animals, *Adv. Agron.* 19:107–149.

Kawano, K., and Tanaka, A. 1967. Studies on the competitive ability of rice plant in population. *J. Fac. Agr. Hokkaido Univ.* 55:339–362.

Kumura, A. 1965. Studies on dry matter production of soybean plant. 2. Influence of light intensity on the photosynthesis of the population. Part 1. Relation between photosynthesis and light receiving aspect of the population in case where light intensity varies with weather condition, *Proc. Crop Sci. Soc. Jap.* 33:473–481.

Loneragan, J. F., and Asher, C. J. 1967. Response of plants to phosphate concentration in solution culture. II. Rate of phosphate absorption and its relation to growth, *Soil Sci.* 103:311–318.

Marshall, B. and Willey, R. W. 1983. Radiation interception and growth in an intercrop of pearl millet/groundnut, *Field Crops Res.* 7:141–160.

Monsi, M. 1968. Mathematical models of plant communities, in: *Functioning of Terrestrial Ecosystems at the Primary Production Level, Proc. Copenhagen Symp., 1965,* pp. 131–149, UNESCO, Paris.

Monsi, M., and Saeki, T. 1953. Über den Lichtfaktor in den Pflanzengesellschaften und seine Bedeutung für die Stoffproduktion, *Jap. J. Bot.* 14:22–52.

Monteith, J. L. 1972. Solar radiation and productivity in tropical ecosystems, *J. Appl. Ecol.* 9:747–766.

Nair, P. K. R., and Balakrishnan, T. K. 1976. Pattern of light interception by canopies in a coconut-cacao crop combination, *Indian J. Agric. Sci.* 46:453–462.

Natarajan, M., and Willey, R. W. 1980. Sorghum-pigeon pea intercropping and the effects of plant population density. I. Growth and yield, *J. Agric. Sci. Camb.* 95:51–58.

Nelliat, E. V., Bavappa, K. V. A., and Nair, P. K. R. 1974. Multistoreyed cropping: A new dimension in multiple cropping for coconut plantations, *World Crops* 26:262–266.

Newman, S. M. 1984a. The design and testing of a system for monitoring the availability of solar radiation for interculture, *Agroforestry Syst.* 2:43–47.

———. 1984b. The use of vegetable phytometers in the evaluation of the potential response of understorey crops to the aerial environment in an interculture system, *Agroforestry Syst.* 2:49–56.

Oikawa, T. 1977. Light regime in relation to plant population geometry. II. Light penetration in a square-planted population, *Bot. Mag. Tokyo* 90:11–22.

Oikawa, T., and Saeki, T. 1977. Light regime in relation to plant population geometry. I. A Monte Carlo simulation of light microclimates within a random distribution foliage, *Bot. Mag. Tokyo* 90:1–10.

Okigbo, B. N. 1981. Evaluation of plant interactions and productivity in complex mixtures as a basis for improved cropping system design, in: *Proc. Int. Workshop on Intercropping,* (R. W. Willey, ed.), International Crops Research Institute for the Semi-Arid Tropics (ICRISAT), Hyderabad, India, 10–13 January 1979, pp. 155–179.

Osiru, D. S. O., and Kibira, G. R. 1981. Sorghum/pigeon pea and finger millet/groundnut mixtures with special reference to plant population and crop arrangement, in: *Proc. Int. Workshop on Intercropping,* (R. W. Willey, ed.), ICRISAT, Hyderabad, India, 10–13 January 1979, pp. 78–85.

Palmer, J. W. 1977. Diurnal light interception and a computer model of light interception by hedgerow apple orchards, *J. Appl. Ecol.* 14:601–614.

Puckridge, D. W., and Donald, C. M. 1967. Competition among wheat plants sown at a wide range of densities, *Aust. J. Agric. Res.* 18:193–211.

Rao, M. R., and Willey, R. W. 1983. Effects of pigeon pea plant population and row arrangement in sorghum/pigeon pea intercropping, *Field Crops Res.* 7:203–212.

Reddy, M. S., and Willey, R. W. 1981. Growth and resource use studies in an intercrop of pearl millet/groundnut, *Field Crops Res.* 4:13–24.

Rorison, I. H. 1968. The response to phosphorus of some ecologically distinct plant species. I. Growth rates and phosphorus absorption, *New Phytol.* 67:913–923.

Saeki, T. 1960. Interrelationships between leaf amount, light distribution and total photosynthesis in a plant community, *Bot. Mag. Tokyo* 73:55–63.

Santhirasegaram, K., and Black, J. N. 1968. The relationship between light beneath wheat crops and growth of undersown clover, *J. Brit. Grassl. Soc.* 23:234–239.

Smart, R. E. 1973. Sunlight interception by vineyards, *Amer. J. Enol. Viticult.* 24:141–147.

Stern, W. R., and Donald, C. M. 1962. Light relationships in grass-clover swards, *Aust. J. Agric. Res.* 13:599–614.

Tooming, H. G., 1977. *Solar radiation and the production of yield,* Gidrometeoizdat, Leningrad, U.S.S.R.

Trenbath, B. R. 1972. The productivity of varietal mixtures of wheat, Ph.D. thesis, University of Adelaide.

———. 1974. Application of a growth model to problems of the productivity and stability of mixed stands, in: *Proc. 12th Int. Grasslands Congr.*, vol. 1, Izdatel'stvo Mir, Moscow, pp. 546–558.

———. 1976. Plant interactions in mixed crop communities, in: *Multiple Cropping,* (R. I. Papendick, P. A. Sanchez, and G. B. Triplett, eds.), Amer. Soc. Agron., Madison, Wisconsin, pp. 129–170.

———. 1978. Models and the interpretation of mixture experiments, in: *Plant Relations in Pastures,* (J. R. Wilson, ed.), CSIRO, Melbourne, Australia, pp. 145–162.

———. 1981. Light-use efficiency of crops and the potential for improvement through intercropping, in: *Proc. Int. Workshop on Intercropping,* (R. W. Willey, ed.), ICRISAT, Hyderabad, India, 10–13 January 1979, pp. 141–154.

———. 1983. The dynamic properties of mixed crops, in: *Frontiers of Research in Agriculture,* (S. K. Roy, ed.), Indian Statistical Institute, Calcutta, India, pp. 265–286.

Trenbath, B. R., and Angus, J. F. 1975. Leaf inclination and crop production, *Field Crop Abstr.* 28:231–244.

Trenbath, B. R., Hartley, P. R., and MacPherson, D. K. 1977. Plant distribution and individual plant photosynthesis, 4th Int. Congr. on Photosynthesis, Abstract, U.K. Science Committee, London, U.K., pp. 384–385.

Willey, R. W., Natarajan, M., Reddy, M. S., Rao, M. R., Nambiar, P. T. C., Kannaiyan, J., and Bhatnagar, V. S. 1983. Intercropping studies with annual crops, in: *Better Crops for Food,* Pitman, London, U.K., pp. 83–100.

Wilson, L. A. 1977. Root crops, in: *Ecophysiology of Tropical Crops* (P. de T. Alvim and T. T. Kozlowski, eds.), Academic Press, New York, pp. 187–236.

Chapter 5

Plant Interactions in Multiple Cropping Systems

Stephen R. Gliessman

In much of conventional agriculture where optimal densities of single crop plantings are the rule, the strict management of the physical factors of the crop ecosystem usually suffices in order to obtain maximum yields. This includes optimum fertilizer levels, water availability, light capture, and other factors that can be manipulated to the benefit of crop output. With additional weed, insect, and disease control, the monoculture cropping system can lead to substantial harvests. But as the advantages of multiple cropping become more recognized (ASA, 1976; Kass, 1978; Harwood, 1979; Willey, 1979a, 1979b; ICRISAT, 1981; Gliessman et al., 1981; Vandermeer et al., 1983), and the move is made toward intercropping two or more different plant species on the same land at the same time, the need for a more in-depth understanding of the biotic components of plant interactions becomes very important. Such information is key to designing and managing complex multiple cropping systems capable of advantageous and consistent results.

Most of our knowledge about the ecological basis for plant interactions comes from the study of natural ecosystems and plant ecology (Daubenmire, 1974). A great variety of classification schemes has been presented in order to better understand the diversity of types of biotic interactions between species, and one I feel best represents the broadest range has been summarized by Odum

(1971, Chap. 7). As can be seen in Table 5.1, the type of interaction will determine the resulting shift in population levels of the species involved. This classification scheme is very broad, and covers plant/animal and animal/animal interactions as well. It focuses on the end result of the interactions as expressed by population growth and survival. The actual mechanisms of interaction are not emphasized. In the design and management of multiple cropping systems, a knowledge of these mechanisms can be very important.

Harper (1977, Chap. 11) most thoroughly examined the importance of understanding the mechanisms of plant interactions. When describing the nature of interactions, he focuses on the fact that a plant may influence its neighbors by changing their environment. These changes may be by means of an addition or removal reaction, and there has been much controversy recently as to which is most important. Many indirect effects on the environment can also affect neighboring species, not by addition or removal of some factor, but by affecting conditions such as temperature, soil insulation, or wind movement, and by altering the balance between beneficial and harmful insects. It can be extremely difficult to separate these factors, but it becomes very important to do so as our desire to intercrop different crops increases. Proper management of such systems

Table 5.1 Analysis of Two-species Population Interactions
(Effect on Population Growth and Survival of Two Populations, A and B)

	When not interacting		When interacting		
Type of interaction	A	B	A	B	General result of interaction
Neutralism (A and B independent)	0	0	0	0	Neither population affects the other
Competition (A and B competitors)	0	0	–	–	Population most affected eliminated from niche
Mutualism (A and B partners or symbionts)	–	–	+	+	Interaction obligatory for both
Protocooperation (A and B cooperators)	0	0	+	+	Interaction favorable to both, but not obligatory
Commensalism (A commensal; B host)	–	0	+	0	Obligatory for A; B not affected
Amensalism (A amensal; B inhibitor or antibiotic)	0	0	–	0	A inhibited; B not affected
Parasitism (A parasite; B host)	–	0	+	–	Obligatory for A; B inhibited
Predation (A predator; B prey)	–	0	+	–	Obligatory for A; B inhibited

Note: + Population growth increased (positive term added to growth equation).
– Population growth decreased (negative term added to growth equation).
0 Population growth not affected (no additional term in growth equation).
Source: Odum, 1971, Chap. 7, by permission of Saunders Publ. Co.

can come only from detailed ecological research. As a guide for such studies, I propose the concept of interference interactions, permitting a mechanistic approach to understanding how the interactions function. For the purpose of this discussion, I break such interactions initially into the reactions the plant has on the environment. This then forms a basis on which to understand the responses observed in crop mixtures.

INTERFERENCE INTERACTIONS

Based on concepts developed by Harper (1961, 1964) and expanded by Muller (1969), interference can be divided into removal reactions of one plant on its environment, and additive reactions where something is added. When some factor is removed from the environment, the resulting response of neighboring species can be negative, positive, or neutral. Examples are competition or herbivory. Where some factor is added to the environment with the same range of responses, we get additive reactions such as allelopathy or symbiosis. Examination of both types of reactions should aid in illustrating such mechanisms.

Removal Interactions

It is reasonable to think that when the resources required for crop growth and yield maintenance are in limited supply, the crop yield will drop. If the resources are limited in an intercrop system, one species of the mixture may be able to remove the needed resource better or sooner than the other. The resulting depression of yield of the other species can be caused by competition. Examples of competition in plant mixtures have been reviewed by many authors, especially in crop mixtures (Trenbath, 1976; Harper, 1977) and weed/crop combinations (Zimdahl, 1980; Radosevich and Holt, 1984). Agronomic journals abound with cases where crop response is improved by the addition of some "limiting" resource, and after such a growth response, the prior limitation is most often ascribed to competition. When an investigation is carried out in order to understand the mechanisms of the competition, most often the result is to encounter a more complex interaction of factors.

The unequal capture of light by one crop over another seems to account for part of the dominance in mixtures. Competition for limited light under the canopy of a crop mixture seems to take place. Stern and Donald (1962) studied canopy development over time in a grass and clover mixture (Fig. 5.1) with four levels of nitrogen application. The effect of the nitrogen levels was to change the balance of species enough so that by the last sample date the mixture at low nitrogen levels was dominated by clover, but at high nitrogen levels the grass was dominant. Clover and grass react differentially to nitrogen supply. Although the application of nitrogen increases early leaf production by clover in the mixture it also increases the height of the grass canopy. The grass increases its advantage over clover when adequately supplied with nitrogen, finally reaching almost total suppression of the clover. An initial look at these results might imply that nitrogen competition was the principal problem. When no additional nitrogen was sup-

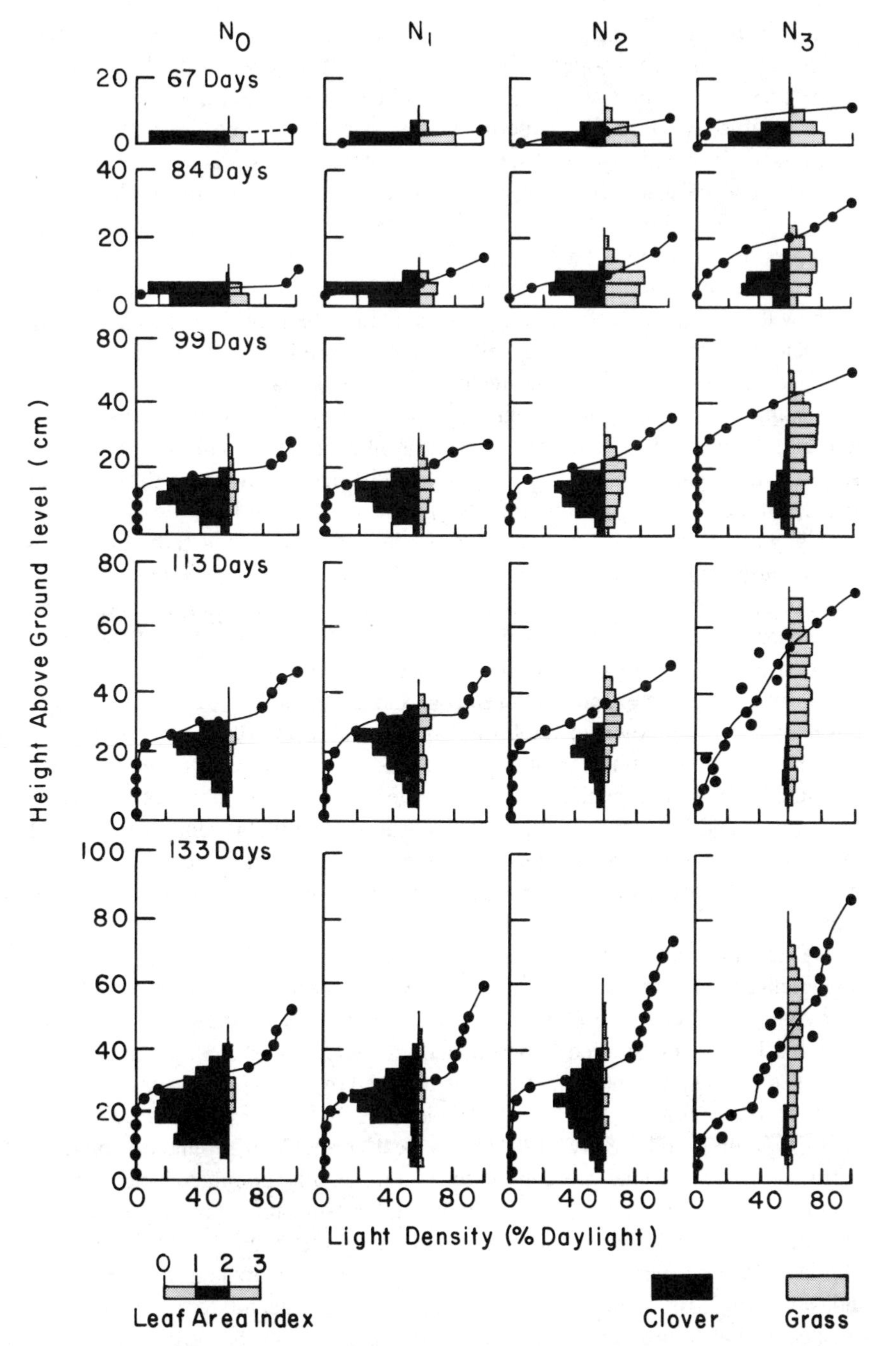

Figure 5.1 Vertical distribution of the leaf area index of grass and clover under four nitrogen treatments together with the profile of light density relative to daylight. (Adapted from Stern and Donald, 1961.)

plied, the nitrogen-fixing legume was at an advantage and able to avoid competition for a limited resource. But the addition of nitrogen shifted dominance to the grass. In the end result, it is impossible to separate nitrogen availability from variable light use. An experiment that started with the examination of a single factor resulted in a complex study of interaction between factors.

The use of water by a crop mixture provides a condition under which competition for a limited resource might come about. There are many reported examples in which one component of a mixture suffers yield depression when water is limiting (Trenbath, 1976; Harper, 1977). However, it is quite difficult to show that competition for water is the actual mechanism. Water as a resource interacts with many other resources. Lowered assimilation rates during water shortage may slow plant canopy development, thus making a crop susceptible to shading. Or, nitrate uptake during water shortage may manifest itself as a nutrient deficiency by the crop. A reduction in root extension during an earlier stage of development by a crop puts it at a distinct disadvantage during water stress, although the reduction initially may have had little to do with water supply. Factors such as interference zones between individual roots, depletion zones caused by root activity, root elongation rates, and root/shoot ratios are some of the characteristics that need to be examined in order to fully understand the nature of the interactions. This is true not only of water, but of the broad range of plant nutrients taken in through the soil solution.

A more specific example comes from a study of rooting depths of different cover-crop grasses sown in apple orchards (Milthorpe, 1961). By comparing moisture availability at different depths, it was found that under *Lolium perenne* moisture levels were reduced to a much greater depth than with two other species (*Poa annua* and *Phleum pratense*), and that subsequent yield reductions occurred in the apples undersown with *Lolium*. Without studies of simultaneous nutrient uptake rates, it would be risky indeed to say that "water competition" was the only factor.

There are enough cases to the contrary that show yield increases, or at least no significant yield reductions, in crop mixtures, despite the fact that removal of resources is occurring (Trenbath, 1974). Very few of these more positive mixtures have been studied experimentally. Yield advantages are most often attributed to complementary interactions between the component crops, with the result of more efficient use of environmental resources. One such study was done by Willey and Reddy (1981) on a pearl millet and groundnut mixture. Partitions in the ground were used to experimentally eliminate belowground interactions between the crops. By calculating LERs for final yields, it was found that the control and dug but nonpartitioned intercrops had a final LER of 1.22 (Table 5.2). The partitioned intercrop suffered a slight reduction in its LER, indicating the loss of some advantage gained from the mixture. The authors concluded that millet benefits from nitrogen available in the soil in groundnut rows; without it, the plants were reportedly pale from nitrogen stress. The belowground interaction could possibly be important for maintaining a competitive balance between the two crops. However, it was concluded that aboveground

Table 5.2 Land-equivalent Ratios of Total Dry Matter Yields throughout the Season and of Final Grain and Pod Yields[a]

	Intercrop control			Intercrop dug (no partitions)			Intercrop partitioned		
Days after sowing	Millet	Groundnut	Total	Millet	Groundnut	Total	Millet	Groundnut	Total
35	0.45	0.72	1.17	0.46	0.72	1.18	0.36	0.77	1.13
50	0.50	0.69	1.19	0.49	0.71	1.20	0.38	0.78	1.16
65	0.52	0.71	1.23	0.53	0.70	1.23	0.44	0.77	1.21
80	0.54	0.71	1.25	0.54	0.72	1.26	0.44	0.80	1.24
95	—	0.68	1.22	—	0.70	1.24	—	0.78	1.22
110	—	0.70	1.24	—	0.71	1.25	—	0.79	1.23
Grain and pod yields	0.50	0.72	1.22	0.52	0.70	1.22	0.43	0.76	1.19

[a]"Expected" LER (i.e., where the degree of competition in intercropping is the same as in sole cropping) is 0.25 for millet and 0.75 for groundnut.
Source: Willey and Reddy, 1981, by permission of Cambridge University Press.

canopy interactions were of greater overall importance, presumably through partitioning of available light (Willey and Reddy, 1981).

Broad application of agricultural inputs usually can overcome any deficiency and eliminate the result of the removal reaction. By being able to determine the specific resource that is limiting, as well as the ecological basis of the interaction, it might be possible to alter the timing, spacing, or even composition of the mixture so that the resource is no longer the limiting factor, avoiding competition and thus yield reduction. Mixtures can be designed more on the basis of the particular ecological requirements of the crop components and by striving for mixtures for which a particular limiting factor is not contested. Such an approach will become increasingly important, either as outside inputs to agriculture become more scarce and/or costly, or as the environmental effects of their more intensive use become better known. This is especially true for smaller-scale systems or those with limited economic resources.

Addition Reactions

From an ecological point of view it is an entirely different situation in a crop mixture when plants are adding to rather than removing materials from the environment. The production of secondary chemicals by plants, followed by their release into the environment and subsequent effects on associated plant species, a concept known as allelopathy, has been recently reviewed in great detail (Rice, 1984). Yet there are many difficulties involved when trying to separate allelopathy from other forms of interference, especially competition (Putnam and Duke, 1978). These include a general lack of nomenclature to describe adequately plant responses that occur in this way, a lack of techniques to separate allelopathy from competition, and the difficulty of proving the existence of direct versus indirect influences from other organisms or microenvironmental modification. Nevertheless, a considerable body of experimental information has been gathered that implicates allelopathy as an important form of interference. Crops that produce allelopathic compounds can have important effects when planted in mixtures, either on the other crops or on weeds. Allelopathy has been demonstrated to play an important role in reducing yields in crops that follow another in rotation, usually due to decomposition products of the residues of the previous crop, or a crop that is inhibited by its own toxins with continual sole cropping of the same soils (see Rice, 1984).

Yet very few studies have been performed in mixed crop systems with the specific intention of looking for the effects of compounds actively being added to the environment by the different members of the mixture. Reports from Russian scientists show both beneficial and detrimental effects from phytotoxins produced in mixtures. In one case, several varieties of legumes were interplanted with corn (*Zea mays*) (Lykhvar and Nazarova, 1970). Many varieties produced yield reductions for corn, and as a consequence, new varieties were selected specifically for use in the mixed cultures. Pronin and Yakovlev (1970) reported that yields of corn and fodder beans increased in mixed plantings, and imply that the influence was associated with a favorable increase of root excretions of each

Table 5.3 Allelopathic Potential of Water Extracts of Air-dried Squash Leaves (*Cucurbita pepo*) in Laboratory Bioassays, Tabasco, Mexico

Test plant	Germination	Radicle length
Corn (*Zea mays*)	100%	85.4%[a]
Beans (*Vigna sinensis*)	96	79.0[b]
Cabbage (*Brassica oleracea*)	92	43.0[b]
Control	100	100.00

[a]*T* test significant at 5 percent.
[b]*T* test significant at 1 percent.
Source: Gliessman, 1983, by permission of *J. Chem. Ecology.*

plant on the other, rather than merely an improvement of nitrogen nutrition by the legume. The interplanting of squash (*Cucurbita pepo*) in corn/bean (*Phaseolus vulgaris*) polycultures in southeastern Mexico aids in weed control (Chacon and Gliessman, 1982; Gliessman, 1983) and produces increases in corn yields and elevated LERs (Amador and Gliessman, 1982). Water extracts of air-dried squash leaves demonstrate a strong allelopathic potential (Table 5.3), especially against the test species that is a recent introduction to the New World. Local farmers describe the use of squash for weed control (Fig. 5.2); in fact, they claimed to plant it for that purpose, and any harvest of fruit was an added bonus (Chacon and Gliessman, 1982). The squash forms a continuous cover over the low-lying weedy species, combining shade with phytotoxins, as well as returning

Figure 5.2 Squash (*Cucurbita pepo*) cover crop dominance following harvest of maize (*Zea mays*); both crops planted simultaneously, near Cárdenas, Tabasco, Mexico. (Photo by Dr. Stephen Gliessman.)

large amounts of biomass to the soil (8 to 10 t dry matter per hectare) (Amador and Gliessman, 1982).

Mutualisms

In contrast to purely additive or removal interactions, mutualisms can often combine several components of interference. The benefits gained by each partner link them into mutual, physiological interdependence. When one component species is absent, the others suffer, and in some cases cannot even exist as free-living organisms. It is difficult to separate mutualisms from the benefits found in beneficial intercropping systems. Yield advantages often come about from the avoidance of direct interference through competition for limited resources or production of phytotoxins which eliminate other competitors. It is much more difficult to demonstrate that benefits are directly derived from the interference. Symbiosis is another term that refers to mutualistic interactions. Ecologically, symbiosis is defined as the permanent, intimate association of two or more dissimilar organisms (Whittaker, 1975). The two terms are frequently interchanged.

Symbiotic nitrogen fixation is the most commonly known mutualistic interaction. Legumes with their accompanying *Rhizobium* bacteria have played important roles in agriculture (Phillips, 1980). The benefits of the mixture of legumes with other crops stem from interactions such as the excretion of nitrogen by the legume for use by the nonlegume, stimulation of soil microorganisms, and the return of nitrogen to the soil (Wilson, 1940). Either through legume/nonlegume mixture or legume/nonlegume rotations, many intercropping systems with this mutualism are practiced today.

The much more widespread importance of mutualisms in ecology are currently being discussed (Boucher et al., 1982). In an evolutionary sense, the benefits gained from mutualisms may tend to be favored over competitive interactions. For example, success of a plant at low levels of available nutrients may come about more through a mutualistic relationship with other species requiring the nutrients, rather than by out-competing them. Resources can be partitioned rather than competed for. Mutual defense from predators, herbivores, or disease organisms can become possible. The long-term benefits to be gained from such an approach to multiple cropping becomes important, then, as we learn to exploit the positive interactions between crop species in mixtures, as well as try to overcome the negative ones.

INTERACTION OF MECHANISMS

In any agroecosystem, the component plant species and the factors of the environment are going to be in constant interaction. The resultant environmental complex changes both temporally and spatially as the crops develop. An excellent example of such a dynamic mixed crop system is the traditional corn/bean/squash polyculture used extensively by farmers throughout mesoamerica (Fig. 5.3). Studies in Tabasco, Mexico, have provided much ecological information on how

Figure 5.3 Interior view of a traditional maize (*Zea mays*), cowpea (*Vigna unguiculata*), and squash (*Cucurbita pepo*) intercrop, near Cárdenas, Tabasco, Mexico. (Photo by Dr. Stephen Gliessman.)

such a system functions (Amador, 1980; Vandermeer et al., 1983; Gliessman, 1984). Corn yields could be stimulated as much as 50 percent beyond monoculture yields when planted in the beans and squash (Table 5.4). Despite yield reductions for beans and squash, the LER of the system was very high. The mechanisms of the stimulation of corn yields have been the subject of further study. On the one hand, beans planted with corn nodulate more and potentially are more active in fixing nitrogen which could become directly available to the corn (Boucher and Espinosa, 1982). This led to observations of net gains of nitrogen in the agroecosystem biomass despite the removal of harvest yields (Gliessman, 1982). This helps ensure better sustainability of the system. Squash fruit yields were low, but foliar biomass was high, contributing to weed control through both removal (shade) and addition (allelopathy) reactions described previously. Noncrop weedy species can impact beneficially on soil resources and insect populations (Gliessman and Altieri, 1982; Altieri and Liebman, Chap 9). Damaging insects are at a disadvantage (Risch, 1980), and the presence of beneficial insects is promoted (Letourneau, 1983). Every component of the system plays an ecological role in maintaining productivity, the combined effects of which have been selected by local farmers for many generations. A cultural and ecological mutualism has been selected for by humans, and an understanding of the intricacies of the interactions teach us a great deal which can be of use for improving present-day cropping systems.

Table 5.4 Yields of Polyculture Corn/Bean/Squash Compared to Monocultures Planted at Four Different Densities, Cárdenas, Tabasco, Mexico

	Monoculture densities				
	Very low	Low	High	Very high	Polyculture
Densities of corn[a]	33,000	40,000	66,000	100,000	50,000
Yield (kg/ha)[b]	990	1,150	1,230	1,170	1,720
Densities of beans	56,800	64,000	100,000	133,200	40,000
Yield (kg/ha)	425	740	610	695	110
Densities of squash	1,200	1,875	7,500	30,000	3,330
Yield (kg/ha)	15	250	430	225	80

[a]Densities expressed as number of plants per hectare.
[b]Yields for corn and beans expressed as dried grain, squash as fresh fruits.
Source: Amador, 1980, by permission of the author.

CONCLUSIONS

Most plants occur in association with other species in natural ecosystems. Our understanding of the complexity of interactions involved in establishing and maintaining such associations has increased greatly (Harper, 1977). Tools exist, then, to better understand the manner in which individual plants in crop mixtures interact with their neighbors. The term interference is used to describe the effect that the presence of one plant has on the environment of another. Addition and removal reactions on the environment are the basis of interference interactions.

The most commonly invoked explanation for yield reduction in mixtures is the removal reaction through competition for a limited resource. Density, the number of plants per unit of area, is the primary component of competition, and agronomists have perfected knowledge of optimal density plantings for sole-crop plantings in order to maximize yields. Very little work has been done on optimizing densities for intercrop systems. In order to accomplish this, our understanding of species proportions and arrangements must be increased. Additive versus substitution arrangements need to be examined from the perspective of interference interactions and resource use in the agroecosystem. Complementary crop mixtures based on different resource needs, either physiologically or temporally, permit crop mixtures capable of over-yielding. Addition reactions, either through allelopathic interference or mutualisms, alter the crop environment and can be of benefit in crop mixtures. The potential role of such mixtures for weed or pest management needs considerably more research. Expanding our view of intercrop systems, then, to include such aspects as pest management, soil conservation, and optimal resource utilization on the one hand, and over-yielding, crop diversification, and multiple-use benefits on the other, can contribute to more resource-conserving and sustainable agroecosystems.

Figure 5.4 Complex, multistoried home garden agroforestry system combining annuals, shrubby perennials, and trees with different ecological requirements and cultural uses, near Cañas, Guanacaste, Costa Rica. (Photo by Dr. Stephen Gliessman.)

From an ecological perspective, all plants in the agroecosystem can be shown to play a role. This includes what are considered to be weeds, perennial shrub and tree components, and even more complex agroforestry combinations (Fig. 5.4). As the need for resource use efficiency combines with the demand for agricultural outputs, being able to employ multiple cropping techniques will become more essential. A theoretical basis for intercropping is being developed (Vandermeer, 1984; other chapters in this volume), and this must be supported by an understanding of the dynamics and complexity of plant interactions involved. This process can contribute greatly to successful multiple crop system design and management.

REFERENCES

Amador, M. F. 1980. Comportamiento de tres especies (Maiz, Frijol, Calabaza) en policultivos en la Chontalpa, Tabasco, México, Tesis Profesional, CSAT, Cárdenas, Tabasco, México.

Amador, M. F., and Gliessman, S. R. 1982. Response of three species (corn, beans, and squash) in polyculture in the Chontalpa, Tabasco, Mexico (unpublished manuscript), 22 pp.

ASA (American Society of Agronomy). 1976. *Multiple Cropping,* Amer. Soc. Agron. Spec. Publ. 27, Madison, Wisconsin, 378 pp.

Boucher, D., and Espinosa, J. 1982. Cropping systems and growth and nodulation responses of beans to nitrogen in Tabasco, Mexico, *Trop. Agric.* 59:279–282.

Boucher, D. H., James, S., and Keeler, K. H. 1982. The ecology of mutualism, *Ann. Rev. Ecol. Syst.* 13:315–348.

Chacon, J. C., and Gliessman, S. R. 1982. Use of the "non-weed" concept in traditional tropical agroecosystems of south-eastern Mexico, *Agro-ecosystems* 8:1–11.

Daubenmire, R. F. 1974. *Plants and environment,* 3rd ed. John Wiley & Sons, New York, 422 pp.

Gliessman, S. R. 1982. Nitrogen distribution in several traditional agro-ecosystems in the humid tropical lowlands of southeastern Mexico, *Plant Soil* 67:105–117.

———. 1983. Allelopathic interactions in crop/weed mixtures: Applications for weed management, *J. Chem Ecol.* 9:991–999.

———. 1984. An agroecological approach to sustainable agriculture, in: *To Meet the Expectations of the Land,* (W. Jackson, W. Berry, and B. Colman, eds.), Northpoint Press, Berkeley, California, pp. 160–171.

Gliessman, S. R., and Altieri, M. A. 1982. Polyculture cropping has advantages, *Calif. Agric.* 36:14–17.

Gliessman, S. R., Garcia, R., and Amador, M. 1981. The ecological basis for the application of traditional agricultural technology in the management of topical agro-ecosystems, *Agro-ecosystems* 7:173–185.

Harper, J. L. 1961. Approaches to the study of plant competition, in *Mechanisms in Biological Competition,* (F. L. Milthorte, ed.), *Symp. Soc. Exp. Biol.* 15:1–39.

———. 1964. The nature and consequences of interference amongst plants, in: *Genetics Today, Proc. XI Int. Cong. Genet.* 2:465–482.

———. 1977. *Population Biology of Plants,* Academic Press, Orlando, Florida, 892 pp.

Harwood, R. R. 1979. *Small Farm Development,* Westview Press, Boulder, Colorado, 160 pp.

ICRISAT (International Crops Research Institute for the Semi-Arid Tropics). 1981. *Proc. Int. Workshop on Intercropping,* 10–13 January 1979, Hyderabad, India, 401 pp.

Kass, D. C. L. !978. Polyculture cropping systems: Review and analysis, *Cornell Int. Agric. Bull.* Ithaca, New York. 69 pp.

Letourneau, D. 1983. Population dynamics of insect pests and natural control in traditional agroecosystems in tropical Mexico, Ph.D. dissertation, University of California, Berkeley, California.

Lykhvar, D. F., and Nazarova, N. S. 1970. On importance of legume varieties in mixed cultures with maize, in: *Physiological Biochemical Basis of Plant Interactions in Phytocenoses,* vol. 1, (A. M. Grodzinsky, ed.), pp. 83–88 (in Russian, English summary).

Milthorpe, F. L. 1961. The nature and analysis of competition between plants of different species, in: *Mechanisms of Biological Competition,* (F. L. Milthorpe, ed.), *Symp. Soc. Exp. Biol.* 15:330–355.

Muller, C. H. 1969. Allelopathy as a factor in ecological process, *Vegetation* 18:348–357.

Odum, E. P. 1971. *Fundamentals of Ecology,* 2nd ed., Saunders, Philadelphia, Pennsylvania, 546 pp.

Phillips, D. A. 1980. Efficiency of symbiotic nitrogen fixation in legumes, *Annu. Rev. Physiol.* 31:29–49.

Pronin, V. A., and Yakovlev, A. A. 1970. Influence of nitrogen conditions and rhizospheric microorganisms on the interrelations of maize and fodder beans in mixed culture, in: *Physiological Biochemical Basis of Plant Interactions in Phytocenoses,* vol. 1, (A. M. Grodzinsky, ed.), pp. 93–101 (in Russian, English summary).

Putnam, A. R., and Duke, W. B. 1978. Allelopathy in agroecosystems, *Annu. Rev. Phytopathol.* 16:431–451.

Radosevich, S. R., and Holt, J. S. 1984. *Weed Ecology, Implications for Vegetation Management,* John Wiley & Sons, New York, 265 pp.

Rice, E. L. 1984. *Allelopathy,* 2nd ed., Academic Press, New York, 422 pp.

Risch, S. 1980. The population dynamics of several herbivorous beetles in a tropical agro-ecosystem: the effect of intercropping corn, beans and squash in Costa Rica, *J. Appl. Ecol.* 17:593–612.

Stern, W. R., and Donald, C. M. 1962. Light relationships in grass-clover swards, *Aust. J. Agric. Res.* 13:599–614.

Trenbath, B. R. 1974. Biomass production in mixtures, *Adv. Agron.* 26:177–210.

———. 1976. Plant interactions in mixed crop communities, in: *Multiple Cropping,* Amer. Soc. Agron. Spec. Publ. 27, pp. 129–169.

Vandermeer, J. 1984. The interpretations and design of intercrop systems involving environmental modification by one of the components: A theoretical framework, in: *Biol. Agric. Hortic.* 2:135–156.

Vandermeer, J., Gliessman, S. R., Yih, K., and Amador, M. 1983. Overyielding in a corn-cowpea system in southern Mexico, *Biol. Agric. Hort.* 1:83–96.

Whittaker, R. H. 1975. *Communities and Ecosystems,* 2nd ed., Macmillan Publishing Co., Inc., New York.

Willey, R. W. 1979a. Intercropping—its importance and research needs. Part 1. Competition and yield advantages. *Field Crop Abstr.* 32:1–10.

———. 1979b. Intercropping—its importance and research needs. Part 2. Agronomy and research approaches. *Field Crop Abstr.* 32:73–85.

Willey, R. W., and M. S. Reddy 1981. A field technique for separating above-and below-ground interactions in intercropping: An experiment with pearl millet/groundnut, *Exp. Agric.* 17:257–264.

Wilson, P. W. 1940. *The Biochemistry of Symbiotic Nitrogen Fixation,* University of Wisconsin Press, Madison, Wisconsin.

Zimdahl, R. L. 1980. *Weed-Crop Competition: A Review,* Int. Plant Protection Center, Oregon State University, Corvallis, Oregon, 196 pp.

Chapter 6

Cereals in Multiple Cropping

M. R. Rao

Cereals are grown throughout the world to provide food for human consumption and feed and fodder for livestock. They are grown on 73.5 percent of the world's arable land and contribute 74.5 percent of the global calorie production (FAO, 1982). Wheat (*Triticum aestivum*) and rice (*Oryza sativa*) are known as fine grains whereas maize (*Zea mays*), sorghum (*Sorghum bicolor*), and millets (*Pennisetum* spp. and others) are referred to as coarse grains. The geographic importance of any particular cereal varies with climate as well as the food habits and living standards of people. Most rice and two-thirds of the wheat produced is used for human consumption but only 25 percent of the coarse grains is consumed directly as human food. The developing nations of semiarid Asia, Africa, and Latin America produce more than 90 percent of the coarse grains, most of which is utilized for human food (Kanwar and Ryan, 1976). The bulk of the coarse grains in developed nations is used as animal feed.

Demand for food is growing with the ever-increasing world population. Compared to the present production at about 1.7 billion tons, the demand for cereals is likely to go up to 2.4 billion tons by the year 2000 (Aziz, 1977). While demand for wheat and rice may increase in the next two decades by 31 and 53 percent, respectively, the demand for coarse grains may double. The developed nations might be able to meet their cereal demand by increasing production at 1.8 percent per annum, but most of the developing nations, with a growth rate of 2.5 percent per annum in cereal production, fall short of their requirement which is increasing at a rate of 3.3 percent per annum due to a high population growth rate.

Cereals play an important role in sustaining livestock which exceed the human population by 2 to 3 times. Cattle thrive principally on cereal straw in the arid and semiarid regions. Fodder scarcity affects milk cattle as well as draft animals, which are important for agricultural operations in developing countries. Food shortage can be corrected by imports, but the shortage of fodder/straw cannot be easily corrected by imports. Thus, large-scale livestock loss in the frequently drought-prone Sahel of Africa, northeast Brazil, and arid and semiarid India is a common phenomenon. This emphasizes the need for ensuring local availability of fodder and the importance of cereal components in cropping systems.

To increase cereal production, the options considered hitherto were to increase the area under cultivation as well as yield per hectare. Both options are possible, but most economically arable land is already cultivated and increasing yield is difficult because (1) yield per hectare of cereals is fairly high compared to other arable crops and further improvement is relatively difficult, and (2) intensive modern cultivation techniques and high inputs are required which in many developing countries are not available to small farmers. However, multiple cropping by sequential planting or intercropping provides a suitable mechanism for increased crop production per unit area and time. While the improved, dwarf, and fertilizer-responsive genotypes paved the way for the green revolution in some cereals in the mid-1960s and 1970s (Dalrymple, 1975), the practice of multiple cropping has the potential to usher in another green revolution in the tropics (Greenland, 1975). The favorable temperature and light regimes in the tropics enable a range of crops to grow over a greater part of the year. However, the length of the cropping period is largely determined by rainfall and/or irrigation and soil moisture holding capacity. Multiple cropping is not new and has been in vogue for centuries in Asia under reservoir and river valley projects. Innovations today include newer genotypes, greater understanding of soil-water-plant systems, and improved management. This chapter discusses cereal-based multiple cropping systems in relation to the management of cereals for improved yields.

CEREAL COMPONENT IN MULTIPLE CROPPING SYSTEMS

Cereals constitute an important component of many cropping systems. As food crops they often receive greater attention than other crops, evident from the more frequent irrigation and higher inputs devoted to them. In India, cereals occupy 74 percent of the irrigated area, rice and wheat together account for 66 percent (SAI, 1979). Whenever new irrigation schemes were introduced, farmers tended to switch from traditional intercropping to rice- or wheat-based cropping systems. Even with dryland cropping, they preferred to invest meager resources in a small area of cereals under assured irrigation (Jodha, 1980). Cereals generally are characterized by fast growth and early maturity, though genotypes in some traditional systems are photosensitive, slow growing, and late maturing. They rapidly cover the ground, which suppresses weeds and reduces runoff. The early maturing cereals enter mostly into sequential multiple cropping systems, while photosensitive long-duration genotypes are intercropped. Warm season C_4 cereals are efficient plants, and their inclusion in the cropping systems enhances the efficiency of resource utilization. Cereals such as rice, sorghum, maize, or millets can be grown throughout the year in most parts of the tropics, provided water is not limiting. Most cropping systems involve at least one cereal crop, and many systems are based solely on cereals. Examples are two or three crops of rice per year in the lowlands of the tropics, two crops of maize or sorghum per year in the uplands, and crop sequences such as maize-wheat and rice-wheat where wheat can be grown. Intercropping with two or more cereals is widespread in the rainfed semiarid tropics (Norman, 1974; Pinchinat et al., 1976; Aiyer, 1949).

Information on the total area use and production of cereals in multiple cropping systems is not available because countries generally do not collect data based on cropping systems. However, surveys conducted in some countries indicate the extent of intercropping and the socioeconomic factors that are involved. Village-level studies conducted by ICRISAT in rainfed, semiarid India indicate that more than 90 percent of the sorghum and all of the pearl millet is grown in intercropping (Jodha, 1980). This practice seemed to be discouraged by high-yielding cultivars and large holdings. Norman (1974) also observed in northern Nigeria that 83 percent of the total cropped area was devoted to crop mixtures. Though the mixtures included two to six crops, the two-crop systems were more predominant. A majority of the systems contained one or more cereals. A high proportion of cereal area is similarly under intercropping in other west African countries (Table 6.1). Steiner (1982) reported that 80 percent of food production in this region might be coming from intercropped fields. In some Central American countries more than 85 percent of the sorghum, mostly photosensitive types produced by small farmers, is intercropped with maize. Maize is also intercropped in Central and South American countries, for example, 56 percent in Brazil (Rao and Morgado, 1984).

Table 6.1 Extent of Mixed Cropping with Cereals in Some African Countries

	Percent mixed cropping			
Cereal	**Nigeria**	**Uganda**	**Ghana**	**Ivory Coast**
Rice	58	—	29	75
Maize	76	84	84	80
Sorghum	80	46	95	72
Millet	90	—	87	81
Eleusine	—	48	—	—

Source: Data on Nigeria and Uganda from Okigbo and Greenland, 1976. Data on Ghana and Ivory Coast from Steiner, 1982.

In traditional systems farmers plant a greater proportion to their staple cereal, ignoring the productivity of the other component. Thus we find in most systems several rows of the cereal alternating with only a few of the other crop (Aiyer, 1949). It is important to produce enough food for subsistence needs, especially when the low-yielding, traditional genotypes are used. However, with present-day, high-yielding genotypes the subsistence criteria need no longer restrict farmers from adopting alternative productive arrangements.

TYPES OF MULTIPLE CROPPING SYSTEMS WITH CEREALS

Multiple cropping practices vary widely from the simplest system of two crops a year in sequence to complex intercropping with many crops. Multiple-cropped lands can be categorized into lowlands, irrigated uplands, and uplands dependent solely on rainfall. Rice-based systems predominate in the lowlands (IRRI, 1982; Gomez and Gomez, 1983). The number of crops harvested per year and the crops that follow or precede rice depend on the period of water availability and the degree of control of water (Fig. 6.1). Where irrigation or rainfall (>200 mm per month) extends over 9 to 10 months the systems could be rice-rice-rice, rice–rice–upland crop, or upland crop–rice–rice. When this period is limited to 6 to 8 months, another system (upland crop–rice–upland crop) may be appropriate. However, because of the reduced growing period, the rice genotypes should be early maturing. Dry seeding rice and use of minimum tillage techniques may help establish crops quickly. When water is available for only 4 to 5 months, only one rice crop is grown. Multiple cropping can be attempted by growing

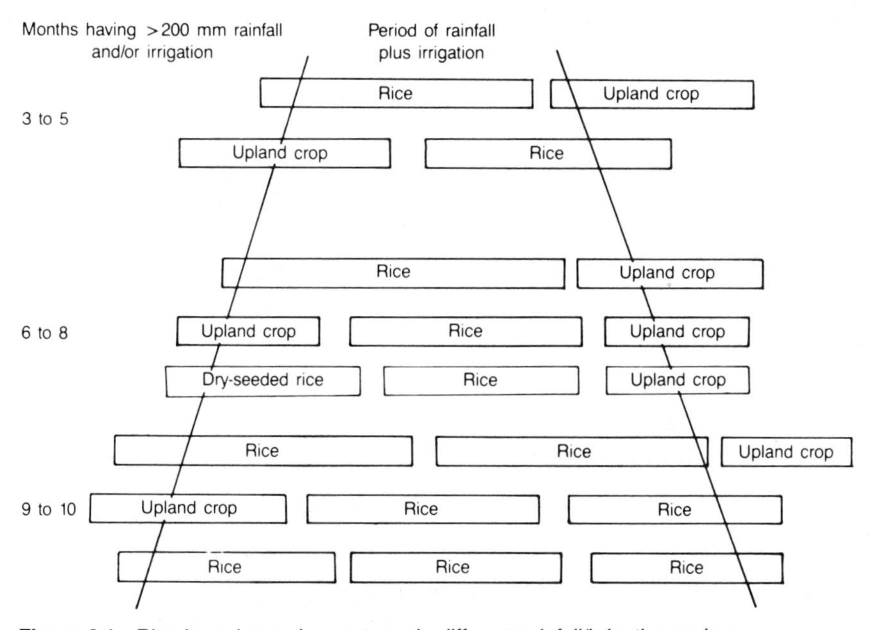

Figure 6.1 Rice-based cropping systems in different rainfall/irrigation regimes.

upland crops before or after rice depending on the early season rainfall and residual moisture. The upland crops that can follow rice include legumes such as mung bean (*Vigna radiata*), black gram (*Vigna mungo*), soybean (*Glycine max*), and groundnut (*Arachys* sp.), cereals such as maize and sorghum, and other crops such as cotton (*Gossypium* sp.) and vegetables. Where the winter is cool (approximately above 25°N or S latitude) and there is good control of water, then wheat, barley (*Avena sativa*), mustard (*Brassica* spp.), chickpea (*Cicer arietinum*), and potato (*Solanum* spp.) can follow rice (Fig. 6.2). Upland crops

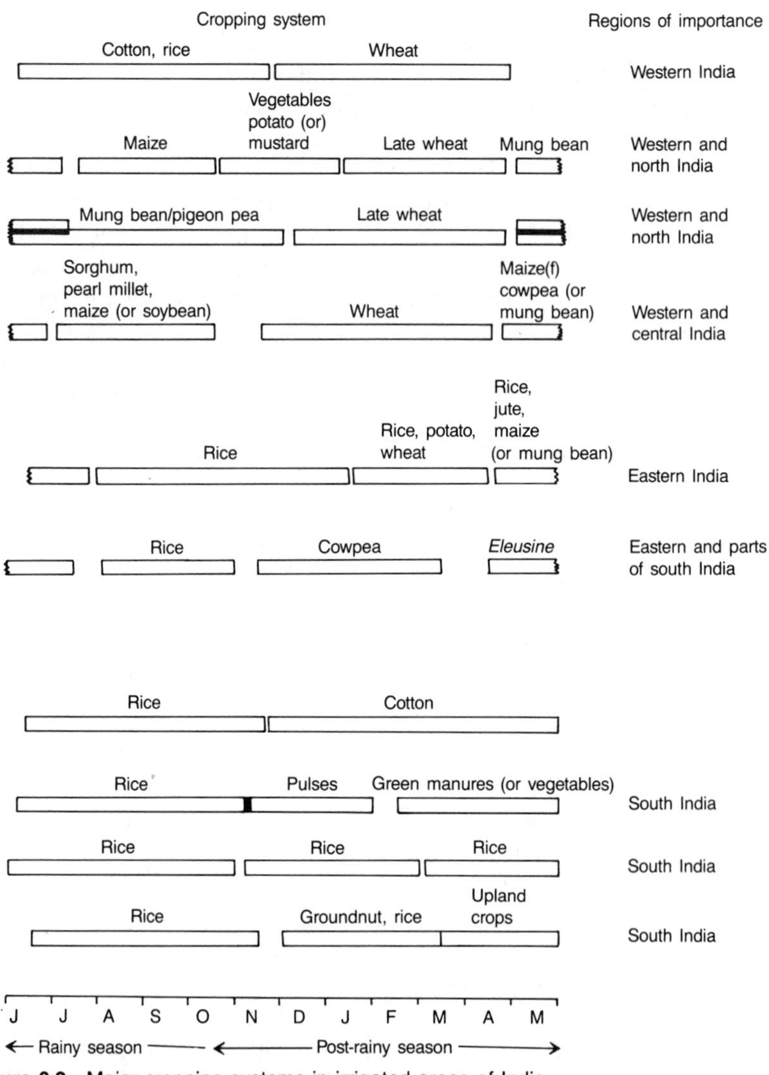

Figure 6.2 Major cropping systems in irrigated areas of India.

can be planted before rice only where the rains start slowly, but they should be early and withstand wet weather at harvest (e.g., cowpea, *Vigna unguiculata*, or maize for green cobs).

On irrigable and high rainfall uplands, sequential multicropping with a wide range of crops is possible. The systems could be cereal-cereal and cereal-legume, oil seeds, or other cash crops. Additional crops are introduced into the system depending on the flexibility of managing the main crops. For example, it has

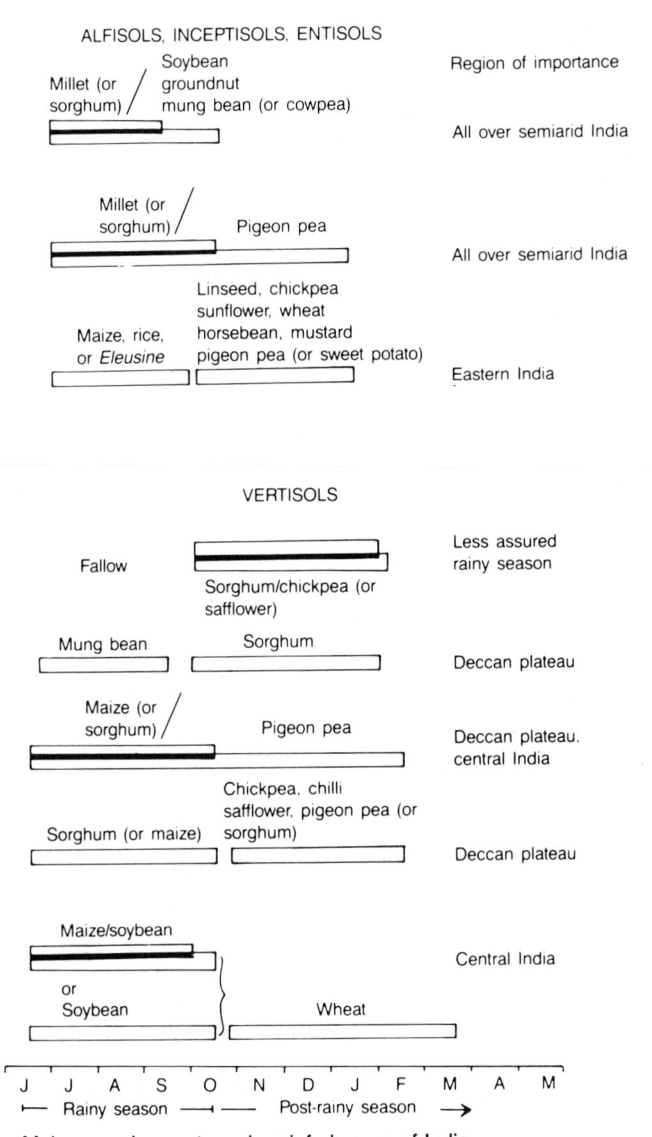

Figure 6.3 Major cropping systems in rainfed areas of India.

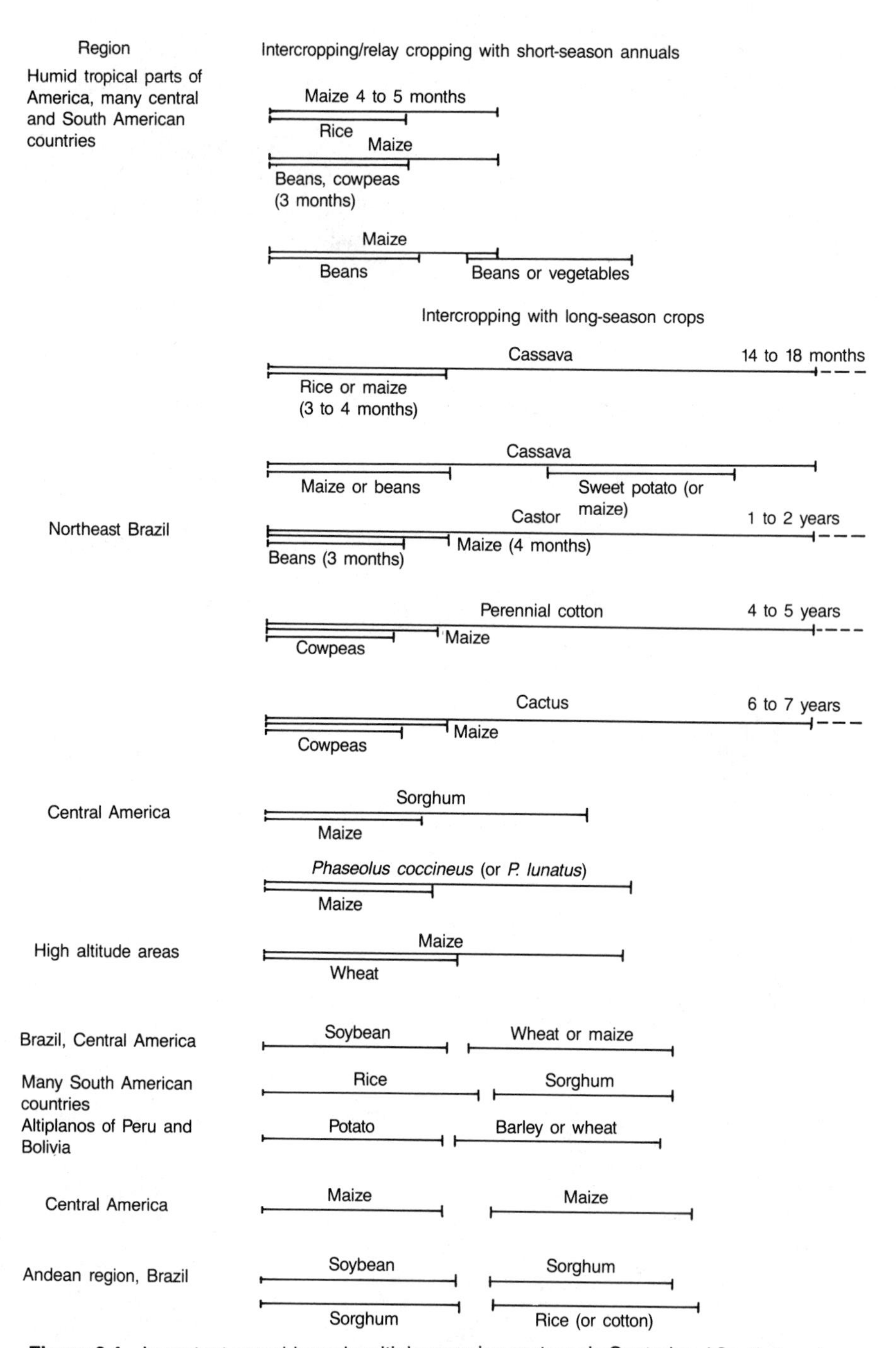

Figure 6.4 Important cereal-based multiple cropping systems in Central and South America.

been shown in northern India that potato or mustard can be added to maize-wheat by relay planting either of these in the standing maize and delaying wheat by 2 months. Short season mung bean or fodder crops can be grown after the harvest of wheat in summer. In the seasonally dry, humid, and semiarid tropics, multiple cropping predominantly takes the form of intercropping. If the cropping season is longer (6 to 9 months) intercropping is mostly with long-season crops such as cassava (*Manihot esculenta*), pigeon pea (*Cajanus cajan*), yams (*Dioscorea* spp.), cotton, and photosensitive sorghum (Figs. 6.3 to 6.5). Although double cropping with sequential planting of 3- to 4-month long crops can be practiced, farmers prefer intercropping for a number of reasons (Norman, 1974; Jodha, 1980). When the cropping period is limited to 4 to 5 months or less, as in most parts of semiarid to arid environments, intercropping with short-season annual crops is the only way to achieve multiple cropping. The following sections describe in detail different intercropping systems.

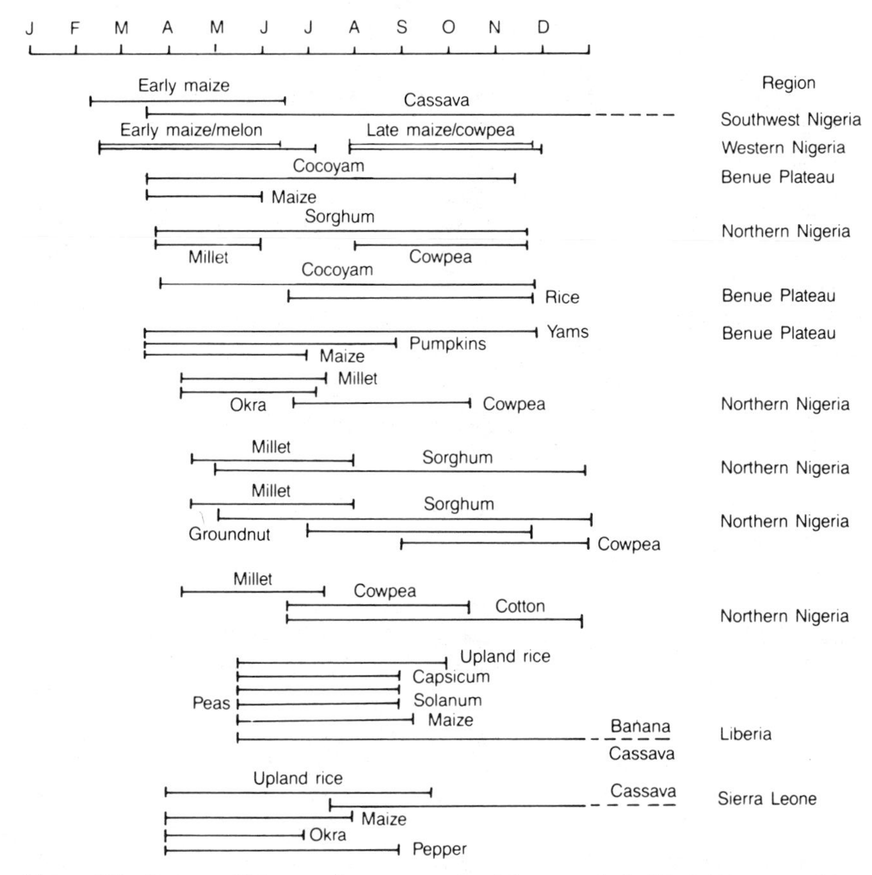

Figure 6.5 Some multiple cropping systems involving cereals in West African countries (Okigbo and Greenland, 1976).

INTERCROPPING SYSTEMS OF CEREALS WITH LONG-DURATION CROPS

Long-season crops initially grow slowly, and require wide spacing as they produce large canopies in the later growth stages. Intercropping with cereals is an excellent way of improving resource utilization because the cereal utilizes the rainy season resources while the later-maturing crop exploits the post-rainy season resources such as residual moisture. The cereal in traditional systems is meant for the subsistence needs of small farmers whereas the other species are grown to provide protein supplement (e.g., pigeon pea, *Dolichos*), additional calories (e.g., cassava), or marketable cash product (e.g., cotton, castor, *Ricinus communis*). The larger temporal difference between species has an imporant bearing on their management.

Cereal/Pigeon Pea Systems

Cereal/pigeon pea systems are popular in India (Aiyer, 1949), East Africa (Enyi, 1973), and the Caribbean (Dalal, 1974). The amount of rainfall determines the types of cereal associated with pigeon pea: rice in regions of 1000 to 1500 mm annual rainfall; maize, around 750 to 1000 mm; sorghum in 500 to 750 mm; and millets in the drier areas having 400 to 600 mm rainfall. Most cereals, depending on their growth duration and height, reduce the growth of pigeon pea and can be ranked for competitiveness: maize > sorghum > pearl millet > setaria (Rao and Willey, 1980a).

Sorghum is the most common cereal intercropped with pigeon pea on a variety of soils in the Deccan Plateau in India. Both crops are planted in the beginning of rains (June to July), with more sorghum in the mixture than pigeon pea. Sorghum is harvested after $3\frac{1}{2}$ to $4\frac{1}{2}$ months while pigeon pea matures in about 6 to 9 months, depending on genotype. Several studies showed that a higher pigeon pea proportion would markedly improve the performance of the system by increasing the pigeon pea yield with little or no reduction in sorghum yield (Willey et al., 1981).

Figure 6.6 shows the growth pattern of crops in sole cropping and in an improved intercrop of one row pigeon pea to two rows sorghum at 45-cm row spacing using the respective sole crop optimum populations, i.e., 180,000 plants per hectare of sorghum and 40,000 to 50,000 plants per hectare of pigeon pea (Willey et al., 1981). On the Vertisols which hold 200 mm or more of available water, the intercropped sorghum accumulated dry matter at a slightly lower rate than the sole crop and produced almost as much yield (94 percent) as sole sorghum (4500 kg/ha). Thus, the higher proportion of pigeon pea had little effect on sorghum yield. Although pigeon pea growth was greatly reduced at first, it compensated in growth after the harvest of sorghum and finally produced 53 percent of the sole crop dry matter. Due to an improved harvest index (30 percent), the intercropped pigeon pea produced a substantial additional yield of 945 kg/ha or 72 percent of sole crop yield. The total productivity of the system was thus 66 percent higher than that of sole cropping. Similarly, planting crops in alternate rows or changes in sorghum population affected only the proportional

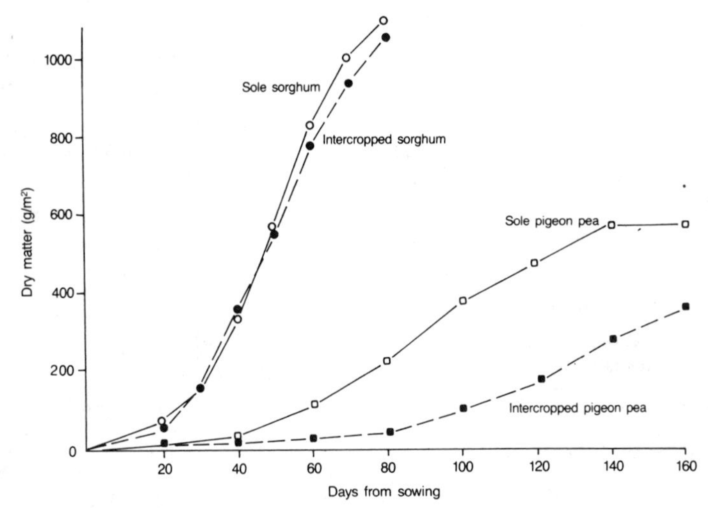

Figure 6.6 Dry-matter production by sorghum and pigeon pea in sole cropping and 2:1 intercropping.

yields and not the overall advantage of the system (ICRISAT, 1980). This observation typifies the wide adaptability and potential benefits of cereal/pigeon pea systems.

On light-textured Alfisols and shallow Inceptisols the advantage of intercropping is less than on Vertisols because of less residual moisture in these soils in the post-rainy period and consequently less compensation from pigeon pea. Whatever the yield advantage (usually 30 to 50 percent), intercropping represents a major benefit because of too short a season for sequential cropping on these soils (ICRISAT, 1980).

Studies at ICRISAT indicated that maize is easy to manage in intercropping with pigeon pea on Vertisols in the high (>750 mm) and assured rainfall areas (Krantz, 1981). The maize/pigeon pea can be managed similarly to sorghum/pigeon pea to realize yield advantages from 44 to 80 percent (Chaudhury, 1981; Rao and Willey, 1980a). Dalal (1974) observed in the Caribbean a 56 percent advantage for mixed planting of maize and pigeon pea, but an 83 percent advantage for alternate row intercropping over sole cropping.

In high rainfall areas of eastern India, pigeon pea is intercropped with upland rice. The system produced 80 to 90 percent of the respective sole crop yields with about 70 percent advantage for intercrop over sole cropping (Sen et al., 1966; Chaudhury, 1981). In situations where high moisture is likely to hamper pigeon pea growth, it may be relay planted into rice after the peak rains.

Millets are associated with pigeon pea in low and unreliable rainfall areas and on shallow, medium- to light-textured soils. Pearl millet is more important than the *Setaria* millet because of its widespread use and higher productivity. Since both millets mature early, they escape competition from pigeon pea and provide the least competition to pigeon pea. Consequently the LER advantage of intercropping pigeon pea with pearl millet was 78 percent (Rao and Willey, 1983b), and for intercropping with the shorter *Setaria* millet it was as high as 90 to 100 percent (Rao and Willey, 1980a). Intercropping with finger millet is carried out to a limited extent on red soils in Karnataka state, India, but because of its heavy tillering and lateral spread it was more competitive to pigeon pea and offered less advantage compared to other millets (ICRISAT, 1974).

The large temporal difference between component species in cereal/pigeon pea systems can be further exploited by introducing another crop that matures earlier than the cereal or by planting after the cereal is harvested. Planting two or more crops with pigeon pea is common in India. In studies at ICRISAT where a number of crops were added between 150 cm rows of pigeon pea, the addition of groundnut alone gave a yield advantage of 41 percent, but the further addition of a setaria millet increased this to 66 percent (Rao et al., 1977).

Cereal/Cassava Systems

Maize and rice are the principal cereals associated with cassava in the humid tropics (Okigbo, 1981; Moreno and Hart, 1978; Mattos and Souza, 1981). Intercropping with sorghum is observed in the semiarid tropics. These cereals ranked for their competitiveness as sorghum > maize > rice, which corresponds to the height of the crops (Zandstra, 1978). Both cassava and maize are planted either at the same time in the beginning of rains, or cassava is relay planted in maize from 4 weeks after planting to as late as after physiological maturity, the stage with maximum dry weight (Moreno and Hart, 1978). In studies at the International Institute for Tropical Agriculture (IITA), Okigbo (1981) observed that relay planting was advantageous only if cassava was planted within 2 months of maize planting. Maize matures in 100 to 150 days, depending on the genotype, and cassava is harvested 10 to 18 months after planting.

Row arrangements vary widely; both crops may be planted in the same row, in alternate rows, or with two to three rows of the cereal alternating with paired rows of cassava. In the humid tropics where moisture is not the major limiting factor and cassava grows for a longer period, the yield advantage of intercropping is very high. Maize yields are unaffected, and although cassava received setback initially, it later produced a high proportion of its sole crop yield (Moreno and Hart, 1978; Sinthuprama, 1978).

In trials at CATIE (Centro Agronómico Tropical de Investigación e Enzeñanza, Costa Rica, 1977–1978) intercropped maize gave 93 percent of the sole crop yields and cassava produced up to 78 percent of the sole crop giving a LER of 1.71. In some situations the total LER was as high as 2.00 (Moreno and Hart, 1978). Similarly, yield advantages of 58 to 77 percent were noted in southern Nigeria (IITA, 1982). In the Amazon region, Marques Porto et al. (1978) observed that a maize/cassava system was more productive than rice/cassava. Of all the cereal/cassava systems, sorghum/cassava gave the lowest advantage. In the semiarid northeast of Thailand and in Brazil intercropping cassava with sorghum gave little or no advantage because in those limited-moisture environments sorghum was very competitive with cassava (Patanothai, 1983; Rao, 1984).

Additional crops can be included in the cereal/cassava systems particularly where moisture is not limiting and cassava grows for a longer period. Systems that included a third, low-canopy type crop such as beans (*Phaseolus vulgaris*), snap beans, lima beans (*Phaseolus lunatus*), and sweet potato (*Ipomoea* sp.) gave higher returns compared to two-crop systems (Hart, 1975; Dos Santos, 1978), but not with the addition of melon (*Cucumis melo*) to a maize/cassava system (Okigbo, 1981). Additional examples are discussed in Chap. 7.

Cereal/Cotton Systems

Cotton is intercropped with sorghum or *Setaria* in India (Aiyer, 1949), with maize or sorghum in West Africa (Baker, 1979a), and with maize in northeast Brazil (Rao, 1984). The most common practice in the Deccan Plateau, India, is to plant several rows of cotton with a strip of sorghum and pigeon pea, either in distinct rows or mixed. With the introduction of improved genotypes of cotton there has been a shift to sole cropping but studies have still shown advantages of intercropping. Thus with Lakshmi cotton and CSH6 sorghum, Prithviraj et al. (1972) showed intercropping advantages of more than 30 percent for row arrangements of two sorghum/two cotton or three sorghum/two cotton. Moreover, both these intercrop systems gave higher monetary returns than either sole crop.

In southern Nigeria where the growing period is longer, cotton is usually relay interplanted into maize but in northern Guinea or Sudan zone delayed planting severely reduces cotton yield. Baker (1979a) showed that the cereal/cotton intercrop can be improved by planting both crops simultaneously at the beginning of rains. Cotton yield decreases with increased maturity of the cereal. Therefore, LER advantage is 62 percent with the earliest Ex-Ghana millet, 28 percent with Ex-Bornu millet, 16 percent with maize, and 21 percent with the traditional tall sorghum. In this 700- to 1000-mm rainfall zone, the better choice for intercropping cotton is millet or early maturing genotypes of other cereals.

Perennial cotton (*Gossypium hirsutum* var. Marie-galante) cultivated for 4 to 5 years in the arid zone of northeast Brazil is invariably intercropped in the first year with maize and/or cowpea (Rao, 1984). Because of low and unreliable rainfall (400 to 600 mm), maize performs poorly in this region and competes with cotton for moisture. A better approach is to substitute cowpea, sorghum, or millet for all or a part of the maize in the intercrop system.

Cereal/Castor Systems

Cereals intercropped with castor include sorghum and millets in India (Aiyer, 1949) and maize with or without beans in northeast Brazil (Rao, 1984). Castor bean is generally sensitive to the cereal competition and shows limited compensation after cereal harvest. Hence no appreciable advantage was observed for intercropping castor with cereals in India, particularly high yielding dwarf cultivars of castor (Freyman and Venkateswarlu, 1977; Reddy et al., 1965; Chaudhury, 1981). Only a few studies reported worthwhile advantage of intercropping castor with sorghum (Rao and Willey, 1980a; Tarhalkar and Rao, 1981; Chinnappan and Paliniappan, 1980).

Cereals with Other Long-Cycle Crops

Eleusine millet is widely intercropped with field beans (*Dolichos lablab*) in the Karnataka State, India, where 4 to 12 rows of the millet alternate with 1 row of beans (Aiyer, 1949). In the improved system two or four rows of millet were planted to every row of beans, which produced 70 to 100 percent millet and 20 to 50 percent bean yields. Grouping the millet rows to as close as 20 cm to create more space for the legume component has improved the legume yields with little affect on cereal yields (Hegde and Havangi, 1974).

Spineless cactus (*Opuntia ficus* var. *indica*) is cultivated for fodder in the *sertão* of northeast Brazil. Initially it grows slowly and is harvested first in the third year after planting. It is invariably intercropped with maize with or without cowpeas in the first year. Maize hardly affects cactus yield and provides a bonus crop during the nonproductive phase of cactus. Maize can be substituted with sorghum or millet (Rao, 1984).

In southern Nigeria, yam (*Dioscorea rotundata* Poir) is intercropped with maize, the principal cereal of the region. This two-crop system showed about 30 percent yield advantage which increased to 49 percent when erect cowpea and melon (*Colocynthis vulgaris* Schrad) were added to the system (Okigbo, 1981).

Cereals can also be intercropped in the early years of orchard crops such as citrus or mango, or cultivated as understory crops in coconuts. The period during which cereals can be associated with perennials is limited because of their relatively poor shade tolerance.

INTERCROPPING SYSTEMS OF CEREALS WITH LOW-CANOPY LEGUMES

Intercropping cereals with low-canopy legumes is widely practiced throughout the tropical world and to a limited extent in some temperate regions. The notable associations of summer cereals are those of maize, sorghum, pearl millet or *Eleusine* millet with legumes such as soybean, groundnut, beans (*Phaseolus vulgaris*), mung bean, cowpea, and black gram. Winter cereals such as wheat and barley are intercropped with chickpea, lentil (*Lens esculentis*), or peas (*Pisum sativum*) in the post rainy season in the Indian subcontinent (Aiyer, 1949), while barley (*Hordeum vulgare*), oats (*Avena sativa*), and rye (*Secale cereale*) are associated with peas or a variety of fodder legumes in temperate countries. A contrasting feature of these systems from the ones described earlier is that complementary effects between species are more likely due to spatial differences in canopy heights and rooting patterns rather than due to temporal differences. Some researchers have postulated that there is a transfer of nitrogen fixed by legumes to the associated cereal (Willey, 1979). However, since both crops mature with little time difference, yield advantages are low. These systems usually occupy the land for $3\frac{1}{2}$ to 5 months, but where the growing period is longer other crops may be relay planted following the harvest of the early maturing component or planted sequentially after the harvest of both crops.

Cereal/Bean Systems

Maize/bean is a predominant cropping system in Central and South America and parts of East Africa where rainfall ranges from 700 to 1500 mm spread over 4 to 5 months. Maize is intercropped with bush beans in the lowlands or with climbing beans in the highlands (Francis et al., 1982a). Both crops are generally planted with the beginning of the rainy season, and where the cropping period is longer another crop may be planted with residual moisture. Beans mature in 90 days and maize is harvested in about 120 to 150 days.

Willey and Osiru (1972) observed in Uganda that maize/bean intercropping was 38 percent more productive than a combination of sole crops. The higher productivity of the intercrop was attributed to better utilization of growth resources, particularly light. They also showed the need for a higher total intercrop population relative to the sole crops. Francis et al. (1982a) reported a yield advantage of 30 percent for the maize/bean intercrop in the dry season and 39 percent in the wet season. Though the total yield advantage remained unaffected, the contribution of the components varied with the bean genotype, row arrangements, and the relative planting time of the component crops. In favorable climates a second bean crop is relay planted into standing maize after harvest of the first bean crop. In such a case maize is doubled over at physiological maturity to minimize competition for light to the young beans. In southeast Brazil, a maize/bean intercrop in the rainy season yielded 3.2 t/ha of maize, and 0.56 t/ha of beans, giving a LER advantage of 45 percent. When a second crop of beans was added to maize in the following dry season the total bean yield increased to 1.5 t/ha and the LER advantage increased to 71 percent (Serpa et al., 1981). Rao and Morgado (1984) found in a review of 51 trials in northeast Brazil that the system produced 70 percent sole maize and 62 percent sole bean yields. However, when the seasonal rainfall was lower than 400 mm, the intercrop did not offer any advantage over sole cropping (Stewart, 1982). Rao and Morgado (1984) showed that sorghum was a better alternative to maize in the semiarid regions to improve and stabilize cereal production without foregoing other benefits of traditional intercropping.

Row arrangements and cereal population vary depending on the farmer's objective and agroclimatic conditions. Where the cereal is most important and is expected to produce high yield (3 to 5 t/ha) maize is planted at the sole crop optimum density and beans are considered as a bonus crop. Thus in favorable climates such as in south Brazil a full density of maize (40,000 to 50,000 plants per hectare) is interplanted with beans (100,000 to 150,000 plants per hectare) (Portes and Caravalho, 1983; Chagas et al., 1983). Maize yields were unaffected (94 percent of sole crop yield) but beans produced 40 percent of sole crop yield, with a system advantage of 34 percent over sole cropping (Serpa et al., 1981).

In unfavorable climates such as the semiarid northeast Brazil, where maize performs poorly and both components are equally important to the farmer, the intercrop is planted with only half the sole crop population of maize (Rao and Morgado, 1984). Other examples are given in Chap. 7.

Cereal/Groundnut Systems

Groundnut is intercropped with sorghum or millets in India and West Africa (Aiyer, 1949; Kassam, 1976), and with maize in Southeast Asia and to a limited extent in Africa (Kassam, 1976). Traditional systems generally contain relatively more groundnut than cereal because groundnut suffers heavily from cereal competition and is more valuable (Kassam, 1976; Mutsaers, 1978). Bodade (1964) observed that two sorghum to eight groundnut was the most profitable system and Lingegowda et al. (1972) reported that one sorghum to three groundnut was most profitable. In Nigeria farmers plant a cereal with wide spacing at the onset of rains and plant groundnut later only if the rains proceed normally. Baker (1978) showed that the cereal can be planted closer (1.37 m compared to farmer practice of 2.7 m apart) to increase the intercrop return to 35 percent over sole cropping, compared to only 15 percent advantage with the wide cereal spacing. With a competitive cereal such as maize, close spacing was detrimental to groundnut, particularly when moisture conditions favored the cereal (Rao et al., 1979).

Results of a detailed pearl millet/groundnut study at ICRISAT are typical of replacement cereal/legume systems (Fig. 6.7; Reddy and Willey, 1981). The intercrop was planted with one millet to three groundnut in 30-cm rows. Sole millet showed a rapid growth rate, producing 8134 kg/ha of dry matter in 85 days (Fig. 6.7). Sole groundnut was slower to develop and yielded 4938 kg dry matter in 105 days. Intercropped groundnut produced 75 percent of the sole crop biomass, while millet produced 50 percent of the sole crop biomass. At harvest the final yields were 55 percent of millet and 71 percent of groundnut, with an intercrop advantage of 26 percent over sole crops. The advantage in intercropping was primarily from increased compensation by millet.

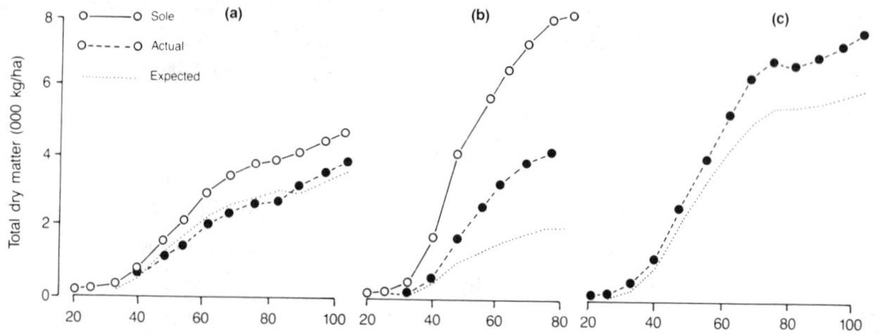

Figure 6.7 Sole crop yields and actual and expected intercrop yields of groundnut and millet. (a) Groundnut, (b) pearl millet, and (c) combined intercrop.

Other Cereal/Legume Systems

Intercropping soybean with maize is found throughout the world, while soybean/sorghum is seen mostly in the semiarid tropics. Whereas most people observed substantial yield advantages of intercropping (Ahmed and Rao, 1982), there have been cases with no worthwhile advantage (Crookston and Hill, 1979; Wahua and Miller, 1978).

In combinations of cereals with cowpea, mung bean, or black gram, the legume matures earlier than the cereal, thus providing some temporal complementarity. When the cereal was planted at sole-crop optimum population, the legume growth was severely reduced, yet the yield advantages of these systems ranged from 20 to 35 percent (Stoop, 1981; Chaudhury, 1981; Rao and Morgado, 1984) and differences between various row arrangements were small with respect to total yields. Reducing cereal population to half the sole crop optimum sometimes improved legume yields and total yield advantage (Rao and Morgado, 1984). Such a practice would be favored only where some sacrifice in cereal yield is acceptable.

On Vertisols in the undependable rainfall areas of south India, only a single post-rainy season crop is grown on the conserved moisture after rainy season fallow (Kanitkar, 1960). Since the growing period in this post-rainy season (October to January) is limited by the fixed stored moisture (200 to 300 mm), the only way that farmers can increase cropping intensity is by intercropping. In this region sorghum is usually intercropped with chickpea or safflower (*Carthamus tinctorius*). On Vertisols in high rainfall areas of central India (1000 to 1400 mm) and on Entisols in northern India, wheat and barley are intercropped with chickpea, mustard (*Brassica juncea*), or safflower following rainy season crops. Because of less temporal difference between the crops and limited moisture, yield advantages for intercropping in the post-rainy season were low (10 to 20 percent) (Saxena et al., 1978; ICRISAT, 1977, 1980). However, wheat/chickpea at Jhansi and barley/chickpea at Agra gave 40 percent advantage over sole cropping (Chaudhury, 1981). The intercrops at these locations were more remunerative than sole cereal, but not always better than sole legume.

INTERCROPPING OF CEREALS WITH CEREALS

Intercropping cereals with cereals may sound counterintuitive, but is common in many parts of the tropics. Since most cereals have similar rooting patterns and have heavy nutrient requirements, the potential for complementary effects of these mixtures with respect to belowground resources is less than with unlike species. However, where large temporal differences exist among cereals, the components may complement each other over time to produce higher yields than sole cropping. Baker (1979b) observed that there was a gain in mixing cereals when the components were separated by at least 43 days. In mixtures of sorghum cultivars, he further observed that components increased their yield over sole cropping if the varieties differed in height by more than 0.6 m and in age at maturity by more than 51 days. Willey and Rao (1981) found that maturity and height differences did not explain much of the variation in total productivity of sorghum/millet intercrops. Thus with some genotypes there may be complementarity for both aboveground and belowground resources.

Intercrop Systems with Long-Cycle Cereals

In the northern Guinea and Sudanian savanna of West Africa, photosensitive long-season sorghum (about 6 months) is intercropped with a number of crops, but intercropping with early (Gero) or late (Maiwa) millet is most important. For example, Norman (1974) observed that sorghum/millet accounted for 26 percent of the total intercropped area in northern Nigeria. Although two crops can be planted sequentially to cover the potential cropping season, this has not attracted the attention of farmers because planting the second crop is difficult in the middle of the rains and crop pests are also severe (Kassam, 1976). Yields in traditional systems are poor, but studies have shown that they can be improved by agronomic management (Andrews, 1972). A 50 : 50 proportion of sorghum and millet using improved genotypes increased yields to over 3 t/ha and showed that the intercrop was 53 percent more productive and 63 percent more profitable than sole cropping. However, an arrangement of one sorghum to two millet and relay planting later with cowpea in the space vacated by millet increased the profitability of the system to 73 percent over sole cropping (Andrews, 1972).

Photosensitive sorghum intercropped with maize is an important cropping system in Honduras, Guatemala, El Salvador, and Nicaragua occupying the 6 to 7 months of the rainy season (Fig. 6.4) on marginal and sloping lands without inputs (House and Guiragosian, 1978; Hawkins, 1984). Experimental data on yield advantages of the system are limited, but it appears that both the components would yield 80 to 100 percent of their respective sole crops (Hawkins, 1984). With improved genotypes it would be possible to harvest yields of 3 t/ha (Clara, 1980). Maize/sorghum utilizes the rainy season more fully than any sole crop system. It further ensures stability in food supply to small farmers, since if maize fails in a drought year sorghum may give some yield.

Sorghum/maize intercropping with or without beans is common in the highlands of Ethiopia where rainfall ranges from 800 to 1000 mm in a bimodal distribution (Gebrekidan, 1977). Sowing follows the onset of short rains in April. Maize is harvested first for green cobs and is followed by beans; sorghum matures in December after the long rains. No yield data have been reported for this three-crop system, although Gebrekidan (1977) showed advantages for the sorghum/bean system.

Long-season sorghum is also intercropped with other crops, such as chat (*Catha edulis*) and sweet potato (*Ipomoea batatas*), in Ethiopia (Gebrekidan, 1977), benniseed, roselle (*Hibiscus sabdariffa*), groundnut, niger (*Guizotia abyssinica*), and okra in Nigeria (Yayok, 1981), and cowpea in other West African countries (Stoop, 1981). While a low sorghum population is used in intercropping with a cash crop such as groundnut, a fairly high population is used in association with cowpea.

Intercrop Systems with Similar Maturity Cereals

Maize/rice intercrops are grown on residual moisture in the *fadamas* (low-lying, seasonally flooded and high water table areas which can be cropped in the dry period on residual moisture) of West African Guinea savanna (Kassam, 1976), and in high rainfall upland humid tropics (CRIA, 1978; Benites, 1981). If rainfall is adequate, rice performs well, but when rains are poor maize is expected to give good yields and safeguard the farmer against risk of total crop failure. Studies have shown that increasing the maize proportion in the traditional system improves the advantage of intercropping, and that high yielding genotypes and fertilization increase yields substantially (Matos and Mota, 1983; IRRI, 1977).

In some parts of India sorghum in the rainy season is mixed with pearl millet, rice, and *Setaria* or *Eleusine* millet. In the post-rainy period sorghum mixed with *Panicum miliaceum* is planted in south India while mixtures of wheat and barley are grown in central and north India (Aiyer, 1949). Mixed cropping of different varieties of a particular cereal is also found (Aiyer, 1949; Mercer-Quarshie, 1979). The advantage of intercropping in these systems was not consistent, but often was in the range of 10 to 30 percent (Rao and Willey, 1980a). Mixtures of oats and barley are also planted in temperate countries, but little advantage over the highest yielding cereal was found (Noworolnik et al., 1984).

MANAGEMENT OF CEREALS IN MULTIPLE CROPPING SYSTEMS

Cereals in sequential cropping systems, whether they precede or follow another crop, can be managed in a similar way as sole crops. For intercropping and relay cropping, sole crop technology may need to be modified depending on the crop associated with the cereal and the degree of overlap.

Plant Population and Row Arrangement

Most cereals show a wide plateau in their yield response to plant population (Fig. 6.8). This enables choice of a density that is less competitive to the associated crop in intercropping while maintaining a high proportion of the potential yield. The requirement of the cereal population is determined by the associated species and the temporal difference between the two crops. In systems where the temporal difference is wide (e.g., cereal/pigeon pea, cereal/cassava) the early cereal should be planted at 100 percent of the sole crop optimum population (IITA, 1982; Natarajan and Willey, 1980). The later-maturing component may require 100 percent or more of its sole crop population depending on competitive ability of the early cereal and its own ability to compensate later in yield (IITA, 1982; Willey et al., 1981). Conversely, the systems with closely maturing crops (e.g., cereal/low canopy legumes) may not require additive populations, though the total population of both crops may be higher than for either

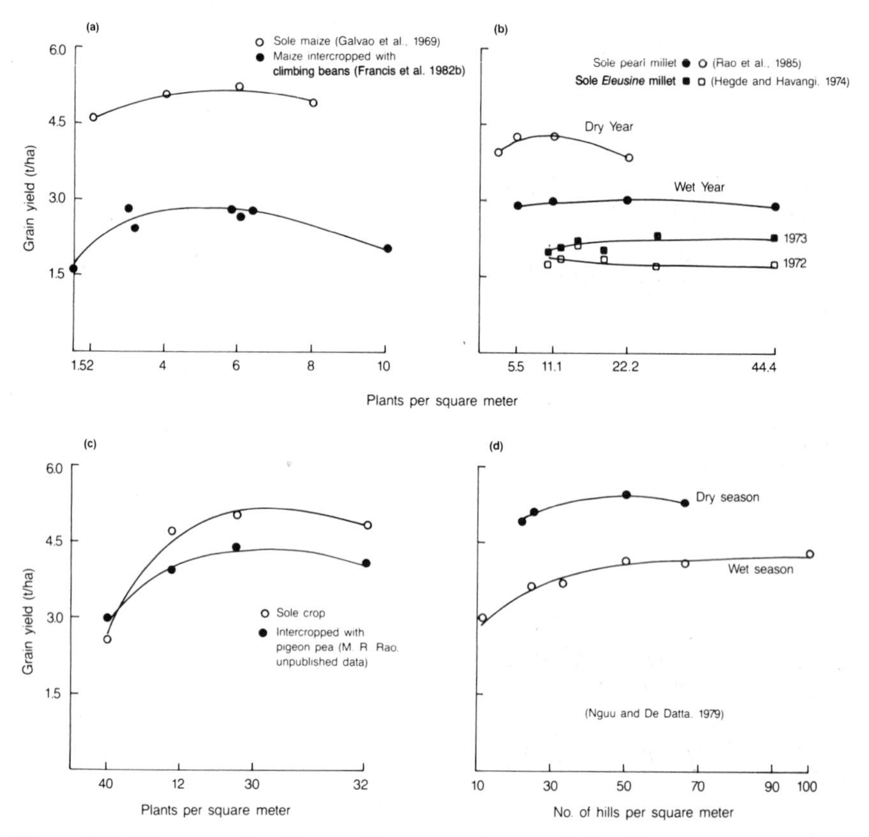

Figure 6.8 Response of major cereal crops to plant population. (a) Maize, (b) pearl and *Eleusine* millet, (c) sorghum, and (d) rice.

of the sole crops (Willey and Osiru, 1972; Osiru and Willey, 1972, Rao et al., 1985). Yield advantage of intercropping may not vary much over a limited range of row arrangements, though the proportional yields may change (Rao and Willey, 1980b, 1983a; Rao and Morgado, 1984). In such cases the choice of any particular row arrangement depends on the requirement of the proportional yields or relative values of the component.

Optimum row arrangement in systems with temporal species differences can be worked out rather easily. The arrangement that utilizes a high proportion of the early crop to maximize its yield and allows the late-maturing component to fully cover the ground in the post-rainy season should normally give the highest productivity. However, in systems based on spatial differences where the competitive balance of the components is critical for yield advantage, a number of factors, such as plant populations, genotypes, moisture, fertilization, canopy size of the two components, and height and maturity of the cereal, needs to be considered. One must seek an arrangement that maximizes the yield per plant of the components. Generally the dominant cereal has to be spaced sufficiently wide enough to minimize competition between species.

Time of Planting

Planting crops at the earliest opportunity is important to allow growth during the most favorable period under rainfed conditions and to gain time for multiple cropping. Timely operations are all the more important in rainfed agriculture where the farmer does not have access to irrigation. All crops in intercropping may be planted at the same time, or planting may be staggered (1) as a safeguard against drought in uncertain rainfall regimes, and (2) to reduce competition between components. Due to unreliable rainfall in the beginning of the season, farmers in northern Nigeria plant first an early maturing millet and later add intercrops such as groundnut or sorghum, depending on whether the rains progress normally or not (Baker, 1978). However, where rains arrive abruptly and are assured, as in many Asian countries, all crops are planted simultaneously at the start of rains; otherwise planting crops later in the middle of rains would be difficult (Willey, 1979).

If dominant cereals are planted some time after establishment of the dominated species such as legumes, the yields of the latter would be improved in intercropping. Delayed planting of cereals may reduce their yield but this can be minimized in some cereals (e.g., millets) by transplanting seedlings. Planting millet 1 to 3 weeks after the establishment of mung bean improved the yields of mung bean and total LER in a mung bean/pearl millet intercrop (De et al., 1978; May, 1982). In Colombia, beans planted 10 days earlier than maize in a maize/bean system produced 83 percent of sole crop bean yield compared to only 40 to 50 percent generally observed in simultaneous planting with maize; the staggered planting did not improve the overall advantage because of a reduction in maize yield (Francis et al., 1982a).

Yields of cotton, beans or cowpea intercropped with maize or sorghum in northeast Brazil could be improved by planting the cereal 2 to 4 weeks later than the other crop (Rao and Morgado, 1983). Sorghum in the maize/sorghum intercrop system in Central America is planted 3 to 4 weeks after maize to ensure 100 percent maize yield (Hawkins, 1984). Limitations of staggered planting in the semiarid tropics are (1) the risk of not getting an opportunity for planting the second crop later because of uncertainty of rainfall, and (2) the difficulty of using mechanical equipment for planting the second crop, making this practice highly labor intensive.

Fertilization

Cereals usually require judicious fertilization to realize their potential yields (Table 6.2). When fertilizing multiple crop systems, the residual effects of previous applications should be considered. Nitrogen fertilizers generally produce little residual effect, but continuous application of even moderate doses of phosphorus (10 to 15 kg/ha) builds up significant residual soil phosphorus (Tiwari

Table 6.2 Crop Yields in Cereal-Based Cropping Systems at Two Fertility Levels on Two Soil Types in Semiarid India

Vertisols			Alfisols		
	Yield (kg/ha)			Yield (kg/ha)	
System	N-P-K 0-0-0	N-P-K 60-12-0	System	N-P-K 0-0-0	N-P-K 60-12-0
Maize/	835	2270	Sorghum/	890	1725
pigeon pea	1050	970	pigeon pea	580	630
Maize/	980	2400	Pearl millet/	1200	2060
chickpea	1100	1330	horsegram	640	745
Sorghum/	1670	3060	Sorghum/	1050	2010
safflower	660	820	ratoon	290	480

Note: Mean of 3 years, 1981 to 1983. The legume in the sequential systems received only a basal dose of 75 kg of diammonium phosphate (18 N–20 P–0 K).
Source: M. S. Reddy, ICRISAT (unpublished data).

et al., 1980). Although specific guidelines cannot be given for fertilization of multiple cropping systems because of the effect of variables (such as crops, system, residual fertility, season, and water availability), generally the nutrient accumulation and consequently the nutrient requirement follows closely the dry matter accumulation (McCollum, 1983).

The fertilizer requirement of two-crop systems is greatly increased compared to a single season cropping, especially if both crops are cereals. A general guideline for fertilizing sequential systems could be to fertilize those crops that respond most in the system and allow others to make use of residual fertility. On that basis, nitrogen and phosphorus should be applied to cereals (ICRISAT, 1981), phosphorus to short-season legumes, potassium to cereals and tuber crops (Goswami et al., 1976), zinc to the rainy season cereals (Tiwari et al., 1980), and sulfur to the oil seed crops (Singh and Sahu, 1981). The cereal following a cereal should be fertilized normally. For example, the response of wheat to nitrogen following different rainy season cereals was similar (Fig. 6.9). Jones (1974) observed in Nigeria that groundnut benefitted the following sorghum to the extent of 60 kg N/ha; grain cowpea did not show such effects. In northwestern India, Lal et al. (1978) observed that wheat following rainy season fodder cowpea required 40 kg less nitrogen per hectare to produce similar yields as that following a cereal (Fig. 6.9). Several other studies indicated that nitrogen could be economized for cereals that follow legumes in sequential systems to the extent of 30 to 80 kg N/ha (Tiwari et al., 1980; Singh, 1983).

Goswami and Singh (1976) suggested that phosphorus could be efficiently managed in double-crop systems by applying it to only one of the crops. In

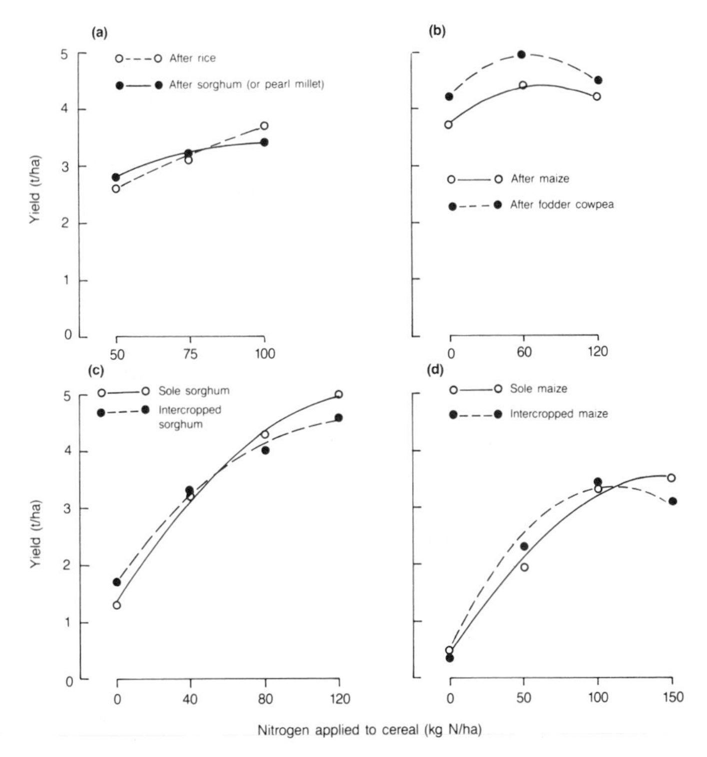

Figure 6.9 Cereal responses to nitrogen in sequential and intercropped systems: (a) wheat following cereals (Bains and Sadaphal, 1971); (b) wheat following a legume or maize (Lal et al., 1978); (c) sorghum in sole vs. intercropped with pigeon pea (ICRISAT, 1981); and (d) maize in sole vs. intercropped with groundnut (ICRISAT, 1981).

wheat-based systems, phosphorus was suggested to be applied to the post-rainy season wheat (Rao and Bhardwaj, 1981; Goswami and Singh, 1976). The residual effect of 26 kg P/ha applied to wheat was sufficient to produce high yields of the following rice. Deep rooted, hardy crops such as pigeon pea, chickpea, and castor were efficient in utilizing the native phosphorus. Therefore, they should be allowed to thrive on residual fertility of applications made to the previous season's crops (Tiwari et al., 1980).

The fertilizer needs of an intercropped cereal may be increased, unaltered, or reduced compared to those of the sole crop depending on the crops involved. When both components respond to a particular nutrient, the intercrop system requirement may be higher than the sole crop needs, e.g., higher nitrogen requirement in the case of cereal/cereal intercrops (Kassam and Stockinger, 1973), and higher phosphorus for legume/legume intercrops. Hedge (1977) observed that phosphorus needs of pigeon pea intercropped with mung bean or cowpea was 32 kg P compared to 24 kg P required for sole pigeon pea. Cereals intercropped with legumes at full population responded to nitrogen in a similar way to sole cropped cereals, though the response in terms of kilogram grain produced per kilogram of applied nitrogen might be less in intercropping because of competition from the legume (ICRISAT, 1981; Ahmed and Rao, 1982). In replacement mixtures with less than 100 percent cereal, it should be fertilized proportionately (ICRISAT, 1981). Nitrogen needs of a cereal intercropped with legumes were reported to be less than for sole cropping due to transfer of some of the fixed nitrogen by legume to the associated cereal during the growing season (Willey, 1979). However, very few intercrop studies have actually estimated the beneficial effect of legumes to the cereal without confounding the effect of plant population in the field (Ahmed and Rao, 1982).

Method of fertilization is important where the components have different requirements, as with nitrogen in cereal/legume systems. Nitrogen should be applied to the cereal as far away from the legume as possible so that nitrogen fixation of the legume is not affected. If both components require the same nutrient, as with nitrogen in cereal/cereal systems and phosphorus in cereal/legume systems, the nutrient can be applied in one application to both.

Improved fertilization generally reduced the relative advantage of intercropping (Ahmed and Rao, 1982). This leads to the suggestion that intercropping may not be worth practicing with high input technology. However, we must draw a distinction between the types of systems. In temporal systems (e.g., sorghum/pigeon pea), the decrease in the relative advantage was minimal because the later-maturing dominated component had enough time to compensate for the earlier competition from the cereal. But in spatial systems (e.g., cereal/low canopy legumes) the advantage decreased sharply at high level of fertilization because of overdominance of the cereal (ICRISAT, 1981, 1982). It must be recognized that although the relative advantage decreased, the absolute returns increased up to a moderate level of fertilization because of increased yields. Other aspects of fertility relations in multiple cropping are discussed in Chap. 8.

Dry Seeding of Rice

Dry seeding of the first rice crop in rice-based cropping systems has been advocated to (1) avoid problems of excess or insufficient water, (2) to overcome labor scarcity for transplanting, and (3) to clear the field of the first crop early enough to permit multiple cropping with upland crops or transplanted rice (Morris, 1982). Dry seeding requires good cultivation during the dry season, which increases power requirement. Other limitations of the practice are increased weeds, uneven germination, and failure of crops whenever long dry spells occur. The greater tillage requirement and the necessity of herbicide use may at times increase the operational expenses, but increased income from the subsequent crops might offset these additional costs. Dry seeding is more appropriate on light- to medium-textured soils than on heavy soils because of the difficulty in plowing the heavy soils in dry conditions and the susceptibility of heavy soils to occasional flooding. Studies showed that dry-seeded rice performs as well as any other method of establishment (Bradfield, 1971; Gomez and Gomez, 1983).

Zero or Minimum Tillage

Establishing crops after cereals is somewhat difficult without adequate land preparation because of large quantities of residues that the cereals leave behind, quick regrowth from the stubble of some cereals, and, specifically after rice, because of the reduced soil chemical conditions under which it is grown. Thorough land preparation entails long turnaround time and reduces soil moisture which is lost to subsequent crops and limits the potential for multiple cropping. No-till or minimum tillage methods which have become easy and effective to practice with the advent of modern herbicides are suggested to reduce the turnaround time between crops.

Different techniques may have to be adopted for different crops that follow rice. Small-seeded legumes such as mung bean, black gram, and sunhemp can be established, especially on heavy soils, by broadcasting presoaked seeds 1 week to 10 days before rice harvest. Crops requiring low populations, such as maize and cotton, can be dibbled by hand. Sorghum or large-seeded legumes need to be drilled. This may be easy on light-textured soils with minimum cultivation. When the soil is too wet for dibbling or drilling, opening small drainage channels at 4 to 5 m may help reduce the turnaround period. Herbicides to kill the rice stubble and rice straw mulch to suppress weeds and conserve moisture have been used effectively for raising upland crops after rice (Syarifuddin, 1982). Though crop yields with minimum tillage were sometimes lower than those after complete tillage, the net returns were comparable or even higher due to reduced operational expenses.

Minimum tillage in rain-fed cropping systems has the added advantage of conserving limited moisture. In studies at ICRISAT, one application of Paraquat [1 kg active ingredient (ai) per hectare] plus Prometryn (0.75 kg ai/ha) eliminated the need for cultivation in establishing a post-rainy season chickpea crop following sorghum or maize (ICRISAT, 1982). No-till systems have become imperative for the success of double cropping in the southern United States. Paraquat to kill the stubble of the previous crop and an appropriate herbicide for controlling weeds in the current crop enabled double cropping with small grains or soybean following winter cereals (Nelson et al., 1977; Sanford et al., 1973; Lewis, 1972).

Relay Planting

Relay planting is practiced primarily (1) to gain time for multiple cropping, (2) to plant the subsequent crops at their optimum planting date when the current crop harvest is delayed (Reddy and Willey, 1982), (3) to avoid moisture stress in the post-rainy season crops, and (4) to avoid labor peaks at the harvest of the first crop and planting of the second crop. It is a labor-intensive operation feasible only on small holdings where labor is not a constraint. Relay cropping is a practical approach in the case of crops established by hand with seedlings or sets of seed material such as vegetables, chili (*Capsicum annum*), sweet potato, sugarcane, and cassava.

In small-plot experiments at ICRISAT, Reddy and Willey (1982) observed that sorghum and pigeon pea in the post-rainy season benefitted from relay planting in maize. The practice could not be carried out successfully on an operation scale (Krantz, 1981). However, there would be no difficulty where seed is broadcast as in the case of small-seeded legumes in wetland rice. Sugarcane is relay planted in rice in Bangladesh, where short-season upland crops are relay planted in sugarcane following rice harvest (Rahman et al., 1982). On Alfisols and lateritic soils in northeast India only a single upland crop is grown in spite of ample rainfall. Studies have shown that cropping can be extended by relay planting pigeon pea in the upland crop of maize or rice (AICRPDA, 1978). However, its adoption will depend on the development of suitable equipment for mechanization.

Relay planting is difficult in the closely spaced and dense canopy cereals such as sorghum or millets unless some rows are skipped. When early maturing crops associated with these cereals in intercropping are harvested, relay crops can be planted (Andrews, 1972). In northern Karnataka, India, cotton is relay planted in the widely spaced pearl millet 7 weeks after it is planted (Koraddi et al., 1970).

Ratooning

Regrowth from sorghum, pearl millet, and rice stubble can be managed to yield another grain or fodder crop (Plucknett et al., 1970). The chief advantages of ratooning are to avoid planting of another crop and save time and costs for seed and land preparation. It also reduces labor inputs during the postharvest period when demand for labor generally is high. The potential for ratooning under assured irrigation should be judged carefully compared to alternative crops that can be grown. In rainfed areas where the potential for planting sequential crops is limited, ratoon crops may provide a partial second crop.

Ratooning sorghum for fodder and grain is more popular than other cereals. Harvest time of the plant crop, cutting height of stubble, tillers per culm, moisture, and nutrients affect the ratoon growth (Plucknett et al., 1970). In India, systems have been described where sorghum was grown for fodder under irrigation in summer and the ratoon was maintained for grain in the following rainy season. The earlier maturing ratoon grain crop gave greater opportunity to plant on time the traditional post-rainy season crop of cotton (Mandal et al., 1965; Shanmugasundaram et al., 1967) or of wheat (Pal and Kaushik, 1969). Another system that utilized the ratoon concept was harvest of rainy season sorghum for fodder at 40 to 50 days and then cultivation of ratoon crop for grain (Willey et al., 1982). This has the advantage of reducing the effect of midseason drought and prolonging the cropping season to produce good quality grain in dry weather. A major drawback of ratoon sorghum crops in India is the severe incidence of shootfly (*Atherigona soccata*) which attacks the young shoots.

Ratoon systems with pearl millet are not as widely used as with sorghum because millet is an early maturing crop which is removed from the field early to allow planting of alternate crops and because the regrowth from stubble is less vigorous and less productive. Of the two different ratoon systems examined with improved millet cultivars at ICRISAT, the system with the plant crop for grain and the ratoon crop for fodder was most productive (ICRISAT, 1975).

When there is not sufficient water to plant two rice crops (Medappa and Mahadevappa, 1976), or when water stagnation hampers freshly planted rice in the wet season (Roy et al., 1983), the ratoon of the previous rice can be managed to produce an additional crop in a shorter period. To be economical, the genotypes should produce adequate ratoon growth. In a study in West Bengal, India, the ratoon crops of several photosensitive cultivars gave similar or better yields than the normally sown *aman* rice, indicating the scope of selecting genotypes for ratoon cropping (Roy et al., 1983).

ROTATIONAL ASPECTS OF CEREALS IN MULTIPLE CROPPING

Although crop rotation has lost prominence in modern agriculture since the advent of chemical fertilizers, there are still situations where the residual effects of some crops on the succeeding crops are of practical importance. The residual effects could be physical due to changes in the soil structure, as in the case of puddling for rice culture, could be chemical due to nutritional imbalances (Tiwari et al., 1980; Flinn and De Datta, 1984), or could be allelopathic caused by root exudates or the buildup of pests and diseases.

Puddling for lowland rice destroys soil structure and leaves the soil in anaerobic condition, making the establishment of subsequent upland crops difficult (Syarifuddin, 1982). The soil immediately after rice harvest is too sticky to cultivate, and if allowed to dry, it becomes hard and cracked. If planted in the wet soil, upland crop germination is impaired due to the lack of aeration, and when the soil becomes hard, root growth may also be affected (Zandstra, 1982). The availability of nitrogen and phosphorus to the following upland crop is affected by transformations that occur under reduced soil conditions of rice paddies. To bring the soil from the reduced state back to a favorable state for arable crops requires repeated cultivations at an appropriate moisture condition. For crops such as groundnut and potato, good tilth is particularly essential. Good soil preparation takes out the ill effects of puddling, but it demands sufficient time which in many instances becomes critical for multiple cropping. However, one advantage of puddling is that it reduces weed problems.

Laboratory and biotests have indicated the existence of toxic (allelopathic) substances that prevent the growth of other plants in cereals such as wheat, oats, barley, and maize. Water extracts of the residues of these crops showed varying degrees of toxicity, but the effects were only for a short period (Yankov, 1984; Guenzi et al., 1967). However, the residues from sorghum roots exhibited the toxic effects for a longer period in laboratory and in field tests (Ruiz-Sifre and Ries, 1983).

Several reports pointed out the poor performance of crops following sorghum. The detrimental effect of sorghum was noticed on succeeding crops of sorghum, wheat, oats, barley, chickpea, pigeon pea, safflower, and groundnut. Wheat yields after sorghum were poor compared to those following maize (Myers and

Hallsted, 1942; Singh and Singh, 1966) and the yields were particularly reduced in the crop planted immediately after sorghum (Myers and Hallsted, 1942). In south India, cotton yields following sorghum were reported to be lower than after pearl millet (Aiyer and Sundaram, 1941) or, in Sudan, after fallow (Burhan and Mansi, 1967; Roy and Kardofani, 1961). In studies at ICRISAT the deleterious effect of rainy season sorghum on the post-rainy season crops was greater in a low-rainfall year (49 to 87 percent) than in a normal rainfall year (10 to 14 percent; Fig. 6.10). The residual effect was much greater following a full season crop than after a crop cut at flowering. Legumes were generally less affected than other crops.

Some possible reasons suggested to explain the deleterious effect of sorghum were that (1) sorghum depletes the nutrients and/or moisture, (2) applied nutrients may be locked up while large quantities of sorghum root residues are decomposed by microbial activity, (3) toxic substances are formed during the decomposition of residues, (4) secretion of toxic root exudates is harmful to crops, and (5) unfavorable soil conditions may exist, e.g., increased replaceable sodium. Burgos-Leon et al. (1980) observed that P-coumaric, O-hydroxybenzoic, and protocatechuic acids extractable from sorghum residues probably contributed to the phytotoxicity after sorghum. While no particular factor could be identified as the sole cause for the sorghum effect in a given environment, one or all of them might be important. The suggested measures for obviating this deleterious effect include mixed cropping of sorghum with legumes (Singh and Singh, 1966), balanced fertilization of sorghum, growing legumes after sorghum with proper *Rhizobium* inoculation (Singh and Singh, 1966; Ruiz-Sifre and Ries, 1983), application of nitrogenous fertilizers to nonlegumes, and removal of sorghum stalks.

Some cereals may promote or reduce pests, diseases, or weeds in cropping systems. Wilt caused by *Fusarium udum* (Butler) is a serious disease on pigeon pea. The disease is aggravated if a sole crop of pigeon pea is grown continuously.

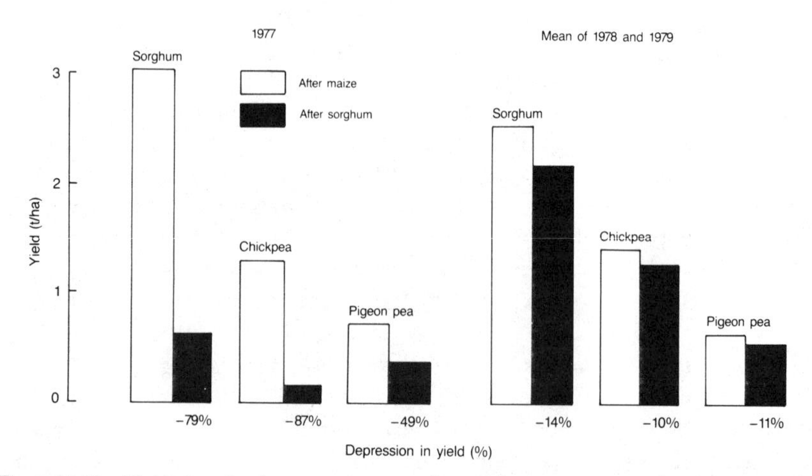

Figure 6.10 Yield of post-rainy season crops after a rainy season crop of maize or sorghum (Willey et al., 1982).

Some reports indicated that pigeon pea wilt was reduced by intercropping pigeon pea with sorghum (Dey, 1948; Gupta, 1961). In a study conducted at ICRISAT on a plot with wilt, several rotational systems in which pigeon pea alternated with 1- to 2-year breaks of sorghum, fallow, tobacco, or cotton were compared. A 2-year break with sorghum was most effective, and pigeon pea that followed showed only 24 percent wilt incidence compared to 80 to 90 percent incidence in the continuous pigeon pea. Intercropping of pigeon pea with sorghum reduced the wilt incidence in the first year to 55 percent and stabilized the level thereafter at about 20 to 30 percent (Natarajan et al., 1985). Such an effect was not observed with another cereal such as maize, and the specific effect of sorghum could be attributed to the secretion of fungi-toxic root exudates (Rangaswami and Balasubramanyam, 1963; Odunga, 1978).

Another cropping system where a cereal influenced diseases was intercropping of groundnut with pearl millet or sorghum. While the early leaf spot (*Cercospora arachidicola*) and late leaf spot (*Phaeoisariopsis personata*) were relatively unaffected by the cereal intercrop, the incidence of rust (*Puccinia arachidicola*) and bud necrosis were less prevalent on intercropped groundnut compared to a sole crop (ICRISAT, unpublished data). The tall cereal may have acted as a barrier for the spread of the airborne rust spores and thrips, the vectors of bud necrosis.

Continued practice of specific cereal-based cropping systems may lead to buildup of problem weeds, such as *Striga,* or nematodes. The parasitic weed striga infests sorghum, pearl millet, and maize. Although there is some host specificity, cross infestation may still occur (Rao and Musselman, 1984). Striga infestation may also occur when new cropping systems with striga-susceptible improved cereal genotypes are introduced. For example, in an ICRISAT on-farm study on Vertisols at Taddanpally, Andhra Pradesh, India, improved cereal cultivars in sorghum/pigeon pea intercrops and maize-chickpea sequential systems were severely affected by striga, reducing the potential benefits of a new double cropping technology. These systems were introduced for efficient utilization of the rainy and post-rainy season in the place of the traditional single post-rainy season cropping with sole sorghum. In another study on Alfisols at Aurupalli, Andhra Pradesh, improved sorghum in a sorghum/pigeon pea intercrop was similarly affected (Shetty and Sharma, 1982).

AT ICRISAT, a research plot used regularly for experiments in which sorghum alternated with a cover crop of maize developed a serious nematode (*Pratylenchus* sp.) problem over a 5-year period that severely reduced the growth of maize and sorghum. Cropping pearl millet in the summer following tobacco has become popular in Gujarat, India, but this sequence has been observed to increase the incidence of root-knot nematode on tobacco (Patel et al., 1979).

DIRECTIONS FOR FUTURE RESEARCH

It is evident from this and other chapters that multiple cropping is complex and labor intensive, requires high inputs, but leads to increased yields. If we accept higher yields as essential to meet the challenge of food supply, we must search for ways of overcoming the limitations of intercropping.

The realization of potential benefits of multiple cropping depends on how effectively the systems are managed at the farm level. The location specificity of cropping systems technologies, imposed by differences in agroclimatic and socioeconomic characteristics, demands multilocation on-farm testing to adapt recommendations of on-station research. Multiple cropping in formerly single-cropped areas involves many changes. Farmers usually will not accept the new technology unless they are convinced of its potential. On-farm research is crucial to involve farmers in the development and demonstration of the potential of new systems. By working closely with the farmer, the researcher gets a clearer understanding of problems to suggest practical solutions. Further discussion on testing is presented in Chap. 13.

It must be recognized that because of socioeconomic differences among farmers, no single system is likely to fulfill the objectives of all farmers in a given environment. Future research should concentrate on developing an array of cropping system options so that the farmer can choose what is most appropriate to each situation. Consistency in the performance of a new system is vital, and the stability of the cereal component is particularly important for subsistence farmers. Farmers would be reluctant to accept new practices and systems unless these show equal or improved stability. Efforts must be made to collect the required data for stability evaluation while testing new cropping systems.

The success of multiple cropping depends on the availability of appropriate genotypes. This calls for determining breeding objectives to develop the genotypes required for different cropping systems. While aiming for good agronomic and grain quality characteristics, fodder yield and quality should not be ignored wherever livestock is an important component of the farming enterprise. The continued practice of multiple cropping may create new problems with soil fertility, pests, diseases, and weeds. Some of these may be recognized only by long-term monitoring. New genotypes should be tolerant to pests and diseases. Appropriate tillage systems to reduce turnaround time between crops and methods of managing crop residues for sustaining soil productivity need immediate attention. The rotational aspect of cropping systems, particularly the role of legumes, and long-term productivity of soils should receive attention. Where allelopathic effects are recognized, methods of using or avoiding these effects should be developed as appropriate. It is worth examining whether there is any potential for identifying sorghum genotypes having less toxic effects on the succeeding crops.

Multiple cropping is a dynamic practice that needs continuing research. As new genotypes keep coming from breeding programs, suitable changes may have to follow in agronomic practices such as plant population and fertilization. Future research should examine new crops that can be associated with cereals, the environments in which intercropping offers or does not offer advantage, interactions of plant population with moisture, fertility, and genotypes, and contingency plans for midseason droughts and crop failures.

Multiple cropping in rainfed areas should receive major attention in the future. If rainfall is the only source of moisture, its management on a watershed basis through suitable land and water management practices holds the key to the success of cropping systems. Provision for farm ponds in watersheds may help recycle runoff water for supplemental irrigation during drought or for the establishment of new crops. By adopting integrated land, water, and crop management techniques, ICRISAT demonstrated that the traditionally single season cropped Vertisols in the assured rainfall zone (750 to 1400 mm per year) can be double cropped (Virmani et al., 1981). These issues are not specific to cereals in multiple cropping. There should be simultaneous improvement in the other components to assure progress in total cropping systems.

REFERENCES

Ahmed, S., and Rao, M. R. 1982. Performance of maize-soybean intercrop combination in the tropics: Results of a multilocation study, *Field Crops Res.* 5:147–161.

AICRPDA (All India Coordinated Research Project for Dryland Agriculture). 1978. *Annual Report, 1978,* AICRPDA, Santoshnagar, Hyderabad, India.

Aiyer, A. K. Y. N. 1949. Mixed cropping in India, *Indian J. Agric. Sci.* 19:439–543.

Aiyer, V. R., and Sundaram, S. 1941. A brief account of the studies on the harmful after-effects of cholam crop on cotton, *Indian J. Agric. Sci.* 11:37–52.

Andrews, D. J. 1972. Intercropping with sorghum in Nigeria, *Exp. Agric.* 8:139–150.

Aziz, S. 1977. The world food situation—today and in the year 2000, in: *Proc. World Food Conference of 1976,* Iowa State University, Ames, Iowa.

Bains, S. S., and Sadaphal, M. N. 1971. Fertilizer application in a multiple cropping system, *Fert. News* 16(5):16–18.

Baker, E. F. I. 1978. Mixed cropping in northern Nigeria. I. Cereals and groundnuts, *Exp. Agric.* 14:293–298.

———. 1979a. Mixed cropping in Northern Nigeria. II. Cereals and cotton, *Exp. Agric.* 15:33–40.

———. 1979b. Mixed cropping in Northern Nigeria. III. Mixtures of cereals, *Exp. Agric.* 15:41–48.

Benites, J. R. 1981. *Nitrogen response and cultural practices for corn-based cropping systems in the Peruvian Amazon,* Ph.D. thesis, North Carolina State University, Raleigh, North Carolina.

Bodade, V. N. 1964. Mixed cropping of groundnut and jowar, *Indian Oil Seeds J.* 8:297–301.

Bradfield, R. 1971. Mechanized maximum cropping systems for the small farms of the rice belt of tropical Asia, *Agric. Mechan. Asia* 1:55–57.

Burgos-Leon, W., Ganry, F., Nicou, R., Chopart, J. L., and Dommergues, Y. 1980. Un cas de fatique des sols induite par la culture du sorgho, *Agron. Trop.* 35:319–334.

Burhan, H. O., and Mansi, M. G. 1967. Rotation responses of cotton in the Sudan Gezira. 1. The effect of crop rotation on cotton yields, *J. Agric. Sci. Camb.* 68:255–261.

Chagas, J. M., Vieira, C., Ramalho, M. A. P., and Pereira Filho, I. A. 1983. Efeitos do intervalo entre fileiras de milho sobre o consorcio com a cultura do feijao, *Pesq. Agropec. Bras. Brasilia* 18:879–885.

Chinnappan, K., and Palaniappan, S. P. 1980. Multi-tier cropping in castor, *Indian J. Agric. Sci.* 50:342–345.

Chaudhury, S. L. 1981. Recent trends in intercropping systems on the drylands of India: Some thoughts, some results, in: *Proc. International Workshop on Intercropping,* 10–13 Jan. 1979, Hyderabad, India, ICRISAT, Patancheru, India, pp. 299–305.

Clara, R. 1980. *Sorghum improvement in El-Salvador* (mimeo), Training Program, ICRISAT, Patancheru, A.P., India.

CRIA (Central Research Institute for Agriculture). 1978. *Cropping Systems Research (Indramayu and Lampung), Annual Report, 1976–77,* Cooperative CRIA-IRRI Program, Bogor, Indonesia.

Crookston, R. K., and Hill, D. S. 1979. Grain yields and land equivalent ratios from intercropping corn and soybeans in Minnesota, *Agron. J.* 71:41–44.

Dalal, R. C. 1974. Effects of intercropping maize with pigeon peas on grain yield and nutrient uptake, *Exp. Agric.* 10:219–224.

Dalrymple, D. G. 1975. Measuring the green revolution: The impact of research on wheat and rice production. U.S. Dep. Agric., Foreign Agric. Econ. Rep. 106, Washington, D.C.

De, R., Gupta, R.S., Singh, S. P., Pal, M., Singh, S. N., Sharma, R. N., and Kaushik, S. K. 1978. Interplanting maize, sorghum and pearl millet with short-duration grain legumes, *Indian J. Agric. Sci.* 48:132–137.

Dey, P. K. 1948. *Plant pathology,* Administrative Rep. Agric. Dep. Uttar Pradesh, 1946–1947, pp. 39–42.

Dos Santos, M. 1978. *Producción biológica y económica de agrosistemas basados en el cultivo de la yuca* (*Manihot esculenta* Crantz). M.Sc. thesis, CATIE-UCR, Turrialba, Costa Rica (in Spanish).

Enyi, B. A. C. 1973. Effects of intercropping maize or sorghum with cowpeas, pigeonpeas or beans, *Exp. Agric.* 9:83–90.

FAO. 1982. *Production year book 1981,* vol. 35, FAO, United Nations, Rome, Italy, pp. 93–108.

Flinn, J. C., and De Datta, S. K. 1984. Trends in irrigated rice yields under intensive cropping at Philippine research stations. *Field Crops Res.* 9:1–15.

Francis, C. A., Prager, M., and Tejada, G. 1982a. Effects of relative planting dates in bean (*Phaseolus vulgaris* L.) and maize (*Zea mays* L.) intercropping patterns, *Field Crops Res.* 5:45–54.

———. 1982b. Density interactions in tropical intercropping 1. Maize (*Zea mays* L.) and climbing beans (*Phaseolus vulgaris* L.), *Field Crops Res.* 5:163–176.

Freyman, S., and Venkateswarlu, J. 1977. Intercropping on the rainfed red soils of the Deccan Plateau, India, *Can. J. Plant Sci.* 57:697–705.

Galvão, J. D., Brandão, S., and Gomez, F. R., 1969. Efeito do população de plantas e niveis de nitrogenio sobre a produção de grãos e sobre o peso medio das espigas de milho. *Experimentae* 9(2):39–82.

Gebrekidan, B. 1977. Sorghum-legume intercropping in the Chercher highlands of Ethiopia, *AAASA J.* 4(2):39–46.

Gomez, A. A., and Gomez, K. A. 1983. Multiple cropping in the humid tropics of Asia, IDRC-176e, Ottawa, Canada, 248 pp.

Goswami, N. N., Leelavathi, C. R., and Singh, R. N. 1976. Potassium in soils, crops and fertilizers, *Bull. Indian Soil Sci.* 10:186–194.

Goswami, N. N., and Singh, M. 1976. Management of fertilizer phosphorus in cropping systems, *Fert. News* 21(9):56–59, 63.

Greenland, D. J. 1975. Bringing the revolution to the shifting cultivators, *Science* 190:841–844.

Guenzi, W. D., McCalla, T. M., and Norstadt, F. A. 1967. Presence and persistence of phytotoxic substances in wheat, oat, corn, and sorghum residues, *Agron. J.*, 59:163–165.

Gupta, S. L. 1961. The effect of mixed cropping of *arhar* (*Cajanus indica* Spreng.) with jowar (*Sorghum vulgare* Pers.) on incidence of *arhar* wilt, *Agric. Anim. Husb. Uttar Pradesh* 3:31–35.

Hart, R. D. 1975. A bean, corn and manioc polyculture cropping system. II. A comparison between the yield and economic return from monoculture and polyculture cropping system. *Turrialba Costa Rica* 25:377–384.

Hawkins, R. 1984. Intercropping maize with sorghum in central America: A cropping system case study, *Agric. Syst.* 15:79–100.

Hedge, B. R., and Havangi, G. V. 1974. Intercropping ragi in Bangalore region, *Intercropping for Increasing Crop Intensity in the Red Soils of the Semi-arid Tropics* (seminar), All India Coordinated Research Project for Dryland Agriculture (AICRPDA), Hyderabad, India. 7 pp.

Hedge, D. M. 1977. Phosphorus management in intercropping with pigeonpea (*Cajanus cajan* (L.) Millsp.) under dryland conditions, Ph.D. thesis, Indian Agricultural Research Institute, New Delhi, India.

House, L. R., and Guiragosian, V. 1978. *Sorghum in Central and South America* (mimeo), ICRISAT, Patancheru, A.P., India.

IARI (Indian Agricultural Research Institute). 1972. Recent research on multiple cropping, *IARI Res. Bull.* (New Series) No. 8. New Delhi. 148 pp.

ICRISAT (International Crops Research Institute for the Semi-Arid Tropics). 1974, 1975, 1977, 1980–1983. *Annual Reports for 1973–74, 1974–75, 1976–77, 1978–79, 1979–80, 1981 and 1982*, ICRISAT, Patancheru, A.P., India.

IITA (International Institute for Tropical Agriculture). 1982. *Annual Report for 1981,* Ibadan, Nigeria.

IRRI 1977. *Annual Report for 1976,* Los Baños, Philippines.

———. 1982. *Report of a Workshop on Cropping Systems Research in Asia,* International Rice Research Institute, Los Baños, Philippines.

Jodha, N. S. 1980. Intercropping in traditional farming systems, *J. Develop. Studies* 16:427–442.

Jones, M. J. 1974. Effects of previous crops on yield and nitrogen response of maize at Samaru, Nigeria, *Exp. Agric.* 10:278–279.

Kanitkar, N. V. 1960. *Dry Farming in India,* 2nd ed., Indian Council of Agricultural Research, New Delhi, India, 470 pp.

Kanwar, J. S., and Ryan, J. G. 1976. Recent trends in world sorghum and millet production and some possible future developments, in: *Symposium on Production, Processing and Utilization of Maize, Sorghum and Millets,* Central Food Technological Research Institute, Mysore, India, December 1976. 52 pp.

Kassam, A. H. 1976. *Crops of the West African Semi-arid Tropics,* ICRISAT, Patancheru-502 324, India, 154 pp.

Kassam, A. H., and Stockinger, K. R. 1973. Growth and nitrogen uptake of sorghum and millet in mixed cropping, *Samaru Agric. Newslett.* 15:28–33.

Koraddi, V. R., Kulkarni, M. V., and Kajjari, N. B. 1970. Overlapping cultivation of cotton in hybrid bajra under dry farming conditions, *Indian Farm.* 20(2):22–24.

Krantz, B. A. 1981. Intercropping on an operational scale in an improved cropping system, in: *Proc. International Workshop on Intercropping,* 10–13 January 1979, Hyderabad, India, ICRISAT, Patancheru, A. P., India, pp. 318–327.

Lal, R. B., De, R., and Singh, R. K. 1978. Legume contribution to the fertilizer economy in legume-cereal rotations, *Indian J. Agric. Sci.* 48:419–424.

Lewis, W. M. 1972. No-tillage production systems for double cropping and for cotton and other crops, in: *Proc. Symp. No-Tillage Systems,* Columbus, Ohio, pp. 146–152.

Lingegowda, B. K., Shanti Veerabadraiah, S. M., Inamdar, S. S., Prithvi Raj, and Krishnamurthy, K. 1972. Studies on mixed cropping of groundnut and hybrid sorghum, *Indian J. Agron.* 17:27–29.

Mandal, R. C., Vidyabushanam, R. W., and Shantanam, V. 1965. Ratooning in hybrid sorghum gives more food and fodder, *Indian Farm.* 15(8):30–31.

Marques Porto, M. C., Almeida, P. A. de, Mattos, P. L. P., and Souza, R. F. 1978. Cassava intercropping in Brazil, in: *Proc. Intercropping with Cassava, International Workshop*, 27 November–1 December 1978, (E. Weber, B. Nestel, and M. Campbell, eds.) Central Tuber Crops Research Institute, Trivandrum, India, IDRC-142e, Ottawa, Canada, pp. 25–30.

Matos, M. A. O. de, and Mota, R. V. da 1983. Avaliaçao de sistemas de arroz solteiro e consorciado com milho, in: *Anais 1 Renião Sobre culturas consorciadas no Nordeste Teresina* - PI, 24 to 28 October 1983, UEPAE de Teresina, Brasil, p. 50.

Mattos, P. L. P. de, and Souza, A. S. da 1981. Mandioca em consorciação no Brasil: Problemas, situação atual e resultados de pesquisa. Documento 1 *Centro Nacional de Pesquisa de Mandioca e Fruticultura,* Cruz das Almas, BA, Brazil, 51 pp.

May, K. W. 1982. Effects of planting schedule and intercropping on green gram (*Phaseolus aureus*) and bulrush millet (*Pennisetum americanum*) in Tanzania, *Exp. Agric.* 18:149–156.

McCollum, R. E. 1983. Dynamics of soil nutrients in multiple cropping systems in relation to efficient use of fertilizers, in: *Fertilizer Use under Multiple Cropping Systems, Report of an Expert Consultation, 3–6 February 1982, New Delhi, Fert. Plant Nutrition Bull. (FAO, Rome)* 5:56–67.

Medappa, C. K., and Mahadevappa, M. 1976. A ratoon crop of 'Intan' paddy in Coorg is possible, *Current Res.* 15(9):148–149.

Mercer-Quarshie, H. 1979. Yield of local sorghum (*Sorghum vulgare*) cultivars and their mixtures in northern Ghana, *Trop. Agric. (Trinidad)* 56:125–133.

Moreno, R. A., and Hart, R. D. 1978. Intercropping with cassava in Central America, in: *Proc. Intercropping with Cassava, International Workshop,* 27 November–1 December 1978, (E. Weber, B. Nestel, and M. Campbell, eds.) Central Tuber Crops Research Institute, Trivandrum, India, IDRC-142e, Ottawa, Canada, pp. 17–24.

Morris, R. A. 1982. Tillage and seeding methods for dry-seeded rice, in: *Report of a Workshop on 'Cropping Systems Research in Asia,'* IRRI, Los Baños, Philippines, pp. 117–132.

Mutsaers, H. J. W. 1978. Mixed cropping experiments with maize and groundnuts, *Neth. J. Agric. Sci.* 26:344–353.

Myers, H. E., and Hallsted, A. L. 1942. The comparative effect of corn and sorghum on the yield of succeeding crops, *Proc. Soil Sci. Soc. Amer.* 7:316–321.

Natarajan, M., and Willey, R. W. 1980. Sorghum-pigeonpea intercropping and the effects of plant population density. 1. Growth and yield. *J. Agric. Sci. Camb.* 95:51–58.

Natarajan, M., Kannaiyan, J., Willey, R. W., and Nene, Y. L. 1985. Studies on the effects of cropping system on *Fusarium* wilt of pigeonpea, *Field Crops Res.* 10:333–346.

Nelson, W. R., Gallahar, R. N., Bruce, R. R., and Holmes, M. R. 1977. Production of corn and sorghum grains in double-cropping systems. *Agron. J.* 69:41–45.

Ngue, N. B., and De Datta, S. K. 1979. Increasing efficiency of fertilizer nitrogen in wetland rice by manipulation of plant density and plant geometry, *Field Crops Res.* 2:19–34.

Norman, D. W. 1974. Rationalizing mixed cropping under indigenous conditions: The example of northern Nigeria, *J. Develop. Stud.* 11:3–21.

Noworolnik, K., Polak, E., and Ruszkowska, B. 1984. A comparison of the productivity of barley and oats grown on soils of a weak rye complex, *Field Crops Abstr.* 37(2, 3):104.

Odunga, V. S. A. 1978. Root exudation in cowpea and sorghum and the effect on spore germination and growth of some soil *Fusaria, New Phytol.* 81:607–612.

Okigbo, B. N. 1981. Evaluation of plant interactions and productivity in complex mixtures as a basis for improved cropping systems design, in: *Proc. International Workshop on Intercropping,* 10–13 January 1979, Hyderabad, India. ICRISAT, Patancheru, 502 324, India, pp. 155–179.

Okigbo, B. N., and Greenland, D. J. 1976. Intercropping systems in tropical Africa, in: *Multiple Cropping* (R. I. Papendick, P. A. Sanchez, and G. B. Triplett, eds.), Amer. Soc. Agron. Spec. Publ. 26, Madison, Wisconsin, pp. 66–101.

Osiru, D. S. O., and Willey, R. W. 1972. Studies on mixtures of dwarf sorghum and beans (*Phaseolus vulgaris*) with particular reference to plant population, *J. Agric. Sci., Camb.* 79:531–540.

Pal, M., and Kaushik, S. K. 1969. A note on ratooning of sorghum in northern India, *Indian J. Agron.* 14:296–298.

Patanothai, A. 1983. The KKU-FORD cropping systems project: An overview, Presented at the *Workshop on Human Ecology at Khon Kaen University,* Kohn Kaen, Thailand, April 18–May 2, 21 pp.

Patel, G. J., Patel, S. H., Patel, N. M., and Patel, D. J. 1979. Feasibility of successful cultivation of bidi tobacco after summer bajri, *Tobacco Res.* 5(1):37–42.

Pinchinat, A. M., Soria, J., and Bazan, R. 1976. Multiple cropping in tropical America, in: *Multiple Cropping* (R. I. Papendick, P. A. Sanchez, and G. B. Triplett, eds.), Amer. Soc. Agron. Spec. Publ. 27, Madison, Wisconsin, pp. 51–62.

Plucknett, D. L., Evenson, J. P., and Sanford, W. G. 1970. Ratoon Cropping, *Advan. Agron.* 22:285–330.

Portes, T. A. de, and Caravalho, J. R. P. de 1983. Area foliar, radiação solar, temperatura do ar e rendimentos em consorciação e em monocultivo de diferentes cultivares de milho e feijão, *Pesq. Agropec. Bras. Brasilia* 18:755–762.

Prithviraj, B. K., Gowda, L., Kajjari, N. B., and Patil, S. V. 1972. Mixed cropping of hybrid jowar and cotton gives higher profits, *Current Res.* 1:18–19.

Rahman, M. M., Razzaque, M. A., Imam, S. A., and Oshima, T. 1982. Agro-economic evaluation of rice-sugarcane winter crop patterns, *Bangladesh. J. Agric.* 7(3/4):1–8.

Rangaswami, G., and Balasubramanyam, A. 1963. Release of hydrocyanic acid by sorghum roots and its influence on the rhizosphere microflora and plant pathogen fungi, *Indian J. Exp. Biol.* 1:215–217.

Rao, M. J. V., and Musselman, L. J. 1984. Host specificity in *Striga* spp. and physiological 'strains', in: *The Biology and Control of Parasitic Weeds. 1. Striga,* CRC Press, Boca Raton, Florida, Uniscience Series.

Rao, M. R. 1984. Prospects for sorghum and pearl millet in the cropping systems of northeast Brazil, Presented at the Workshop on Sorghum and Millets in

Latin American Farming Systems, 16–22 September 1984, CIMMYT, El-Baton, Mexico.

Rao, M. R., Ahmed, S., Gunasena, and Alcantara, A. 1979. Multilocational evaluation of productivity and stability of some cereal-legume intercrop systems, A review of INPUTS trial III. in: *Proc. Final INPUTS Review Meeting,* August 20–24 1979, Resource Systems Institute, East-West Center, Honolulu, Hawaii, pp. 123–160.

Rao, M. R., and Bhardwaj, R. B. L. 1981. Direct, residual and cumulative effect of phosphate in *arhar*-wheat cropping system. *Indian J. Agric. Sci.* 52:96–102.

Rao, M. R., and Morgado, L. B. 1983. Consorciacao com as culturas de algodao e mandioca no Nordeste do Brasil—resultados atuais e perspectivas para futuras pesquisas. in: *Anais 1 Reunião Sobre Culturas Consorciadas no Nordeste* 24–28 October 1983, Teresina-Piaui, CPATSA, Petrolina-PE, Brazil, 58 pp.

———. 1984. A review of maize-beans and maize-cowpea intercrop systems in the semi-arid northeast Brazil, *Pesq. Agropec. Bras. Brasilia* 19:179–192.

Rao, M. R., and Willey, R. W. 1980a. Preliminary studies on intercropping combinations based on pigeon pea or sorghum, *Exp. Agric.* 16:29–40.

———. 1980b. Evaluation of yield stability in intercropping: Studies with sorghum/pigeon pea, *Exp. Agric.* 16:105–116.

———. 1983a. Effects of pigeon pea plant population and row arrangement in sorghum/pigeonpea intercropping, *Field Crops Res.* 7:203–212.

———. 1983b. Effects of genotype in cereal/pigeon pea intercropping on the Alfisols of the semi-arid tropics of India, *Exp. Agric.* 19:67–78.

Rao, M. R., Lima, A. F., and Willey, R. W. 1985. Plant population and row arrangement studies on pearl millet/groundnut intercropping, ICRISAT Journal Article 504, ICRISAT, Patanchera, 502 324, India.

Rao, M. R., Rego, T. J., and Willey, R. W. 1977. Plant population and spatial arrangement effects in monocrops and intercrops in rainfed areas, in: *Seminar on Dry Farming,* 16 April 1977, Institute of Agricultural Technologists, Directorate of Agriculture, Karnataka State, 18 pp.

Reddy, G. P., Rao, C. S., and Reddy, P. R. 1965. Mixed cropping in castor, *Indian Oil Seeds J.* 9:310–313.

Reddy, M. S., and Willey, R. W. 1981. Growth and resource use studies in an intercrop of pearl millet/groundnut, *Field Crops Res.* 4:13–24.

———. 1982. Improved cropping systems for the deep Vertisols of the Indian semi-arid tropics, *Exp. Agric.* 18:277–287.

Roy, D. K. D., and Kardofani, A. Y. 1961. A study of long-term rotation effects in the Sudan Gezira, *J. Agric. Sci., Camb.,* 57:387–392.

Roy, S. K. B., Ghosh, R., and Mandal, J. 1983. Comparison of photo-period sensitive ratooned rice and 'aman' rice, *Int. Rice Res. Newslett.* 8(5):22–23.

Ruiz-Sifre, G. V., and Reis, S. K. 1983. Response of crops to sorghum residues, *J. Amer. Soc. Hort. Sci.* 108:262–266.

SAI (Statistical Abstracts India). 1979. *New Series No. 24,* Central Statistical Organization, Department of Statistics, Ministry of Planning, Government of India, pp. 43, 75.

Sanford, J. E., Myhre, D. L., and Merwine, N. C. 1973. Double cropping systems involving no-tillage and conventional tillage, *Agron. J.* 65:978–982.

Saxena, M. C., Singh, H. P., Yadav, D. S., and Sharma, R. P. 1978. Comparative performance of winter cereals, pulses and oil seeds in pure and mixed cropping under upland and lowland rainfed conditions of Tarai, *Pantnager. J. Res.* 3(1):41–45.

Sen, S., Sen Gupta, K., Sur, S. C., and Mukherjee, D. 1966. A study on mixed cropping of arhar (*Cajanus cajan* (L.) Millsp.), *Indian J. Agron.* 11:357–362.

Serpa, J. E. S., Fontes, L. A. N., Galvão, J. D., and Conde, A. R. 1981. Comportamento do milho e do feijão em cultivos exclusivos consorciados e em faixas alternadas, *Revista Ceres.* 228:236–252.

Shanmugasundaram, A., Tangavelu, S., Mylasamy, V., and Suresh, S. 1967. Successful ratooning of hybrid jowar in Madras State, *Indian Farm.* 17(6):29–31.

Shetty, S. V. R., and Sharma, M. M. 1982. Striga experience from farming systems research at ICRISAT, in: *AICSIP/ICRISAT Working Group Meeting on Striga Research,* ICRISAT, September 30–October 1, 1982, 7 pp.

Singh, H. G., and Sahu, M. P. 1981. Fertilizer use in groundnut based cropping systems in different agroclimatic zones, *Fertil. News* 26(9):51–55.

Singh, K., and Singh, A. 1966. Studies on correcting after effects of sorghum, *J. Res. Ludhiana* 3:135–141.

Singh, S. P. 1983. Summer legume intercrop effects on yield and nitrogen economy of wheat in the succeeding season, *J. Agric. Sci. Camb.* 101:401–405.

Sinthuprama, S. 1978. Cassava and cassava based intercrop systems in Thailand, in: *Proc. Intercropping with Cassava, International Workshop,* 27 November–1 December, (E. Weber, B. Nestel, and M. Campbell, eds.), Central Tuber Crops Research Institute, Trivandrum, India, IDRC - 142e, Ottawa, Canada, pp. 57–66.

Steiner, K. G. 1982. Intercropping in tropical small holder agriculture with special reference to West Africa, *Schriftenreihe der GTZ* 137, Eschborn.

Stewart, J. I. 1982. Crop yields and returns under different soil moisture regimes, presented at the *Third FAO/SIDA Seminar on Field Food Crops in Africa and the Near East,* Nairobi, Kenya, 6–12 June 1982, 19 pp.

Stoop, W. A. 1981. Cereal-based intercropping systems for the West African semi-arid tropics, in: *Proc. International Workshop on Intercropping,* 10–13 January 1979, Hyderabad, ICRISAT, Patancheru, A.P., India, pp. 61–68.

Syarifuddin, A. K. 1982. Tillage practices and methods of seeding upland crops after wetland rice, in: *Report of a Workshop on Cropping Systems Research in Asia,* IRRI, Los Baños, Philippines, pp. 33–42.

Tarhalkar, P. P., and Rao, N. G. P. 1981. Genotype-plant density considerations in the development of an efficient intercropping system for sorghum, in: *Proc. International Workshop on Intercropping,* 10–13 January 1979, Hyderabad, India, ICRISAT, Patancheru, Andhra Pradesh, India, pp. 35–40.

Tiwari, K. N., Pathak, A. N., and Tiwari, S. P. 1980. Fertilizer management in cropping systems for increased efficiency, *Fertil. News,* 25(3):3–20.

Virmani, S. M., Willey, R. W., and Reddy, M. S. 1981. Problems, prospects, and technology for increasing cereal and pulse production from deep black soils, in: *Proc. Management of Deep Black Soils for Increased Production of Cereals, Pulses, and Oilseeds,* 21 May 1981, New Delhi, ICRISAT, Patancheru, A.P., India, pp. 21–36.

Wahua, T. A. T., and Miller, D. A. 1978. Relative yield totals and yield components of intercropped sorghum and soybeans, *Agron. J.* 70:287–291.

Willey, R. W. 1979. Intercropping—Its importance and research needs. 1. Competition and yield advantages, and 2. Agronomy and research approaches. *Field Crop Abstr.* 32:1–10, 73–85.

Willey, R. W., and Osiru, D. S. O. 1972. Studies of mixtures of maize and bean (*Phaseolus vulgaris*) with particular reference to plant population, *J. Agric. Sci. Camb.* 79:517–529.

Willey, R. W., and Rao, M. R. 1981. Genotype studies at ICRISAT, in: *Proc. Int. Workshop on Intercropping*, 10-13 Jan. 1979, ICRISAT, Patancheru, Andra Pradesh, India, pp. 117–127.

Willey, R. W., Rao, M. R., and Natarajan, M. 1981. Traditional cropping systems with pigeonpea and their improvement, in: *Proc. Int. Workshop on Pigeonpeas,* 15–19 December 1980, ICRISAT, Patancheru, India, vol. 1., pp. 11–25, ICRISAT, Patancheru, A.P., India.

Willey, R. W., Rao, M. R., Reddy, M. S., and Natarajan, M. 1982. Cropping systems with sorghum, in: *Proc. Sorghum in the Eighties, Int. Symp. on Sorghum,* 2–7 November 1981, ICRISAT, Patancheru, India, pp. 477–490.

Yankov, B. 1984. Effect of a soybean catch crop during wheat and barley monoculture on the toxicity of the soil solution, *Field Crop Abstr.* 37(8):637.

Yayok, J. Y. 1981. Crops and cropping patterns of the Savanna region of Nigeria: The Kaduna situation, in: *Proc. International Workshop on Intercropping,* 10–13 January 1979, Hyderabad, India, ICRISAT, Patancheru, A.P., India, pp. 69–77.

Zandstra, H. G. 1978. Cassava intercropping research: agroclimatic and biological interactions, in: *Proc. Intercropping with Cassava, Int. Workshop,* 27 November–1 December 1978, (E. Weber, B. Nestel, and M. Campbell, eds.) Central Tuber Crops Research Institute, Trivandrum, India, IDRC-142e, Ottawa, Canada, pp. 67–76.

Zandstra, H. G. 1982. Effect of soil moisture and texture on the growth of upland crops after wetland rice, in: *Report of a Workshop on Cropping Systems Research in Asia* IRRI, Los Baños, Philippines, pp. 43–54.

Chapter 7

Multiple Cropping with Legumes and Starchy Roots

Jeremy H. C. Davis
Jonathan N. Woolley
Raul A. Moreno

Multiple cropping systems that involve legumes and root crops are important where traditional farming systems prevail, in the tropics. They are used principally by small farmers, who need intensified crop production systems and who tend to be involved in more labor-intensive agriculture. Limited-resource farmers also are concerned with risk avoidance, and in some parts of the tropics are more oriented toward subsistence than toward markets for selling their products. This is by no means always true, as small farms in parts of South America and Asia demonstrate (Pachico, 1984). As a generalization, the small farms are found more in marginal areas. Multiple cropping with legumes and root crops offers special advantages for farmers in this situation. For example, a number of authors have shown that there is a much lower probability of income falling below a disaster level in intercrops than in the equivalent sole crops (Rao and Willey, 1980; Baker, 1980; Francis and Sanders, 1978). In temperate-climate countries there is also mounting interest in returning to more diversified agricultural production systems to obtain improved crop protection and benefit from the increased productivity offered by many intercropping systems.

Traditional farming systems with legumes and root crops are not necessarily stagnant because they are traditional. Innovation and change are normal features of traditional farming systems (Ruthenberg, 1980), and the challenge is to accelerate this process while maintaining the benefits inherent in stable farming systems. The effort going into breeding improved cultivars of grain legumes and roots in many tropical countries and in the Consultative Group for International Agricultural Research (CGIAR) system is resulting in higher crop productivity, including improved stress tolerance which is relevant to farmers using multiple cropping systems.

Beans (*Phaseolus vulgaris*) are the most important of the grain legumes, with annual world production of over 8 million tons (Laing et al., 1984). Of these, about 4 million tons are produced in Latin America (Sanders and Alvarez, 1978), and 1.4 million tons in Africa (CIAT, 1981a). A definition of the principal cropping systems in which beans are found and their relative importance in different agroecological zones of Latin America were presented by Laing et al. (1984). Francis et al. (1976) presented data showing that 90 percent of the bean crop in Colombia, 73 percent in Guatemala, and 80 percent in Brazil is grown in association with other crops. Cowpeas (*Vigna unguiculata*) in Africa are almost entirely grown with other crops (Arnon, 1972). These percentages may have decreased in recent years in some countries, where the opportunity cost for labor has increased, but it is likely that multiple cropping will persist for the foreseeable future as an important feature of grain legume production.

Grain legumes are sometimes a secondary crop to a major cereal or root crop, where the legume provides groundcover and a bonus to the yield of the primary crop. This is the case of cowpea with sorghum (*Sorghum bicolor*) in northern Nigeria (Andrews, 1972). In other areas the legume may be the major crop, as with beans and maize (*Zea mays*) in eastern Antioquia, Colombia. The importance attached to each crop component will be reflected in its agronomic management, such as relative planting date and plant population.

For legumes, the association with cereals is of major importance. The largest section of this chapter is devoted to this crop association, and an attempt is made to draw conclusions on the plant types required for improving these cropping systems. We also explore the way systems interact with the control of diseases and pests, the management of plant density, fertilization, relative planting dates, and weed control. Appropriate examples have been taken principally from the major grain legumes that are intercropped in the tropics: bean, cowpea, pigeon pea (*Cajanus cajan*), peanut (*Arachis hypogaea*), and chickpea (*Cicer arietinum*). There are too many species and crop combinations involved to treat each separately. Rather, information of a generally applicable nature is included.

Of the starchy root crops grown in the tropics, cassava (*Manihot esculenta*) is the most important, with world production of more than 100 million tons of fresh roots per year (Onwueme, 1978). The proportion of this production that is intercropped is much lower than for grain legumes. On the other hand, one of the major problems of cassava production on sloping land is erosion, and here intercropping has a particularly important role to play.

The sections of this chapter on root crops deal individually with the most important crop associations. Finally, a section of the chapter is devoted to the need to develop and adapt technology for the small farmer growing legumes or root crops in multiple cropping systems.

MULTIPLE CROPPING OF LEGUMES WITH CEREALS

Characteristics of the Plant

Most studies of intercropping legumes with cereals have emphasized agronomic aspects rather than genetic improvement. Where intercropping plays a major part in crop production, for example common beans with maize or pigeon peas with sorghum, potential new cultivars should be tested in these cropping systems before release. The next step is to carry out part of the process of breeding, using the traditional cropping system rather than sole cropping when making selections. The possibility of breeding grain legumes for intercropping was first studied with beans at CIAT (Centro Internacional de Agricultura Tropical) (Francis et al., 1976) and with cowpeas at the International Institute for Tropical Agriculture (IITA, 1976, 1977).

Breeding cultivars specifically for a cropping system is justified if the interaction between genotypes and different cropping systems is highly significant, or if selection is carried out more efficiently in intercropping than in sole cropping. Francis (1981) has reviewed the significance of genotype by cropping system interactions in a number of legume/cereal intercrops. The conclusion was that in most cases there is a significant interaction, and that in some cases this deserved attention by the breeder. This approach may not give the whole answer, however, since it involves trials with finished lines or cultivars. In most cases multiple cropping involves competition, and as such can be considered as a stress. A breeding strategy designed to incorporate tolerance to the stresses involved may result in improved cultivars that are more suitable for intercropping. Selection should be directed at minimizing intercrop competition and maximizing complementary effects (Willey, 1979a). More detailed treatment of breeding decisions is found in Chap. 10.

The other aspect to study is the relative efficiency of selection in intercropping versus sole cropping. In some cases the efficiency may be lower in intercropping, due to reduced yields of the legume and increased coefficients of variation (Francis et al., 1978a, 1978b). These problems can be overcome, however, by increasing plot size or the number of replications, as shown for climbing beans intercropped with maize (Davis et al., 1981). Climbing beans provide an example where it is more efficient to select in intercropping with maize than in sole cropping, because of the much reduced cost of managing the intercrop compared to managing a sole crop that requires a trellis or stakes. In addition, recent evidence indicates that the heritability of yield for beans in the early generations of a breeding program is higher in intercropping with maize than in sole cropping (Zimmermann et al., 1984). The reason for this may be that competitive ability has a higher heritability than yield potential because it is associated with highly heritable characteristics such as plant height and vigor.

Intercrops of legumes with cereals can be broadly divided into those which take advantage of temporal differences between the crops (e.g., sorghum/pigeon pea intercrop), and those which mostly rely on spatial differences (e.g., maize/climbing bean intercrop). The plant characters associated with these two categories are time to maturity (temporal differences) and plant architecture above and below ground (spatial differences). Important characters of the legume expected to influence its suitability for intercropping with cereals are its tolerance of shade and its ability to fix nitrogen. The priorities for breeding resistance to particular diseases and pests may differ for intercropping versus sole cropping, due to the effects of the cropping system on the incidence and severity of these. Integration of the breeding program with an on-farm research program is ideal for setting priorities relevant to the needs of the farmer.

Maturity Differences In the sorghum/pigeon pea intercrop, which is important in India, the sorghum dominates the early stages of growth, maturing after 4 months and leaving the pigeon pea to flower and ripen after harvest of the sorghum (Willey et al., 1983). The slower-growing pigeon pea has virtually no effect on sorghum yield, and is able to grow on residual moisture at the end of the rainy season after the sorghum harvest. In the work of Natarajan and Willey (1979), the competition from the sorghum resulted in a 47 percent reduction in dry matter yield of the pigeon pea, but this was accompanied by a remarkable increase of the harvest index from 22 to 30 percent. This is one of the few cases reported where harvest index is increased by competition from an intercrop, and perhaps only occurs where there are large maturity differences. In most cases harvest index of the legume is reduced by competition, for example, with peanut/pearl millet (*Pennisetum americanum*) (Willey et al., 1983) and with beans/maize (Davis et al., 1984). A relatively high harvest index, in the face of competition, is a desirable characteristic of the plant for intercropping.

Maturity differences between intercropped species have to be quite large to obtain the benefits of temporal separation. For example, although some reports of beans and maize indicate an advantage for early beans associated with late maize (Osiru, 1982), the difference is not usually great enough to compensate for the lack of competitive ability of the early beans, so that the later climbing types usually give the highest yields with maize (Davis et al., 1984).

An important factor in controlling maturity differences is the photoperiod reaction of cultivars. Studies at different latitudes in Nigeria indicate that photoperiod response of cowpeas controls their adaptation to the cereal farming systems with sorghum and pearl millet. Regardless of seasonal variation in sowing date, photosensitive cowpeas tend to flower close to the average date when the rains end, so that the seeds mature into the dry season as the leaf area of the associated cereal begins to decline (Steele and Mehra, 1980). Traditional photosensitive lines yield less than photoinsensitive lines in sole crop in northern Nigeria, but the photosensitive types tend to do better when intercropped with a cereal because they can maximize vegetative growth, delaying flowering until the start of the dry season (Wien and Summerfield, 1980).

Photoperiod insensitivity has been a major breeding objective in most crops where wide adaptation is sought. As the need for specific adaptation to different agroecological zones becomes more widely recognized, photoperiod sensitivity may need to be reexamined as a selection criterion (Hershey et al., 1982).

Plant Architecture Kass (1978) mentioned the need to consider the effect of each crop component on the other components in order to determine the plant types most appropriate for intercropping. If there were no genotype by cropping systems interaction, this would imply either that there were no differences in competitive ability among the genotypes, or that competitive ability was just as favorable for sole cropping as for intercropping. Studies with crops such as barley have indicated that competitive ability is negatively related to yield in sole cropping (Hamblin and Donald, 1974), whereas it is positively related to yield in beans intercropped with maize and not correlated to yield of beans when sole cropped (Davis and Garcia, 1983).

Medium- to short-stature cereal cultivars provide less competition to an intercropped legume than do tall cereals (Andrews, 1972). The value of this depends on the growth habit of the legume, however. In the case of maize/beans, a vigorous climbing type will pull over the stem of a short stature maize (Table 7.1), and the total yield of the lodged crop will suffer accordingly. On the other hand, there is a significant reduction in root lodging of the maize when intercropped with beans, apparently due to improved anchoring of the roots and basal portion of the stem (Table 7.2). The result is that some maize cultivars may yield more in certain intercropping combinations than in sole cropping due to reduced lodging (Davis et al., 1981). This is not a valid selection criterion, however, since it would be preferable to select lodging resistant maize cultivars.

Higher total yields may be obtained from an intercrop due to better interception of light in a stratified crop canopy, consisting of the cereal above and the legume filling the gaps below. The land equivalent ratio (LER) provides measure of the efficiency of particular crop associations relative to the sole crops (Willey, 1979b). For the purpose of selecting genotypes, however, an economic

Table 7.1 Effect of Growth Habit of Indeterminate Beans on Lodging and Yield Reduction in Maize

Bean line	Bean plant height (cm)	Maize lodging		Maize yield (% reduction)
		Stem	Root	
G2801	90	8.5%	2.5%	0.0
G4446	90	9.0	2.1	1.2
G2525	230	16.3	3.9	16.0
G2258	220	20.2	6.3	32.1
G3371	270	24.2	5.1	33.4
LSD, $p = 0.05$	27	4.1	2.3	—

Source: Garcia and Davis, 1985.

Table 7.2 Comparison of the Effect of Intercropping Indeterminate Beans on the Stem and Root Lodging of Maize

Maize cultivar	Maize plant height (cm)	Maize stem lodging		Maize root lodging	
		S%	I%	S%	I%
ICA H-210	191	1	0	2	2
Antig. × Rep. Dom.	246	2	9	81	51
La Posta	256	1	6	43	25
ICA H-209	262	1	4	77	17

Note: S% = Lodging % in maize sole crop.
I% = Lodging % intercropped with beans.

analysis is more useful because it deals with absolute values rather than ratios. When beans are intercropped with maize, climbing ability is found to be the most important character determining competitive ability (Davis et al., 1984). In an economic analysis of combinations of 3 genotypes of maize (short, medium, and tall stature), and 10 genotypes of beans (semiclimbing, medium climbing, and aggressive climbing types), it was found that the best net income was obtained with the medium stature maize (Suwan-1) associated with the aggressive climbing beans (Davis and Garcia, 1983). The yield of beans and maize was negatively correlated when different bean cultivars were planted with the same maize, and between 0.6 and 1.9 kg of maize yield were lost per additional kilogram of bean yield, due to competition. This relationship varied from one location to another, and the factors involved are not yet understood, but may be related to the genotype of maize and the relative importance of aboveground and belowground competition. Maize yield reduction can be used as an independent measure of bean competitive ability, and this was found to be positively correlated with bean yield when intercropped. Bean competitive ability was not correlated with yield when sole cropped. The conclusion is that competitive ability and productivity in sole cropping are unrelated characters, as originally suggested by Donald (1963).

Competitive ability in beans is closely related to plant height and biomass production (vigor). The ideal legume, however, would have a reduced competitive effect on the companion cereal crop, and at the same time show high productivity. An attempt has been made at CIAT, in early generation yield trials, to select breeding lines of beans significantly above the regression line between bean and maize yield with the goal of improving bean yield with a minimum reduction of maize yield. The F_3 families of beans shown in Fig. 7.1 fall into groups according to their growth habit: determinate growth habit I (bush) families caused slight maize yield reduction, but bean yield was poor; indeterminate growth habit IV (climbing) families caused a greater reduction in maize yield, but bean yield was much higher; indeterminate type II and III families fell between the other two groups. Although some families fell above the regression

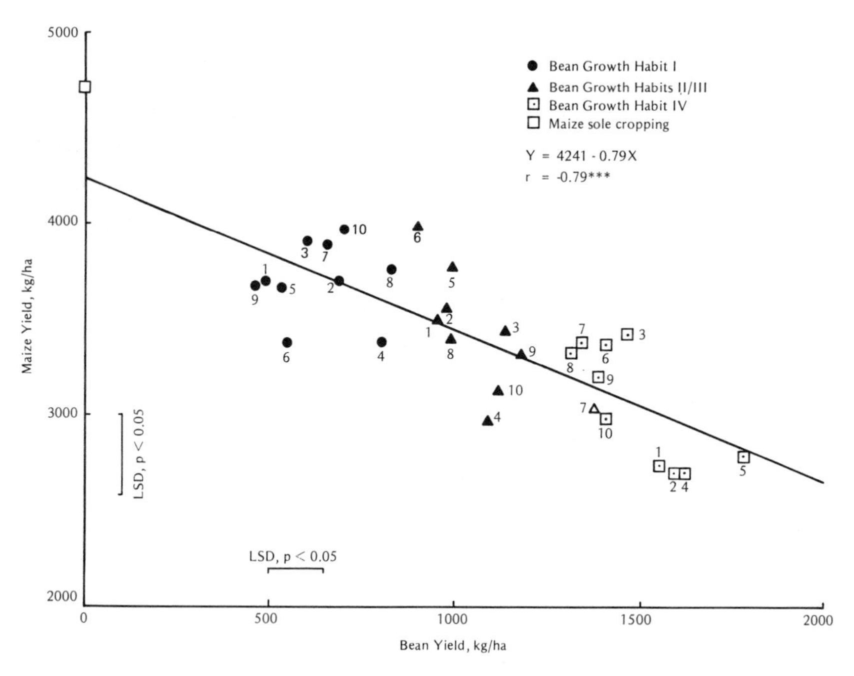

Figure 7.1 Mean yields of beans and maize (kg/ha) for F_3 families (numbered 1 to 10) of beans intercropped with maize (mean of 3 genotypes). (From Davis et al., 1983.)

line, the deviations from regression were not significant in most cases. The results to date suggest that there is little flexibility to improve bean yield without a corresponding reduction in maize yield in conditions where other factors are not limiting.

Climbing in beans is stimulated by the presence of a support, which may be a bamboo stake or a maize plant. Most indeterminate genotypes will react to this stimulus by developing more nodes on the main stem, but this varies according to the growth habit. Type IIa (CIAT, 1981b) has the least ability to respond, and the type IVb has the greatest response. For example, node number in line VRB 81057 varied from 15 nodes without support, to 27 nodes with a 2-m bamboo stake, and to 28 nodes when intercropped with Suwan-1 maize (Davis et al., 1984). Those genotypes that were able to respond to the stimulus to climb suffered less reduction in yield and harvest index due to competition from the maize, and consequently there was a significant interaction of bean lines with cropping system, both for yield and harvest index. The competitive relationship between maize and beans is reflected in a strong negative correlation between biomass production of the two crops. Because the legume normally suffers some loss in harvest index due to competition, increased total production of the intercrop versus the sole crop must be attributed to increased total biomass production.

The quality of light can also affect climbing in certain genotypes of beans

(Kretchmer et al., 1977). When the dark period was interrupted with 15 minutes of red light, indeterminate bush types became climbers. The filtering of light through the canopy of maize may bring about this phytochrome-controlled effect, but the importance of this in the field has yet to be demonstrated. The light intensities needed to trigger the response are very low.

In cowpeas, Wien and Smithson (1981) demonstrated significant interactions of genotypes by cropping systems. Plant size (vigor) in late pod fill was consistently correlated with seed yield in intercropping, demonstrating the need for sufficient vegetative growth to support pod development in stress situations. They concluded that an initial screening could be made in sole cropping, selecting for disease and insect resistance and eliminating plants with low vigor and erect plant type. This would need to be followed by testing the material in the cropping system in which it would ultimately be grown.

Management of Pests and Diseases

In natural plant communities, the diversity of hosts results in spatial isolation, and this, combined with low levels of resistance to pathogens in the population, results in a balance where pathogens and hosts can coexist (Burdon and Shattock, 1980). The same principle applies to crops: diversity within one crop species (cultivar mixtures) and diversity between species (intercropping) are both liable to lead to more stable cropping systems that are less prone to attack from diseases and pests (Suneson, 1960). This may be related to the diluting effect of nonhost plants on the dispersal of pathogens and insects (Trenbath, 1977). In addition, yield may be buffered by compensatory growth of other components of the mixture that are not attacked by the pest or disease.

As with competition, the effect of intercropping on disease and pest control will depend on the temporal and spatial relationship between the components. In young cowpea plants, disease susceptibility was found to be induced by maize pollen (Allen and Skipp, 1982). Where shade favors the pathogen, intercropping may often result in more disease development, and this will be affected by planting density. The plant architecture of the companion crop may affect disease development through differences in humidity and shading. More details are presented in Chap. 9.

There seems to be great variation in the protection against diseases and pests, according to the companion crop. While there is no obvious trend for improved control of diseases resulting from intercropping, it is clear that the spectrum of diseases will be different in intercropping than in sole cropping.

Management of Density and Spatial Arrangement

Intercropping gives a yield advantage when the optimum total plant density is higher than that of either sole crop (Willey, 1979b). It is often stated that the legume should provide an additional yield bonus on top of the normal cereal yield, and the cereal is therefore planted at optimum density for sole cropping. It is assumed that the cereal and the legume are able to exploit different parts of the environment, the legume receiving the light not intercepted by the cereal, and the roots exploiting different strata.

The total density that can be sustained depends on the environmental resources. In stress conditions plant density should be low. When there was severe drought stress, Fisher (1977) indicated that sole cropping beans was preferable to intercropping with maize. On the other hand, Lépiz (1978) demonstrated that intercropping beans and maize resulted in greater stability of production, since any loss of plant density of one crop tended to be compensated by the other crop. For small farmers, this is probably a major factor influencing the decision to intercrop.

Willey and Osiru (1972) used the replacement series concept with beans and maize, where the optimum number of bean plants in sole cropping was twice the number of maize plants. In intercropping, one maize plant was considered equivalent to two bean plants. At low density the replacement of maize by beans gave no advantages, but at higher total density yield advantages of 38 percent were obtained. In mixtures the optimum plant population was higher than in the pure stands as a result of improved utilization of environmental resources.

Intercropping experiments with green gram (*Phaseolus aureus*) and bulrush millet (*Pennisetum americanum*) (May, 1982) indicate that the best combined plant population is somewhere between the optimum for sole cropping, calculated on the basis of replacement, and double these rates (additive intercropping).

Research at CIAT with beans and maize indicates that bush beans can be planted at their optimum sole crop density (approximately 25 plants per square meter) with maize at moderate density (4 plants per square meter), but climbing bean density needs to be reduced (Francis et al., 1982a, 1982b). The average optimum density for climbing beans in sole cropping is 12 plants per square meter and for intercropping is 6 plants per square meter. This is because the climbing bean can pull over the stem of the associated maize if the density is too high, resulting in lodging of the whole crop. This depends on the relative vigor of the two crops. Planting the two crops together in hills (several seeds of each crop planted together in clumps) reduces the amount of stem lodging, when compared with planting in rows (seeds of each crop distributed evenly in separate furrows), and this may partially explain why hill planting is commonly used by farmers in the tropics (Table 7.3). Another advantage of hill planting is that it allows very accurate placing of the fertilizer at the side of the emerging plants.

The benefit of planting in hills was not found in maize/pigeon pea intercrops (Dalal, 1974), probably because pigeon pea does not climb. Competition for light and nutrients was greater for both crops when grown in the same hill compared with alternate rows, and this particularly affected the yield of maize. The orientation of rows in relation to the movement of the sun was studied with a maize/soybean intercrop by Pendleton et al. (1963), but no significant effect was found. On sloping land, the orientation of rows to follow the contours for erosion control is probably more important than its effect on light penetration. Farmers sometimes report that they plant climbing beans upwind of the maize so that the wind blows the young bean plants against the maize and encourages them to climb.

Work conducted in farmers' fields on plant population and spatial arrange-

Table 7.3 Effect of Hill Planting on Lodging in Maize Associated with Climbing Beans at Densities of Three Plants for Maize and Six Plants for Beans, per Square Meter

	Percent lodging[b]		
Planting system[a]	Stem	Root	Total
Rows, 100 cm apart	17.0a	3.2a	20.1a
Hills, 66 × 100 cm	13.0b	3.8a	16.8b
Hills, 100 × 100 cm	9.8c	3.0a	15.2b

[a]The term "hill" is used for clumps of seeds of the two crops planted together in the same hole. Rows consisted of a central row of maize, flanked by a row on either side of beans at 15 cm distance from the maize; the distance between central rows of maize was 100 cm.

[b]Means with the same letter within a column are not significantly different based on the Duncan's multiple range test at the 0.05 probability level.

Source: Data from Garcia and Davis, 1985.

ment is relatively limited in comparison to that conducted on experiment stations. In farmers' cropping systems, which have evolved over a period of time, changes in plant density or spatial arrangement without an accompanying change in cultivar or other practices are unlikely to improve the system. This is because farmers readily experiment with these factors, and information is exchanged between neighbors. It is unlikely that a researcher can improve on this experimentation in a simple and readily varied component. An exception to this rule might be found in settlement schemes, or where a new cropping system has been introduced.

An example of farmer experimentation with spatial arrangements of two crops may be seen in an area of Nariño, Colombia. In an area where beans are row intercropped with maize, farmers use a substitution system, replacing a row of beans at regular intervals by a row of maize. Sampling of farmers' fields shows that the number of bean rows per maize row has increased recently from a mean of three in 1982 to five in 1984. Interrow and interplant distances of maize and beans, however, have not changed. Farmers, when interviewed, usually rationalized this change by stating that the maize provided too much shade for the beans in the earlier system, although an increase in the bean/maize price ratio has also probably encouraged this change.

The use of row substitution suggests that row intercropping developed from an original bean sole cropping system. This is supported by recent observations in another area of Nariño, Colombia. Maize/bean row intercrops have appeared in an area that was almost completely dedicated to bean sole crops. Every second row of beans has been substituted for a row of maize in this case. The evolution of this new system is being studied to determine how farmers vary the proportion of the two crops according to the competition effects.

Changes in crop spatial arrangements may be useful in farmers' fields if they accompany another change. Experiment station results in Antioquia, Colombia, had shown that a change from four maize plants and two climbing bean plants per hill spaced 1 m apart (a typical farmer's planting arrangement), to two maize plants and two bean plants spaced 1 × 0.5 m between hills (row width and maize density unchanged; bean density doubled) increased the yield of new early maturing bean cultivars without affecting maize yield or causing lodging. Preliminary calculations showed that the additional cost of seed and labor was more than compensated by the additional bean yield. Such a change was not suitable for the traditional cultivar since the increased bean density caused maize lodging. This change was tested on farms in Nariño and the highest net benefit (Table 7.4) was obtained by replacing both the local maize and the local bean by an early maize (Cundinamarca 431) and an early bean (ICA-Llanogrande). Whereas the most suitable plant population for local maize/new bean was 4 + 4 plants per square meter, that for the new maize/new bean was 4 + 8 plants per square meter. Thus a change in population and spatial arrangement permitted the proper expression on the farm of the different plant types of maize and beans, which did not appear useful at the farmers' usual population densities.

Fertilization

Soil fertility typically interacts with a number of other agronomic variables such as plant density, weed control, and cultivar. In most crop associations of legumes with cereals, the fertilizer is applied to the cereal and the legume may take advantage of it. In general the cereal demands more nitrogen than the legume. If the legume is fertilized separately, as in a relay system, it receives more phosphorus than nitrogen.

Ahmed and Gunasena (1979) showed that the yield of legumes intercropped with maize always decreased with increasing rates of nitrogen due to increasing competition from the maize. In all experiments the intercrop provided higher economic returns than the sole crops at corresponding nitrogen levels. When the legume is less fertilizer responsive (e.g., soybean), the relative advantage of intercropping may decrease with increasing levels of nitrogen application (Liboon and Harwood, 1975). The same conclusion was reached by Mutsaers (1978) for the peanut/maize association in Cameroon, which was found to be better buffered against low fertility levels than pure maize stands. Intercropping tends to be especially beneficial, therefore, when soil fertility is limiting.

On the other hand, Krijgsman and Damla (1979) demonstrated in 173 farm trials in Nigeria that fertilizer has a beneficial effect on farmers' maize/cowpea and sorghum/cowpea intercrops. Net profits were generally higher for intercrops than for sole crops, and fertilizer alone gave a higher value-to-cost ratio than any other improved practice. Ratios were generally higher for mixed than for sole cropping.

In sequential or relay cropping the first crop can have a beneficial effect on the second crop. Benclove (1970) found that the nitrogen fertilizer for the maize can be reduced following peanuts in a double cropping system.

The effect that an intercropped legume has on the cereal depends on growth and the timing of nutrient requirements. Sole-cropped legumes in rotation with

Table 7.4 Yields, Costs and Benefits in Farm Trials Involving a Change in Spatial Arrangements Accompanying Less Vigorous, Early Varieties of Maize and Beans

Ipiales District, Nariño, Colombia 1982 (Mean of 3 Trials)

Maize variety	Bean variety	Within row spacing (m)	Population (m^{-2})		Yield (kg/ha)		Variable costs[a] ($Col/ha)	Net benefit ($Col/ha)
			Maize	Beans	Maize	Beans		
Local	Mortiño (local)	1.0	4	2	2147	545	2159	43,872[b]
Local	Llanogrande	1.0	4	2	2541	297	1511	37,457
Local	Llanogrande	0.5	4	4	3100	447	4314	51,704
Local	Llanogrande	0.5	4	8	2706	582	6901	46,969
Cundinamarca 431	Llanogrande	1.0	4	2	2284	377	1050	36,186
Cundinamarca 431	Llanogrande	0.5	4	4	2573	813	3392	57,356[b]
Cundinamarca 431	Llanogrande	0.5	4	8	2847	861	5979	62,080[b]
LSD (5%)					935	175	—	—

[a]Fixed costs = 22,759 Colombian pesos per hectare ($Col/ha).
[b]Economically efficient treatments.

cereals may fix more nitrogen than intercrops, and therefore leave more residual nitrogen for the next crop (Nnadi, 1978). There is no evidence yet for a direct transfer of nitrogen from the legume to the cereal in association, and fixation is normally reduced in association due to competition (Searle et al., 1981) since the nitrogen contribution from the legume is related to its vegetative vigor.

Although some of the papers cited above indicate sustained advantage from intercropping even with improved fertilization, it is more common to find the land equivalent ratio (LER) reduced, especially at higher levels of applied nitrogen, demonstrating the value of intercropping cereals and legumes especially for low-input farming systems.

Relative Planting Dates

The effect of competition between crops is greatly alleviated when their maximum demands on the environment occur at different times (Andrews, 1972). Work on the association of cowpeas with maize at IITA was based on prior evidence from Baker (1975) that the maize/cowpea combination offers a more productive alternative to the traditional sorghum/millet/cowpea system. Cowpeas produce 35 percent of their sole crop yield on average when planted simultaneously with maize, declining to very low yields when planted 30 to 40 days after the maize, and recuperating again as the planting date is delayed still further (Wien and Smithson, 1981). As the duration of the overlap between the two crops is reduced, so the yield of the cowpea increased, provided they are not pushed too far into the dry season. Knowing the length of the rainy season at different latitudes in Nigeria, and the effect of different relative planting dates on competition between cowpea and maize, Wein and Smithson (1981) were able to develop a model to predict the yield of cowpea for different times of planting after maize at a given latitude.

In beans the situation is somewhat different, in that simultaneous planting of beans and maize is common in much of the tropics. This results typically in a 50 percent reduction in yield of the beans compared with sole cropping. When relative planting dates of beans and maize were varied by up to 20 days, Francis et al. (1982c) found that maize and bean yields were negatively correlated. Relay intercropping, involving larger differences in relative planting date, does occur in some areas where the rainfall distribution permits, for example in Antioquia and in parts of Central America, Mexico, and Brazil. There is no obvious trend in relative planting dates that can be associated with latitude, however. Agroecological studies of Latin American bean production regions currently underway at CIAT (1980) should lead to a better definition of rainfall distribution and thus of the maturity classes of maize and beans needed for the different relay cropping systems.

A change in relative planting dates may be useful in beans, especially if coupled with a change in plant morphology. As an example, in Central America the bush or semiclimbing beans are often planted at physiological maturity of the preceding maize crop, and thus suffer drought stress if the rainy season ends early. Woolley and Smith (1985a) showed that a potential solution to this problem was to advance the planting date of beans by about 3 weeks. When less leafy maize cultivars were used, bean yield was not affected. This provides an alternative solution to the use of early maize cultivars which yield less than those presently used. There is evidence that maize cultivars with reduced leaf area produce at least as well as the more leafy types. Bean cultivars differ in their adaptation to advancing the planting date (Woolley and Smith, 1985b).

Changing planting dates will require a shift in the times of maximum demand for labor, and this factor needs to be taken into account when designing modifed cropping systems and testing them on farm.

Weed Control

The spreading canopy of cowpeas sown under cereals smothers weeds and makes further weeding unnecessary (Summerfield et al., 1974). Hart (1974) reported that weeds accounted for 20, 25, and 83 percent of the total biomass of maize, cassava, and bean sole crops, and 16 percent when these crops were intercropped. This may be an important factor influencing the farmer's decision to use intercroppping, particularly since chemical weed control is not widely available or practiced by farmers growing cereals and legumes in the tropics. The weed-suppressing ability of intercrops will depend on the component crops and the plant type of the cultivars selected (vigorous, spreading legumes will compete better than erect types), and on spatial arrangement. The enhanced competitive ability of intercropping may be due to high total plant population in some cases.

On the other hand, weeding of intercrops can be more time consuming, and herbicide use may be less effective because of the need to select herbicides that do not damage the cereal or the legume (Moody and Shetty, 1981). Combinations of effective herbicides have been successfully used in maize/bean intercrops at CIAT for a number of years.

MULTIPLE CROPPING OF ROOTS WITH CEREALS

Cassava with Maize

There are several commonly used cropping systems involving these species. The first is simultaneous planting, in which the maize takes advantage of the light and soil that are not totally used by the cassava due to its slow initial growth rate. During the first phase of growth, the relationship between these species is complementary. Afterward, competition occurs (principally for light) as the two species develop a larger leaf area. Maize competes with weeds while the cassava is in its initial growth stages. When the maize reaches physiological maturity it is doubled over (the stem is bent over below the ear, a common practice in Latin America), and the leaf area of the cassava occupies the space left free by the maize. In sedentary agriculture this system is important where weeds are a major problem, and where soil preparation is avoided as much as possible between each crop. In Africa, Unamma and Ene (1984) found that cassava was more affected than maize by competition from weeds. The first 4 to 8 weeks were found to be most critical for weed competition.

Another system is the intercropping of cassava between the maize when the latter is reaching physiological maturity. This arrangement is common in shifting agriculture, based on long periods of fallow, although it is also sometimes used in more sedentary agriculture in the humid tropics. Cassava is one of the last crops established in the succession of species after the cutting and burning of primary forest in shifting agriculture, or after cutting secondary vegetation in agriculture based on long fallow periods.

These intercropping systems have been traditionally practiced in Latin America since the time of the Mayan civilization (Roger, 1965), and more recently in Africa and Asia. When these species are planted together simultaneously, the planting densities used are more critical than in relay planting because of the competition. The maximum density of maize that can be used without adversely affecting the cassava root yield is closely related to the plant type, both of maize and cassava, and the spatial arrangement used (Akobundu, 1981; Kang and Wilson, 1980; Meneses and Moreno, 1983; Meneses et al., 1983). Double rows of cassava have been found to be the best arrangement for intercropping maize in terms of the LER and net income (Mattos et al., 1980; Ternes, 1981).

Maize yield depends on an adequate supply of phosphorus and nitrogen, whereas cassava is more tolerant of low fertility. Fertilizer applications should, therefore, be calculated on the basis of the requirements of the maize. In the case of potassium, the situation is reversed, cassava being more demanding (Meneses and Moreno, 1983). The level of fertilization affects the competitive relationship between these species: the yield of maize is more affected by the level of fertility than the yield of cassava.

In Central America the incidence and severity of *Uromyces manihotis* and *Elsinoe brasiliensis* were found to be reduced in cassava intercropped with maize, but *Oidium manihotis* increased (Moreno, 1977, 1979; Larios and Moreno, 1976, 1977). In Africa, Arene (1976) and Ene (1977) found that the attack of *Xanthosoma manihotis* was reduced in cassava associated with maize and melon, when compared with the sole crop of cassava.

Sweet Potato with Cereals

The sweet potato (*Ipomoea batatas*) forms part of diverse crop associations in America, Asia, and Africa. In Asia it forms part of cropping patterns with rice, although intercropping is becoming less common as agriculture is modernized (Wan, 1982). In America it is grown with several other species, especially maize. Most of the literature indicates that the yield of sweet potato is reduced by intercropping with taller crops, which intercept the light, although it is known that some cultivars are more tolerant of shade (Moreno, 1982; Zara et al., 1982). In general there is a correlation between intercepted radiation and total biomass of sweet potato (Escobar, 1975; Lizarraga, 1976). Probably because of this effect of shade, most farmers reduce the density of maize when they plant it with sweet potato. The same phenomenon was reported in Uganda (Okigbo and Greenland, 1976).

The level of fertilization required by the maize may be too high for an adequate distribution of biomass to the roots of sweet potato. A relationship of 1:2 for nitrogen/potassium was found to be optimal in Central America (Jaramillo, 1977).

Sweet potato is frequently used in sequential cropping. It is a good crop for using up residual fertilizer applied to preceding crops. In Central America good yields (18 t/ha) can be obtained with just 133 kg/ha of potassium, when following an association of maize/beans (Brioso de Leon, 1979). In the Caribbean it is also common to plant sweet potato after other horticultural crops that receive fertilizer.

MULTIPLE CROPPING OF ROOTS WITH LEGUMES

Cassava with Legumes

Root crop associations are common with beans, cowpeas, pigeon peas, and peanuts; occasionally they occur with lima beans (*Phaseolus lunatus*), soybeans (*Glycine max*), and forage legumes.

Intercropping cassava and beans is not as common as cassava and maize, due to the different temperature and soil requirements of these species. Cassava is adapted more to the lowland tropics than beans, though there is considerable overlap in their range; beans are generally adapted to more fertile soils than cassava. Cowpeas, like cassava, are able to adapt to acidic, infertile soils, whereas beans will generally not thrive under these conditions (Leihner, 1978). Intercropping with cowpeas, pigeon peas, and peanuts is common in regions of high temperature with extended dry periods. Intercropping with lima beans, soybeans, and cover crops may be more frequent at the experimental level than on the farm. Simultaneous planting of cassava with beans, cowpeas, or soybeans reduces the yield of both species, but gives the highest values of LER when compared with changing the relative planting dates (Thung and Cock, 1979).

Some cultivars of cassava lose part of their foliage toward the end of the vegetative growth period, and these can be intercropped with legumes toward the end of the cassava crop cycle as well as at the beginning. The highest values of LER have been obtained in this system, since cassava yields are not affected by the late planted legume which itself suffers yield reduction of 17 to 35 percent (Moreno and Meneses, 1980). In general, 240 days after planting the cassava is found to be optimal for intercropping with legumes (Moreno and Meneses, 1980; Castellanos, 1981).

For cassava and beans, optimal densities for sole cropping are found to be adequate for intercropping (Thung and Cock, 1979), and changing the spatial arrangement has not been found to produce significant effects (Leihner, 1978). As long as the canopy of the cassava is above that of the legume, its yield is not significantly affected (Leihner, 1983). Pruning the cassava to reduce light interception, provide forage for animals, and reduce the excess leaf area produced by some cultivars was found to be beneficial by Castellanos (1981).

Cassava cultivars with late branching are less competitive for light with legumes planted simultaneously (Thung and Cock, 1979). On the other hand, those that lose their foliage after 210 days from planting are the best for intercropping with legumes at the end of the cassava crop cycle. In legumes, a spreading plant type that provides rapid groundcover is desirable for erosion control and competition with weeds when the legume is planted simultaneously with cassava. On the other hand, climbing cultivars of legumes take better advantage of the cassava for support when planted at the end of the cassava growth cycle (Moreno and Meneses, 1980).

Sweet Potato with Beans

Intercropping sweet potatoes with beans is not very common in America, but is one of the more important and successful cropping systems in parts of central Africa, especially Rwanda. Sweet potato can be relay cropped with beans 60 to 90 days after planting the beans, taking advantage of residual fertility. There is generally some yield loss of sweet potato. However, values of LER above 1.6 have been obtained with an application of 145:61:45 kg/ha of N/P/K fertilizer (Moreno, 1982).

ROOT CROPS IN ANDEAN AGRICULTURE

In the high Andes of South America, a number of root and tuber crops are important, especially potato (*Solanum* spp.), oca (*Oxalis tuberosa*), olluco (*Ullucus tuberosus*), isaño (*Tropaeolum tuberosum*), and arracacha (*Arracacia xanthorrhyza*). These are intercropped with quinoa (*Chenopodium quinoa*) and/or tarwi (*Lupinus mutabilis*). The latter is often planted around the edges of fields, apparently to discourage animals.

The crop species and the sequence of cropping used depends on the altitude. Between 3400 and 3600 m elevation in southern Peru and Bolivia, maize is the most important crop, and is commonly rotated in one of the following 4-year cycles: maize–maize–maize–potatoes; maize–maize–potatoes–fallow; potatoes or barley–faba beans (*Vicia faba*)–maize–maize; maize–wheat–faba beans–maize.

Between 3600 and 3800 m, the most common crop sequences are: potato–wheat–faba beans–barley; faba beans–wheat–peas–barley; potatoes or quinoa–barley–peas–fallow; faba beans–barley–faba beans–fallow. Above 3800 m, the cropping systems used commonly are potato–oca–isaño–barley, or potato–fallow (6- to 7-year fallow). At high altitudes the bitter potato (*Solanum andigena*) is used (Tapia, 1982; Tapia et al., 1983).

MULTIPLE CROPPING OF LEGUMES AND ROOTS WITH PERENNIAL CROPS

Perennial crops may be divided into three classes (Ruthenberg, 1980): (1) perennial field crops, including sugar cane, pineapple, sisal and bananas; (2) shrub crops, especially coffee and tea; and (3) tree crops, such as cacao, coconut, and forestry trees.

Both roots and legumes are found intercropped with perennial crops. Examples are the production of beans under bananas in Rwanda, and the production of cassava in newly established coconut plantations in Tanzania and South India (Kumar and Krishi, 1979).

Perennial crops may reduce soil erosion between seasons of the annual crop, and may bring up nutrients from deeper soil strata in some cases. The annual crop provides soil cover in the early stages of establishment of a perennial crop. This is particularly the case when intercropping beans or cowpeas with sugarcane or coffee. Cassava is intercropped with bananas and plantains in Central America during their establishment.

Work has been reported by CATIE (1981) on the economic yield of various combinations of perennial crops (cacao and coffee) with shade trees (*Erythrina poeppigiana* and *Cordia alliodora*), pastures, and annual crops (maize and beans).

Research in agroforestry, where annual crops are grown between trees planted for erosion control and fuel, has been reported at CATIE (Heuveldop and Lagermann, 1981) and International Center for Research in Agroforestry (ICRAF) (Torres and Raintree, 1984). There are few research results available yet in any system of perennial crops with legumes or roots, but there is great potential in this area, due to increasing pressure on land resources in the humid tropics where there is a need for cropping systems which provide continuous groundcover.

MULTIPLE CROPPING AND ON-FARM RESEARCH

Most multiple cropping research up to now has been conducted on experiment stations. We believe that much more research will need to be conducted on farmers' fields to realize the benefits of research on improved production techniques. The usefulness of on-farm research (OFR) is not confined to multiple crops. The two areas of study tend to be considered together, however, because a high proportion of small farmers' crops are intercropped. OFR is mostly directed at small-scale farmers. Conceptually, it is helpful to follow the terminology used by Simmonds (1984) who divides farming systems research (FSR) into three types:

1. FSR *sensu stricto,* which is the detailed description of existing farming systems, involving numerical modeling, and often has an academic rather than a practical objective.
2. OFR with a farming systems perspective (OFR/FSP), which seeks stepwise changes within farmers' existing cropping systems while taking into account the relationship between these systems and the rest of the farm. The aim is to rapidly identify appropriate technology for farmers by working closely with them, both in the diagnosis of problems, and in on-farm trials designed to find the solution to these problems.
3. New farming systems development (NFSD), which seeks a radical redesigning of cropping systems rather than stepwise change. NFSD is normally, but not exclusively, conducted on experimental stations. A good

example of this approach is provided by the work of the ICRISAT Farming Systems Program in India (ICRISAT, 1984). Most multiple cropping research conducted up to the present has contributed to NFSD; radical changes in species and relative arrangement in the field have been sought, without particular concern at this stage for their adaptation to "farmers' economics."

NFSD has made useful contributions to understanding cropping systems, and it is vital that alternatives to overstressed tropical farming systems become available. There is, however, a need for more multiple cropping research to be conducted in the style of OFR/FSP, adapting technological components to farmers' cropping systems by direct work in those systems on-farm, so as to achieve stepwise improvements in the short term. Background research in farmers' fields on single technological components is also needed. The latter is not NFSD since it does not imply a radical restructuring of the system, but is not OFR as usually defined. Here it will be called "OFR for technology development."

In the case of legumes, an interesting example of the need for different approaches is provided by beans in Latin America and Africa. In some parts of Latin America diseases are the most serious limiting factor. Research, mostly conducted on experimental stations, has developed technological components, namely disease-resistant cultivars. These cultivars are adapted to local conditions that include the farmers' multiple cropping systems which are simulated on the station. This research is followed by the testing and adaptation of these cultivars, sometimes together with agronomic innovations, on farmers' fields. There has been little need so far of OFR for technology development. In contrast, in eastern Africa, soil fertility problems appear more limiting to bean production than diseases. Research being planned to overcome this constraint includes both the development of tolerant cultivars (on-station and on-farm) and OFR for technology development, including OFR on soil conservation measures. After such components have been developed, they are expected to be tested through more conventional "adaptive OFR."

Since OFR/FSP has the objective of evaluating changes of technological components in existing cropping systems, it is not usual to include sole crops for comparison, nor to estimate the LER, since an economic analysis comparing the new treatment with the farmers' practice is more relevant.

Choice of Components for OFR

Some technological components are, by their nature, very susceptible to local conditions of soils, climate, and farmers' crop management. These cannot be satisfactorily researched on experiment stations and most or all work on them should be conducted in farmers' fields. Clear examples are plant density and the dosage of individual fertilizers (rather than the search for new products). Soil scientists have often carried out farm trials before making recommendations for the fertilization of a particular crop in a region, but more research on fertilizer application in multiple crops needs to be conducted on the farm. Some technology components, such as change in plant population, are unlikely to be successful without introducing some other component, such as a new cultivar.

Farmers accept some types of changes more readily than others. For Latin American farmers with some access to purchased inputs the approximate order of increasing difficulty of adoption is: change of cultivar, change in amount of existing input (including change in planting density), introduction of a previously unknown input, change in plant spacing within the row, change in mechanical land tillage practices, change in interrow spacing, addition of a new crop to the system. For farmers in regions where purchased inputs are little known or beyond their financial reach there would be a different order of difficulty of adoption.

OFR in Practice

One reason for the relatively small proportion of multiple cropping research conducted on farm, in relation to stations, may be that researchers fear failure on-farm, but can guarantee obtaining results on station. In fact, when suitable technological components are available for testing in farmers' systems, OFR is rapid and highly effective. It is effective since farmer participation in the conduct of trials, even if only in maintaining researcher-managed trials, greatly reduces the problems of adoption typically encountered with research results extrapolated directly from experimental stations to a target area as extension recommendations.

An example from southern Nariño, Colombia, described more fully by Woolley (1984), illustrates the use of OFR in practice in a mixed cropping system of maize and climbing beans in the Andean Highlands. Beginning with very few cultivar trials in 1981 to 1982, 56 simple on-farm trials (not more than 500 m^2 in size) were conducted in the following 2 years. The number of trials was greater than is common in OFR since many different trial designs were used each year because of the existence of a number of promising technological components. By the end of the second year of trials, a genotype from the trials was being grown by 60 percent of the collaborators in the trials. This line offered tolerance to foliar diseases and root rots, was early to maturity, and yielded more under farmers' conditions. Another more recent advanced line appeared still more promising but needed more testing. Bean and maize cultivars with sufficient earliness to permit the inclusion of another crop in the yearly cycle had been identified, and other technological components necessary for their use were being incorporated. Inproved chemical disease control and additional fertilization, especially when combined with the new genotype, were less promising because they were variable in effect between sites and years. A change in maize spatial arrangement and bean density, combined with the new genotype, was economically attractive in trials but has been shown to need modification in order for the change in spacing to be acceptable to farmers.

Socioeconomic analysis forms a vital part of on-farm multiple cropping research. It is important both for diagnosis of the principal problems before starting on-farm experimentation (Byerlee and Collinson, 1980), as well as for the analysis of agronomic experiments (Perrin et al., 1976).

In the particular case of legumes and roots, studies of their market acceptability are more important than they are for cereals. The preference of consumers for different grain legumes varies greatly from one region to another, both between but also within species. The most extreme case is provided by beans. International trials coordinated by CIAT are divided into 16 broad grain types for bush and climbing beans (CIAT, 1981b), but consumer preferences within any of these subdivisions still separate many types. The price of nonpreferred grain types may vary depending on the availability of beans in the market.

The value of root crop production of different species also varies greatly depending on local preferences. Cassava, for example, has many different uses: for fresh consumption, for processing as farinha or gari (cassava meal fermented to remove cyanogenic glucosides), or for drying as an animal feed. Fresh root crops suffer greater problems of storage and distribution than grain legumes. Thus, the place where the crop is produced may be as important as the yield in determining the economic value of a farmers' root crop.

Finally, research in multiple cropping may be biased to the benefit of the small farmer. This is because associated crops are more labor intensive and are more difficult to mechanize. The complex problem of small-farmer biases in technology for the bean crop has been discussed by Pachico (1984).

CONCLUSIONS

Although there is still a lack of standardization in the definitions of cropping systems, this is improving. In the past, there has been a tendency to classify crops according to the major crop component, ignoring the intercrops which are often legumes. Extension recommendations often have ignored intercrops, but again there is evidence that research that has demonstrated the advantages of these systems is having an effect. The development of OFR in many countries is leading to more appropriate recommendations and closer integration of research and extension.

There is now a firm background of research in intercropping with legumes and root crops, and in legumes this has developed into a breeding effort to develop improved cultivars for intercropping. Contrary to earlier suggestions that this would not be worthwhile, there is accumulating evidence that selection can be efficiently carried out in cropping systems other than sole cropping. A complementary approach is to consider the stresses involved in multiple cropping, especially those related to competition, and select suitable cultivars accordingly. An example of this is the need for early defoliating cultivars of cassava for relay cropping with legumes, or late branching types of cassava for simultaneous planting with maize. Very little is known about shade tolerance in different genotypes of legumes or roots, and this seems worth studying.

Research on agronomic practices for intercropping with legumes and roots began on the research station, and is now moving to the farm. Most research has been with intercropping (planting in rows), whereas many farmers use systems of mixed cropping (planting in hills, or with no distinct arrangement).

Spatial arrangement is often determined more by cultivation practices and rotations than by differences in productivity, another reason for doing more of the research on the farm.

Yield advantages from multiple cropping are obtained from a better use of growth resources (Willey, 1979a). Improved disease and pest control has been shown to occur in many instances in legumes and roots when intercropped, but examples have also been given of poorer control. The greater the diversity, however, the less likely that total crop loss will occur. In this context, the use of cultivar mixtures, or their more sophisticated counterparts (composites, multilines), will provide an additional line of protection.

Some crop combinations interact more than others. The spatial arrangement of a climbing legume, for example, may be largely controlled by that of the companion crop. Cultural practices which reduce lodging in a cereal (e.g., earthing up around maize stems) will result in a benefit for the associated legume. Planting indeterminate beans with maize reduces root lodging especially in tall cultivars of maize, but highly vigorous climbing beans may cause an increase in the stem lodging of the maize. Planting in hills, rather than in rows, significantly reduces the incidence of stem lodging. It is the complexity of these relationships that has delayed research progress in this area.

There has been little work done on differences in rooting patterns which might lead to exploitation of different soil layers. Nitrogen fixation in the legume probably reduces competition for this element and enriches the soil for sequential crops. Legumes seem to be more successful at this when rotated than when intercropped. In some circumstances, however, the depletion of soil nitrogen by a cereal may stimulate nodulation in the legume. Differences in rooting depth are most significant in associations with perennial crops, for example beans with plantains. More work is also needed on the associations of perennial crops with annual crops, because of the possibility of obtaining sustained groundcover and better nutrient recycling.

It is not clear yet whether multiple cropping is more or less beneficial in stress conditions than in optimum conditions, and this will depend on the crop combination. Most intercropping with legumes suggests a greater benefit in low soil fertility conditions. Differences in the timing of maximum demand for water can also lead to a buffering against unpredictable drought stress.

Future advances are expected to be achieved by an increasing involvement with farmers. The problem of developing cultivars and agronomic practices specifically adapted to different agroecological systems needs to be faced, and an improved definition of cropping regions, based on climatic and socioeconomic data, should help to rationalize the development and testing of new technology.

REFERENCES

Ahmed, S., and Gunasena, H. P. M. 1979. N-Utilization and economics of some intercropped systems in tropical countries, *Trop. Agric. Trinidad* 56(2):115–123.

Akobundu, I. O. 1981. Weed control in maize-cassava intercrop, in: *Triennial Root Crops*

Symp. Int. Soc. for Tropical Root Crops (E. R. Terry, K. A. Oduro, and F. Caveness, eds.), Ibadan, Nigeria, series IDRC-163e, pp. 124–128.

Allen, D. J., and Skipp, R. A. 1982. Maize pollen alters the reaction of cowpea to pathogens, *Field Crops Res. 5:265–269.*

Andrews, D. J. 1972. Intercropping with sorghum in Nigeria, *Exp. Agric.* 8:139–150.

Arene, O. B. 1976. Influence of shades and intercropping on the incidence of cassava bacterial blight, in: *Proc. Workshop on Cassava Bacterial Blight,* (G. Persley, E. R. Terry, and R. MacIntyre, eds.), Ibadan, Nigeria, IDRC.

Arnon, I. 1972. *Crop Production in Dry Regions,* vol. 2, Leonard Hill, London, U.K.

Baker, E. F. I. 1975. Research on mixed cropping with cereals in Nigerian farming systems—a system for improvement, in: *Int. Workshop on Farming Systems,* 18–21 November 1974, ICRISAT, Hyderabad, India, pp. 287–301.

———. 1980. Mixed cropping in northern Nigeria. IV. Extended trial with cereals and groundnuts, *Exp. Agric.* 16:361–369.

Benclove, E. K. 1970. Crop diversification in Malaysia, Incorporated Society of Planters, pp. 61–295.

Brioso de Leon, I. 1979. Fertilización de un sistema de producción de cultivos con granos y raices en una distribución de precipitación con un periodo seco corto, M.S. thesis, CATIE, Turrialba, Costa Rica, 100 pp.

Burdon, J. J., and Shattock, R. C. 1980. Disease in plant communities, *Appl. Biol.* 5:145–219.

Byerlee, D., and Collinson, M. P. 1980. Planning technologies appropriate to farmers, Concepts and procedures, CIMMYT, Mexico.

CATIE (Centro Agronómico Tropical de Investigación y Enseñanza). 1981. *Progress Report,* Turrialba, Costa Rica.

Castellanos, V. H. 1981. Comportamiento de la yuca (*Manihot esculenta* Crantz) sometida a una poda parcial y cultivada en asociación con frijol común arbustivo y voluble (*Phaseolus vulgaris* L.), Tesis Mg. Sci., CATIE, Turrialba, Costa Rica, 104 pp.

CIAT (Centro Internacional de Agricultura Tropical). 1980. *Bean Program Annual Report, 1979.* Cali, Colombia, 109 pp.

———. 1981a. Regional workshop on potential for field beans in Eastern Africa, Lilongwe, Malawi, 1980, Proceedings, Cali, Colombia, CIAT, 226 pp.

———. 1981b. *Bean Program Annual Report, 1980,* Cali, Colombia, 92 pp.

———. 1982. *Bean Program Annual Report, 1981,* Cali, Colombia, 198 pp.

Dalal, R. C. 1974. Effects of intercropping maize with pigeonpeas on grain yield and nutrient uptake, *Exp. Agric.* 10:219–224.

Davis, J. H. C., Amezquita, M. C., and Muñoz, J. E. 1981. Border effects and optimum plot sizes for climbing beans (*Phaseolus vulgaris*) and maize in association and monoculture, *Exp. Agric.* 17:127–135.

Davis, J. H. C., and Garcia, S. 1983. Competitive ability and growth habit of indeterminate beans and maize for intercropping, *Field Crops Res.* 6:59–75.

Davis, J. H. C., Perez, A. V., and Hopmans, S. 1983. Early generation yield testing in beans in monoculture and intercropped with maize, EUCARPIA, Section Vegetables, Meeting on *Phaseolus* bean breeding, Hamburg, 19–21 July 1983, pp. 54–65.

Davis, J. H. C., van Beuningen, L., Ortiz, M. V., and Pino, C. 1984. Effect of growth habit of beans (*Phaseolus vulgaris* L.) on tolerance to competition from maize when intercropped, *Crop Sci.* 24:751–755.

Donald, C. M. 1963. Competition among crop and pasture plants, Advan. Agron. 15:1–118.

Ene, L. S. O. 1977. Control of cassava bacterial blight, *Trop. Root Tuber Crops Newslet.* 10:10–31.

Escobar, R. 1975. Análisis del crecimiento y rendimiento del camote en monocultivo y en asociación con frijol, maiz y yuca, M.S. thesis, CATIE, Turrialba, Costa Rica, 81 pp.

Fisher, N. M. 1977. Studies in mixed cropping. I. Seasonal differences in relative productivity of crop mixtures and pure stands in the Kenya highlands, *Exp. Agric.* 13:177–184.

Francis, C. A. 1981. Development of plant genotypes for multiple cropping systems, in: *Plant Breeding II,* (K. J. Frey, ed.), Plant Breeding Symposium, 2d, Iowa State University, Ames, Iowa.

Francis, C. A., Flor, C. A., and Temple, S. R. 1976. Adapting varieties for intercropping systems in the tropics, in: *Multiple Cropping,* (R. I. Papendick et al., eds.), Amer. Soc. Agron. Spec. Publ. 27, Madison, Wisconsin, pp. 235–253.

Francis, C. A., Prager, M., and Laing, D. R. 1978a. Genotype × environment interactions in climbing bean cultivars in monoculture and associated with maize, *Crop Sci.* 18:242–247.

Francis, C. A., Prager, M., Laing, D. R., and Flor, C. A. 1978b. Genotype × environment interactions in bush bean cultivars in monoculture and associated with maize, *Crop Sci.* 18: 237–242.

Francis, C. A., Prager, M., and Tejada, G. 1982a. Density interactions in tropical intercropping. I. Maize (*Zea mays* L.) and climbing bean (*Phaseolus vulgaris* L.), *Field Crops Res.* 5:163–176.

———. 1982b. Density interactions in tropical intercropping. II. Maize (*Zea mays L.) and bush beans (Phaseolus vulgaris* L.)., *Field Crops Res.* 5:253–264.

———. 1982c. Effects of relative planting dates in bean (*Phaseolus vulgaris* L.) and maize (*Zea mays* L.) intercropping, *Field Crops Res.* 5:45–54.

Francis, C. A., and Sanders, J. H. 1978. Economic analyses of bean and maize systems in monoculture versus associated cropping, *Field Crops Res.* 1:319–335.

Garcia, S., and Davis, J. H. C. 1985. Effects of spatial arrangement and density on yield and lodging of maize associated with indeterminate beans, Draft for CIAT, A.A. 67–13, Cali, Colombia.

Hamblin, J., and Donald, C. M. 1974. The relationship between plant form, competitive ability and grain yield in a barley cross, *Euphytica* 23:535–542.

Hart, R. D. 1974. The design and evaluation of a bean, corn and manioc polyculture cropping system for the humid tropics, Ph.D. thesis, University of Florida, Gainesville, Florida.

Hershey, C., Miles, J., and Davis, J. H. C. 1982. Strategies for genetic improvement of the CIAT commodities: Seeking a balance between broad adaptability and site specificity, Discussion paper prepared for 1982 Annual Review, CIAT, Cali, Colombia, 78 pp.

Heuveldop, J., and Lagermann, J. 1981. *Agroforestry,* Proceedings of a seminar held in CATIE, Turrialba, Costa Rica, DSE, Feldafing, Germany.

ICRISAT (International Crops Research Institute for the Semi-Arid Tropics). 1984. Overview of ICRISAT's FSR program, ICRISAT, Patancheru, Andra Pradesh, India.

IITA (International Institute for Tropical Agriculture). 1976. *Annual Report for 1976,* Ibadan, Nigeria.

———. 1977. *Annual Report for 1977,* Ibadan, Nigeria.

Jaramillo, S. 1977. Absorción de nutrimento por maiz (*Zea mays* L.) y camote (*Ipomoea batatas*) en asociación y su fertilización con nitrógeno y potasio, M.S. thesis, CATIE, Turrialba, Costa Rica, 194 pp.

Kang, B. T., and Wilson, G. F. 1980. Effect of maize population and nitrogen application on maize-cassava intercrop, in: *Triennial Root Crops Symposium*, International Society for Tropical Root Crops (E. R. Terry, K. A. Oduro, and F. Caveness, eds.), Ibadan, Nigeria, series IDRC-163e, pp. 129–133.

Kass, D. C. L. 1978. Polyculture cropping systems: Review and analysis, *Cornell Int. Agric. Bull.* 32:1–69.

Kretchmer, P. J., Ozbun, J. L., Kaplan, S. L., Laing, D. R., and Wallace, D. H. 1977. Red and far-red light effect on climbing in *Phaseolus vulgaris* L., *Crop Sci.* 17:797–799.

Krijgsman, D. W., and Damla, J. D. 1979. Fertilizer demonstration and distribution programme, Plateau State, Nigeria, Interim Report, 1 February 1978 to 31 March 1979, FAO, 1979.

Kumar, R. M., and Krishi, N. 1979. Intercropping systems with cassava in Kerala State, India, in: *Intercropping with Cassava*, (E. Weber, B. Nestel, and M. Campbell, eds.), IDRC, Ottawa, Canada.

Laing, D. R., Jones, P. G., and Davis, J. H. C. 1984. Common Bean (*Phaseolus vulgaris* L.), in: *The Physiology of Tropical Field Crops*, (P. R. Goldsworthy and N. M. Fisher, eds.), John Wiley & Sons Ltd., Chichester, U.K., pp. 305–351.

Larios, J., and Moreno, R. 1976. Epidemiologia de algunas enfermedades foliares de la yuca en diferentes sistemas de cultivo. I. Mildiu polvoroso y roya, *Turrialba* 26:389–398.

———. 1977. Epidemiologia de algunas enfermedades foliares de la yuca en diferentes sistemas de cultivo. II. Roya y muerte descendente, *Turrialba* 27:151–156.

Leihner, D. 1978. Agronomic implications of cassava-legume intercropping systems, in: *Intercropping with Cassava*, (E. Weber, B. Nestel, and M. Campbell, eds.), Proceedings of a workshop held in Trivandrum, India, 27 November to 1 December 1978, series IDRC-142e, Ottawa, Canada, pp. 103–112.

Lépiz, R. 1978. La asociación maiz-frijol y el aprovechamiento de la luz solar, Tesis, Doctor en Ciencias, Especialidad Genetica, Chapingo, Mexico, 304 pp.

Liboon, S. P., and Harwood, R. R. 1975. Nitrogen response in corn/soybean intercropping, *6th Annual Scientific Meeting of the Crop Science Society of the Philippines*, 8 to 10 May 1975, Bacolod City, Philippines.

Lizarraga, N. 1976. Evaluación del crecimiento del camote (*Ipomoea batatas*) y su relación con la radiación solar en monocultivo y en asociación con yuca (*Manihot esculenta*) y maiz (*Zea mays*), M.S. thesis, UCR-CATIE, Turrialba, Costa Rica.

Mattos, P. L. P., de Souza, L., and Correa, R. 1980. Double row planting systems for cassava in Brazil, in: *Workshop on Cassava Cultural Practices*, (E. J. Weber, J. C. Toro, and M. Graham, eds.), Salvador-BA, Brazil, series IDRC-151e, pp. 54–58.

May, K. W. 1982. Effects of planting schedule and intercropping on green gram (*Phaseolus aureus*) and bulrush millet (*Pennisetum americanum*) in Tanzania, *Exp. Agric.* 18: 149–156.

Meneses, R., and Moreno, R. 1983. Efecto de diferentes poblaciones de maiz (*Zea mays*) en la producción de raices de yuca (*Manihot esculenta*) al cultivarlos en asocio. I. Aspectos agronómicos, *Turrialba* 33:109–116.

Meneses, R., Navarro, L. A., and Moreno, R. A. 1983. Efecto de diferentes poblaciones de maiz (*Zea mays*) en la producción de raices de yuca (*Manihot esculenta*) al cultivarlos en asocio. II. Aspectos económicos, *Turrialba* 33:291–296.

Moody, K., and Shetty, S. V. R. 1981. Weed management in intercropping systems, in: *Proc. Int. Workshop on Intercropping,* 10–13 January 1979, ICRISAT, Hyderabad, India, pp. 229–237.

Moreno, R. 1977. Efecto de diferentes sistemas de cultivo sobre la severidad de la mancha angular del frijol (*Phaseolus vulgaris* L.) causada por *Isariopsis griseola* Sacc., *Agron. Costarricense* 1(1):39–42.

———. 1982. Intercropping with sweet potato (*Ipomoea batatas*) in Central America, in: *Sweet potato, Proceedings of the First International Symposium,* (R. L. Villarreal and T. D. Griggs, eds.), AVRDC, Taiwan, pp. 243–254.

Moreno, R. A. 1979. Crop protection implications of cassava intercropping, in: *Intercropping with Cassava,* (E. Weber, B. Nestel, and M. Campbell, eds.), IDRC-142e, Ottawa, Canada, pp. 113–127.

Moreno, R., and Meneses, R. 1980. Rendimiento de algunas leguminosas intercultivadas al final del ciclo de vida de la yuca (*Manihot esculenta* Crantz), *Proceedings of the 26th Annual Meeting of the PCCMCA,* Guatemala.

Mutsaers, H. J. W. 1978. Mixed cropping experiments with maize and groundnuts, *Neth. J. Agric. Sci.* 26:344–353.

Natarajan, M., and Willey, R. W. 1979. Growth studies in sorghum/pigeon pea intercropping with particular emphasis on canopy development and light interception, in: *Proc. Int. Workshop on Intercropping,* 10–13 January 1979, ICRISAT, Hyderabad, India, pp. 183–190.

Nnadi, L. A. 1978. Nitrogen economy in selected farming systems of the savanna region, *SCOPE/UNEP Workshop on Nitrogen Recycling in West African Ecosystems,* 11–15 December 1978, IITA, Ibadan, Nigeria.

Okigbo, B. N., and Greenland, D. J. 1976. Intercropping Systems in Tropical Africa, in: *Multiple Cropping,* Amer. Soc. Agron. Spec. Publ. 27, pp. 63–101.

Onwueme, J. C. 1978. *The Tropical Root Crops: Yams, Cassava, Sweet Potato, Cocoyam,* John Wiley & Sons, Chichester, U.K., 234 pp.

Osiru, D. S. 1982. Genotype identification for intercropping systems (summary), in: *Intercropping,* (C. L. Keswani and B. J. Ndunguru, eds.), Proc. Second Symp. Intercropping in Semi-Arid Areas, Morogoro, Tanzania, 4–7 August 1980, IDRC, Ottawa, Canada, pp. 91–92.

Pachico, D. 1984. Bean technology for small farmers: Biological, economic and policy issues, *Agric. Admin.* 15:71–86.

Pendleton, J. W., Belen, C. D., and Seif, R. D. 1963. Alternating strips of corn and soybean versus solid plantings, *Agron. J.* 55:293–295.

Perrin, R. K., Winkelmann, D. L., Moscardi, E. R., and Anderson, J. R. 1976. *From Agronomic Data to Farmer Recommendations: An Economics Training Manual,* CIMMYT, Mexico.

Rao, M. R., and Willey, R. W. 1980. Evaluation of yield stability in intercropping: Studies on sorghum/pigeon pea, *Expl. Agric. 16:105–116.*

Roger, D. 1965. Some botanical and ethnological consideration of *Manihot esculenta, Econ. Bot.* 19:369–377.

Ruthenberg, H. 1980. *Farming systems in the Tropics,* 3rd ed. Oxford University Press, Oxford, U.K., 424 pp.

Sanders, J. H., and Alvarez, P. C. 1978. Evolución de la producción de frijol en América Latina durante la última década, CIAT, serie 06SB-1, Agosto 1978.

Searle, P. G. E., Comudom, Y., Shedden, D. C., and Nance, R. A. 1981. Effect of maize and legume intercropping systems and fertilizer nitrogen on crop yields and residual nitrogen, *Field Crops Res.* 4:133–145.

Simmonds, N. N. 1984. The state of the art of farming systems research, in: *Agriculture and Rural Development,* World Bank, Washington D.C., p. 135.

Steele, W. M., and Mehra, K. L. 1980. Structure, evolution and adaptation to farming systems and environments, in: *Vigna, Advances in Legume Science,* vol. 1, (R. J. Summerfield and A. H. Bunting, eds.), Proc. Int. Legume Conference, Kew, 31 July to 4 August, 1978.

Summerfield, R. J., Huxley, P. A., and Steele, W. M. 1974. Cowpea (*Vigna unguiculata* (L.) Walp.), *Field Crops Abstr.* 27:301–312.

Suneson, C. A. 1960. Genetic diversity—a protection against plant diseases and insects, *Agron. J.* 52:319–321.

Tapia, M. 1982. El medio, los cultivos y los sistemas agricolas en los Andes del Sur del Peru, IICA-CIID, Cusco, Peru, p. 80.

Tapia, M., Valladolid, J., Blanco, D., and Lescano, L. 1983. *Informe Tecnico 1980–1983,* Proyecto Investigación de los Sistemas Agrícolas Andinos IICA/CIID, Universidades de Ayacucho, Cuzco y Puno, Peru, 35 pp.

Ternes, M. 1981. Análisis agro-económico del sistema maiz-yuca según variaciones de población y arreglo espacial, Tesis Mag. Sci., CATIE, Turrialba, Costa Rica, Universidad de Costa Rica, 118 pp.

Thung, M., and Cock, J. H. 1979. Multiple cropping cassava and field beans: Status of present work at the International Centre of Tropical Agriculture (CIAT), in: *Intercropping with Cassava, Proc. Int. Workshop* (E. Weber, B. Nestel, and M. Campbell eds.), Trivandrum, India, 27 November to 1 December 1978, series IDRC-142e, pp. 7–16.

Torres, F., and Raintree, J. B. 1984. Agroforestry systems for smallholder upland farmers in a land reform area of the Phillipines: The Tabargo case study, Working Paper 18, ICRAF, Nairobi.

Trenbath, B. R. 1977. Interactions among disease hosts and diverse parasites, in: *The Genetic Basis of Epidemics in Agriculture* (P. R. Day, ed.), *Ann. N.Y. Acad. Sci.* 287:124–150.

Unamma, R. P. A., and Ene, L. S. O. 1984. Weed interference in cassava-maize intercrop in the rain forest of Nigeria, in: *Tropical Root Crops: Production and Uses in Africa,* Proc. 2nd Triennial Symp. Int. Soc. Tropical Root Crops, Donala, Cameroon, 14–19 August 1983, IDRC, pp. 59–62.

Wan, H. 1982. Cropping systems involving sweet potato in Taiwan, in: *Sweet Potato, Proc. First Int. Symp.* (R. L. Villarreal and T. D. Griggs, eds.), AVRDC, Taiwan, pp. 223–254.

Wein, H. C., and Smithson, J. B. 1981. The evaluation of genotypes for intercropping, in: *Proc. Int. Workshop on Intercropping,* 10–13 January 1979, ICRISAT, Hyderabad, India.

Wien, H. C., and Summerfield, R. J. 1980. Adaptation of cowpeas in West Africa: Effects of photoperiod and temperature responses in cultivars of diverse origin, *Advances in Legume Science,* vol. 1, (R. J. Summerfield and A. H. Bunting, eds.), Proc. International Legume Conference, Kew, 31 July to 4 August, 1978.

Willey, R. W. 1979a. Intercropping—its importance and research needs. Part I. Competition and yield advantages, *Field Crop Abstr.* 32:1–10.

———. 1979b. Intercropping—its importance and research needs. Part II. Agronomy and research approaches. *Field Crops Abstr.* 32:73–85.

Willey, R. W., and Osiru, D. S. O. 1972. Studies on mixtures of maize and beans (*Phaseolus vulgaris*) with particular reference to plant population, *J. Agric. Sci.* 79:517–529.

Willey, R. W., Natarajan, M., Reddy, M. S., Rao, M. R., Nambiar, P. T. C., Kannaiyan, J., and Bhatnagar, V. S. 1983. Intercropping studies with annual crops, in: *Better Crops for Food,* CIBA Foundation Symposium 97, 1983, Pitman, London, U.K. p. 248.

Woolley, J. N. 1984. Avances en la investigación a nivel de finca en sistemas de cultivos que incluyen frijol, CIAT training document, Cali, Colombia.

Woolley, J. N., and Smith, M. E. 1985a. Determination of the plant type of maize suitable for use with present and possible bean relay systems in Central America, (unpublished manuscript).

———. 1985b. Evaluation of bush and semi-climbing bean genotypes for use in relay systems with maize in Central America, (unpublished manuscript).

Zara, D. L., Cuevas, S. E., and Carlos, J. T. 1982. Performance of sweet potato varieties grown under coconuts, in: *Sweet Potato, Proc. First Int. Symp.* (R. L. Villarreal and T. D. Griggs, eds.), AVRDC, Taiwan, pp. 234–242.

Zimmermann, M. J. O., Rosielle, A. A., and Waines, J. G. 1984. Heritabilities of grain yield of common bean in sole crop and in intercrop with maize, *Crop Sci.* 24:641–644.

Chapter 8

Agronomy of Multiple Cropping Systems

Thomas C. Barker
Charles A. Francis

The agronomy of multiple cropping systems includes all practices controlled by the farmer that contribute to crop growth and production. Both plant and soil resource management decisions are necessary. Many of these choices are strongly influenced by the climate, inherent soil properties, and socioeconomic constraints. Each decision on crop species, land preparation, fertility inputs, and other agronomic practices will have impact on other factors as well.

Initial considerations for the farmer include what species to plant, in what proportions, on what dates, and with what final objectives. Next, the farmer must determine the type and timing of land preparation, application of natural or chemical fertilizers before or during crop growth, and whether or not to irrigate the crop if that option is available. Pest and weed control methods, planting dates of two or more crops, plant densities, spatial arrangements, and physical proximities need to be chosen. If the planting pattern will include a relay planting of another crop, this should be decided prior to initial planting. Finally, the plans for harvest or multiple harvests must be made, including the sale or storage of the products.

These are among the agronomic questions that the farmer considers at the outset of the cropping season. In addition, the amount of land available and the labor needed to perform each of the tasks listed above must be evaluated. This is a critical element of planning, since manual land preparation or weeding may be the single most critical production constraint in some areas. Thus, the agronomy of cropping systems, and especially multiple cropping, includes a long series of interrelated questions and decisions before starting the cropping season.

The reader should refer to the discussion of cereals in multiple cropping systems by Rao (Chap. 6) and the presentation of legumes and starchy roots by Davis, Woolley, and Moreno (Chap. 7). This discussion of agronomy does not overlap with their presentations except where necessary for clarity.

One of the best reviews to date of the agronomy of multiple cropping remains that of Willey (1979a, 1979b). In addition, there are books by Beets (1982), Gomez and Gomez (1983), and Papendick et al. (1976), plus numerous articles in the literature over the past 20 years. Four workshops have placed major emphasis on multiple cropping systems (IDRC, 1976, 1980; IRRI 1977; ICRISAT, 1981). These workshop proceedings should be consulted for more detailed information on the agronomy of cropping systems. There is also a book in preparation by Robert Willey at University of East Anglia (R. Willey, personal communication). His extensive personal field experience and publications assure that this will be a classic in the agronomic aspects of multiple cropping.

This chapter deals with the important variables that are under the control of the farmer in planting and harvesting a successful multiple crop pattern. More emphasis is given to the methods of learning about systems than about specific results which are available in the literature. Methods of analysis, which are important in the evaluation of cultural practices, and how the decisions are made to follow a certain set of practices are discussed. The chapter includes land preparation, planting practices, cultural management of crops, and fertility management. The debate about the cross interactions between cereals and legumes in an intercrop or a multiple crop system is evaluated. Throughout the discussion, there is an attempt to show which monoculture results and experiences are applicable to multiple cropping, or at least how to decide if results can apply. Finally, an attempt is made to evaluate the state of the art in the agronomy of multiple crops, and the needs for future research.

SEEDLING ENVIRONMENT: TILLAGE, CULTIVATION, AND WEED CONTROL

General Principles

The time of seed germination and seedling establishment is one of the most critical stages of the plant's life cycle. After planting, the seed imbibes water and breaks dormancy; then the radicle and shoot establish a seedling. The young plant is subject to stress conditions in this stage, and the major objective of land preparation and primary tillage is to establish a zone for the seedling that minimizes stress. Potential stress conditions include inadequate or excess moisture, soil temperatures too high or too low for a given species, crusting of the soil surface, shading from weeds or another crop, and insect or pathogen attack.

Unger and Stewart (1976) summarize the needs of the seedling as three basic requirements: oxygen, water, and temperature; they list seven secondary requirements that help to assure appropriate conditions for seed germination and establishment:

1. Adequate soil aeration for gaseous exchange in the seed and root zone
2. Adequate seed/soil contact to permit water flow to seeds and seedling roots
3. A noncrusted soil to permit seedling emergence
4. A low density soil that permits root elongation and proliferation
5. An environment that provides adequate light to the seedling
6. An environment that affords protection against wind and water erosion
7. A pest-free or pest-controlled environment

Temperate Systems

Unger and Stewart (1976) discussed these factors in the temperate climate context, where double cropping is becoming a common practice. They provide a review of literature and a comparison of conventional or complete land preparation with no-tillage methods to save time and energy in the turnaround period between crops.

In the high plains and the southwest region of Nebraska, an "ecofallow" system is growing in popularity. This pattern allows three crops in 4 years, or an intensification from the previous wheat (*Triticum aestivum*)-fallow system with two wheat crops every 4 years. Although not technically multiple cropping, the system provides information in zero-tillage that is useful to this discussion. When the goal is to establish a summer cereal such as maize (*Zea mays*) or sorghum (*Sorghum bicolor*), a zero-till planter that tills a narrow strip is used to seed the cereals. Tilling this strip allows the soil in that micro-area to dry more quickly and thus warm more rapidly than a comparable micro-area where seed is planted directly into wheat stubble without a tilled strip. This promotes germination in the cool early season. But there is a trade-off with moisture loss. Where rainfall is only about 400 mm per year, and a part of this is residual from snow fall during winter, any loss of moisture through tillage may result in decreased yields. Maize appears better able to germinate under the cool and moist conditions of zero-till than sorghum, a smaller seeded crop with less energy reserve in the seed. Thus the interaction of crop species with tillage and planting method is important, and genetic variation exists for tolerance to soil conditions within species as well.

The importance of establishing a seedbed or a micro-zone for early seedling development is the same in both conventional monoculture and in multiple cropping; however, complications may occur in more intensive systems. The timing of tillage and planting operations and the possible presence of another crop or its residue complicate the preparation of an optimum seedbed. The many variations in land preparation for double cropping in the temperate zone have been discussed in several recent reviews, including publications by Unger and

Stewart (1976), Lewis and Phillips (1976), Gomm et al. (1976), Gallaher (1980), and Sanford (1982). These reviews will lead the serious reader to greater detail in the literature on tillage, comparisons of conventional versus minimum or zero-till, and weed control options.

An additional tillage system that is receiving increasing attention in the United States for handling tillage in summer cereals interplanted with cool season legumes is "ridge tilling" and the establishment of permanent beds (RAA, 1985). On a number of farms in the maize belt, promising results have been reported on the use of permanent ridges on which to grow the cereal along with an overseeded legume to provide winter cover, erosion control, and additional nitrogen for the summer cereal. In this system, the ridge is scalped off just ahead of the planter units, and the low-growing cool-season legume and weeds on top of the ridge are moved away from the planting zone to give the cereal (usually maize) or soybean (*Glycine max*) a chance to germinate and grow in a weed-free and moist environment. Shortly after germination, and again at about 10 days after emergence, the field is cultivated rapidly with a rotary hoe. This kills small weed seedlings without doing major damage to the growing crop. Later, the row middles are cultivated with wide sweeps that cut weeds between the crop rows and throw soil over weeds within the row, thus re-forming the permanent beds. This system has been used with reduced rates of herbicide banded in the row with the crop, or even without herbicide. Either option reduces farmer production costs.

Tropical Systems

While the principles outlined by Unger and Stewart (1976) and cited above apply to any cropping situation, in tropical multiple cropping systems the constraints to optimum seedbed preparation are often different from those encountered in temperate cropping systems. For example, in tropical areas moisture, rather than temperature, is more often the factor determining planting date and tillage timing.

Since multiple cropping systems are often used by farmers with limited land and power resources, much of the tillage is done by hand or with animal traction. Minimum tillage schemes are common, and may involve cutting the existing plant cover, burning, or less often controlling weeds with herbicide. The result of any of these events is a greater residue cover on the soil surface than with a conventional power tillage operation. This may be advantageous for the seedling. In Nigeria, Rockwood and Lal (1974) demonstrated that no-till planting of sorghum into crop residues maintained the seed at 10°C lower temperature at 5-cm depth than a conventional clean-tilled planting. In the latter tillage system, temperature in the seedling zone reached 41°C. Sorghum yields were about 50 percent higher in the no-till planting due to reduced water stress at the lower temperature. This demonstrates the importance and interaction between temperature and water stress, and the subsequent effect of this interaction on young seedlings.

A challenge unique to the humid tropics is the annual conversion of flooded, puddled soils suitable for irrigated or rainfed rice, to an upland, well-drained soil environment for succeeding dryland crops during the dry season. The problems in converting a paddy field to a well-aerated field are outlined by Gomez and Gomez (1983), who discuss several planting options. In this process it is important to conserve as much moisture as possible, since the second crop(s) will depend on residual moisture plus very limited rainfall to grow and produce seed or forage. The SORJAN system developed in Indonesia combines upland and lowland crops in alternating strips that allow simultaneous cultivation. Rice (*Oryza sativa*) is grown in the flooded strips, while maize, beans (*Phaseolus vulgaris*), mung bean (*Vigna radiata*), sorghum, cassava (*Manihot esculenta*), cowpea (*Vigna unguiculata*), or sweet potato (*Ipomoea batatas*) are grown in the upland strips. During the dry season, cassava or sorghum can be grown in the upland areas on residual moisture, while a catch crop of a legume or maize can be grown where the rice has been harvested. A major constraint in the SORJAN system is the labor required to build and maintain permanent raised planting areas. Specific adaptation of the system depends on the crops to be grown, the land preparation, and the rainfall pattern. A number of current studies on SORJAN and similar cropping systems in Indonesia are summarized by McIntosh (1984). Although unique to areas in Southeast Asia, this cropping system illustrates the challenges of establishing a seedbed in the alternating wet/dry conditions.

Additional research is needed on the methods and tools that will allow rapid planting of a second crop either into a growing first crop (relay planting) or quickly after harvest of the first crop. Zero or minimum tillage has many advantages for moisture control and time savings. Weed control may be a problem unless it is possible to use selective herbicides in the system. The use of chemical herbicides is complicated by the need for tolerance by two or more crops in the pattern (often both a grass species and a legume), the cost of the product and its application, and the complication this introduces into management by the small farmer. Allelopathy may be another problem, and more information is needed on the effects of one species on another, and how this biological process can be used to advantage on the farm.

PLANTING DATES, PATTERNS, AND DENSITIES

The potential number of combinations of planting patterns of two or more crops, coupled with relative densities and dates of planting, borders on the infinite. An early review of cropping systems by Aiyer (1949) and more recent descriptions in the books and symposia previously cited give a catalog of prevalent multiple cropping systems practiced in the world. Several chapters in the book *Multiple Cropping* (Papendick et al., 1976) detail systems in Asia, tropical America, tropical Africa, the Middle East, and the United States. Experimental work on cropping patterns can be gleaned from literature searches using the key words "multiple cropping," "intercropping," "double cropping," and "relay cropping." These articles reveal a modest but growing body of information on specific agronomic aspects of multiple cropping. The work by Bradfield (1966) was among the first to explore the ultimate potential intensity of cropping systems where up to seven crops were planted in relay and sequential patterns through the course of a year in the Philippines. It is crucial to examine the extent of differences between known practices for monoculture, and the additional or different information needed for managing multiple cropping systems. One must build on what is already known in order to search efficiently for improved technology.

Planting Dates

Similar factors determine the date of planting for each crop during the year in multiple cropping and monoculture systems. Dates are adjusted to ensure adequate temperature for germination and growth, avoid extremes in temperature that could cause stress or difficulty in setting and maturing seed, provide adequate moisture for growth and completing the life cycle, and minimize other stress conditions during the growing season. Genetic changes in crop species have allowed new cropping patterns to develop which utilize production resources more efficiently than traditional patterns. For example, improved cold-temperature tolerance in maize has allowed earlier planting in temperate spring conditions. This trait coupled with early maturity could allow planting a second crop after maize. The development of photoperiod-insensitive rice varieties in the tropics has provided shorter cycle maturity in rice, which allows planting of two crops per year, with time in some areas for a third crop of mung bean or another short-cycle legume to be grown. These are examples of increased utilization of the available growing season and rainfall resulting from genetic improvement of crops and the testing of these new cultivars in new patterns with adjusted planting dates.

The potentials of nitrogen fixation by cool-season, cover crop legumes and their contributions to cereal crops or soybeans in the United States have raised new questions about dates of planting. If the legume is established during the late summer and fall, then allowed to grow and produce dry matter and nitrogen in the spring, the optimum date of incorporation for nitrogen production needs to be determined. There is a question of trade-offs in nitrogen production vs. yield loss due to changing maturity of the summer maize or soybean crop. The longer a legume such as red clover (*Trifolium repens*) stays in the field, the more nitrogen will be produced for spring plow-down. This would postpone the planting of maize, for example, with an increasing yield loss due to shortening the potential growing season. Conversely, when the clover is turned under early to allow for optimum maize planting, there is a loss in potential nitrogen production. Thus the optimum balance between biological nitrogen production, the cost of replacing that nitrogen with purchased chemical nitrogen or manure, and the loss of yield by delayed maize planting requires further research. There may also be potential for genetic improvement of the overseeding system. Variation exists among legume species and cultivars in terms of spring growth rate, potential biological nitrogen fixation rate and efficiency, and the sensitivity of forage species to shading by the cereal crop at the time of overseeded legume establishment. Variation also exists in maize with respect to production as a function of planting date.

Management decisions become more complicated as multiple cropping systems become more intensive. For example, the farmer faces difficult decisions on when to plant component crops in a mixed intercrop with several species having overlapping growth cycles and different cultural requirements. The effects of temporal differences in planting and crop life cycles were reviewed by Willey (1979b), who described the concept of *temporal complementarity*. His description may be interpreted as "the greater the difference in maturity and growth factor demands of the crop components, either because of genetic differences or manipulation of planting dates, the more opportunity for greater total exploitation of growth factors and overyielding" (i.e., a total production greater than monoculture of each crop). An example is the intercropping of a long-season crop such as cassava or pigeon pea (*Cajanus cajan*) with a short-season pulse such as cowpeas or mung bean. The low-growing grain legumes mature in less than 80 days at low altitudes. This allows the longer season taller crops to develop with little loss of yield of either component.

Dates of planting may depend on the yield objectives of the farmer for the two or more crops in the cropping pattern. Some of the biological trade-offs in yield are illustrated by an example from Colombia with different relative planting dates of maize and beans (Francis et al., 1982). Table 8.1 illustrates the effects of four different types of beans which were planted (1) with maize on the same date, (2) five and ten days before the maize, and (3) five and ten days after the maize. The effects of the changes in relative planting dates and the effects of

Table 8.1 Intercropped Bean and Maize Yields (t/ha) with Five Relative Planting Dates and Four Bean Plant Types
CIAT, Colombia

Planting dates	P788 (type I) (determinate)		P566 (type II) (semi-indeterminate)		P498 (type III) (indeterminate, nonclimbing)		P589 (type IV) (indeterminate, climbing)	
	Bean	Maize	Bean	Maize	Bean	Maize	Bean	Maize
Beans 10 days before maize	1.4	3.6	1.9	2.8	1.5	3.0	1.4	1.6
Beans 5 days before maize	0.8	5.8	1.4	5.2	1.2	4.8	1.6	3.3
Beans and maize, same day	0.7	6.3	1.0	5.8	0.9	5.3	1.4	4.7
Maize 5 days before beans	0.6	6.5	1.0	5.9	0.7	5.7	1.0	5.3
Maize 10 days before beans	0.5	6.8	0.8	6.2	0.5	5.9	1.1	5.7
Bean monoculture	1.5	—	1.6	—	1.3	—	3.0	—
Maize monoculture	—	6.4	—	6.4	—	6.4	—	6.4

Source: Data from Francis et al., 1982.

the type of bean are apparent from the graphs. Early bean planting clearly favors the beans, while causing reduced yield of the maize. Advanced maize planting causes an even greater proportionate reduction in the bean yield, due to early competition and shading of the beans. This affects the relative yields of the two crops, as well as the total income derived. Income is determined in part by the relative prices of the two crops. An example for climbing beans associated with maize at a range of relative prices for bean/maize from 1:1 to 8:1 is given by Francis and Sanders (1978) with crop data from Colombia. The relative prices affect total income and the farmer's decision about which multiple cropping system to plant. In the bean/maize intercrop in Colombia, it was more advantageous for net income to plant an intercrop compared to either of the monocrops at bean/maize price ratios below 4 : 1. Due to the greater variance in monocrop bean yields, the probability of making any profit at all was highest in intercrop systems up to a 6:1 price ratio. In contrast to the maize/bean data, Misbahulmunir et al. (1984) found no difference among yields or LERs when maize and groundnut were planted over a range of relative planting dates. These are some of the interacting biological and economic factors that influence the farmer's decisions.

Planting Patterns

Chapter 1 listed a number of categories of cropping patterns: mixed cropping, row intercropping, strip cropping, relay cropping, and others. These have a wide range of variants when each possible row width, number of rows, density, and plant species combination is considered. A number of combinations for cereals (Chap. 6) and for legumes and starchy roots (Chap. 7) plus other examples are presented in each chapter. More important here is the discussion of how to decide on planting patterns, and what influence planting patterns have on the component crop yields, total production, and economic return from the multiple cropping system.

In temperate zone double-crop, strip-crop, and relay systems, planting patterns are often determined by the available equipment and labor requirements for planting and harvesting. Double cropping utilizes the same equipment as monoculture, and needs no further discussion. Strip cropping may depend on the four-row or six-row planters available, and needs to coincide with the width of a combine that is used to harvest mechanically. It stands to reason that if interactions between component crops are expected to provide benefits to total production, the narrower the strips, and thus the greater the extent of plant interactions, the better, within the limits of available equipment. Since four-row strips appear to be advantageous in maize/soybean combinations under a fairly wide range of conditions, a four-row planter and combine with 10-foot header and two-row or four-row maize head would be needed to harvest. Two-row strips of maize planted at specified intervals in soybean fields have been shown to increase soybean yield by 10 to 50 percent by reducing wind and thus increasing humidity in the crop canopy; this was in addition to the harvest of the strips of maize (Brown and Rosenberg, 1975).

Relay planting a second crop depends on the row spacing of the first, as well as the opportune time and type of equipment that can be driven through to plant the second crop. A number of research studies and farmer experiences have shown that planting soybeans into a growing winter wheat crop in the spring can be done more easily if every third or every fourth row of wheat had not been seeded the previous year. This does promote weed growth in the missing wheat rows, and the overall benefit may be lost. Patterns of overseeded legumes in small grains or maize are determined by the available seeding method. They may be broadcast if seeded aerially or banded between the maize rows if seeded at the last cultivation.

Most of these relay patterns can be implemented with existing equipment (as is the case with double cropping), or with modifications that farmers can make with the units they already have. Thus it appears that equipment is not a major limitation to introducing multiple cropping systems into mechanized agriculture.

The multiplicity of possible patterns is introduced with the option of hand planting, cultivation, and harvest of multiple cropping systems. One of the simpler systems is the bean/maize intercrop in the Andean zone and Central America, where the two crops may be planted in alternating rows, as in the case of bush beans, or in the same rows, using climbing beans and maize. There is a wide range of combinations even within this one system. At the other end of the spectrum is the mixture of 12 or more crops in a complex pattern (Fig. 10.2) which takes into account micro-relief of the field, proximity of taller and shorter crops, and the range of food products for consumption or sale on the small farm in Nigeria (Okigbo and Greenland, 1976). These patterns may be specific to each farm and the nutritional and economic needs of each family. Barker (1984a) found 12 to 15 sweet potato cultivars (of 45 available cultivars) interplanted in a single field with pigeon peas and other crops in a shifting cultivation area of the Philippines. Such a diverse array of species and cultivars with unique characteristics allows a precise cropping pattern to suit a given microclimate and set of household objectives.

Such systems have developed over centuries in some regions to meet this range of needs, and the design of new technology to improve them is complex (Francis, 1981; Barker, 1983). The more complex patterns such as the one in tropical West Africa show an amazing similarity to the natural ecosystem of the area. The diversity of species and their different growth cycles mimic the natural system and take advantage of some of the same types of biological structuring and buffering that occur in the natural systems.

In the study of spatial organization or planting patterns leading to the design of new alternatives to improve yield, food supply, or income it is important to improve on the complimentarity of the component species or their ability to maximize utilization of available growth factors in the environment. The new dwarf rice varieties with a shorter crop cycle, which permit two crops of rice plus a catch crop of mung bean or sorghum, provide this advantage through temporal efficiency, allowing three crops per year where one was possible before. Each of these varieties may need to be engineered for the specific conditions at the time of planting in the total pattern. A legume for overseeding in the temperate zone may need to establish well in the shaded conditions below the overstory maize crop, then develop well in cool fall conditions when maize is drying down for harvest. It also needs the ability to start fast in the spring and produce as much dry matter and nitrogen as possible before incorporation and planting of the next crop of maize or soybeans.

The potato (*Solanum tuberosum*)/maize/climbing bean relay system in Antioquia, Colombia, traditionally occupied the land for 14 to 15 months, and thus it was not possible to repeat this pattern in successive years. By breeding a shorter-cycle maize with good stalk strength to support the bean, seeking a shorter-cycle bean, and putting these components together with the same potato variety, it was possible to complete this complex cycle in 1 year to ready the field for another cycle.

These examples illustrate the interaction of plant breeding with agronomy in the design of new and improved cropping patterns. In each of the examples, there are differences in the height of crop components, differences in time to maturity, differences in optimum time of planting during the year, or other complementarity that leads to greater total resource use through the year. When growing crops and their residues are in place during most or all of the year, this also ties up as much of the nitrogen and other nutrients as possible and prevents erosion and leaching of valuable nutrients for these and succeeding crops.

Although this discussion gives no clear guidelines for how to conduct research or exactly what the eventual systems might be for a given set of components, it does point out some of the complexities that are operative in the design of planting patterns and the need to consider a wide range of ecological factors and objectives of the farmer. This is necessary as the patterns are put together with different species, varieties of each species, planting date combinations, and densities of each crop component.

Plant Densities

In his review of intercropping, Willey (1979b) described the importance of distinguishing between total plant population of all crops in the pattern and the component plant population of each component. There is obviously not a direct substitution of one plant of maize for one bean plant when mixtures are described and designed. Willey suggested that optimum densities of monocrops be used to calculate component populations on a relative basis. For example, if optimum monocrop density of maize is 50,000 plants per hectare and optimum monocrop density of bean is 200,000 plants per hectare, then a 50:50 mixture of the two crops would contain 25,000 maize and 100,000 bean plants per hectare. Other combinations can be used to determine the optimum for a given set of objectives of the farmer. This is preferred to the replacement series approach (Osiru and Willey, 1972; Willey and Osiru, 1972), since it relates intercrop performance to optimum sole-crop yields of each component. Furthermore, there are situations where a full complement of one species is planted with the addition of a partial density of a second crop (Willey, 1979b). A recent study (Carter et al., 1985) has shown that the equivalent numbers of two species in an intercrop are not one constant value, but vary over a range of possible combinations of the two components.

There are probably as many reports in the literature on component crop density in multiple cropping as on any other single factor. Thus a researcher would be well advised to study previous work carefully before beginning any new studies. Another valuable step would be to study work done on monocultures of each component crop, and to see if the optimum levels on experiment stations are the same as on farms. With this information in hand, it is possible to design new studies that can add to current information and provide new guidelines to farmers. It is important, when considering research goals, to determine if sub-optimum density is a real constraint on the farm, and if it is perceived as such by farmers. This is one innovation that is easily introduced if a demonstrated effect (yield increase from greater crop densities) is present. The only additional cost is for seed to provide the increased density.

Most complex among the decisions for farmers and agronomists alike is the integration of planting patterns, combinations of crop components, and the densities of these component species. With the multiplicity of combinations possible, it is best to concentrate initially on the systems that are already popular with farmers and see if there are weaknesses that can be corrected through some change in these factors. Procedures are presented in Chap. 13 to evaluate constraints to production and to design research to solve them on the research station and on the farm. Statistical methods to help design these studies are presented in Chaps. 13 and 14. There is a forthcoming book by W. T. Federer, Cornell University, Ithaca, New York, which will present detailed statistical procedures for multiple cropping. Care must be taken to evaluate the total systems, and to make sure that evaluation criteria are in agreement with the objectives of farmers. This is a complex area that is not easily understood or subjected to technical scrutiny because of the wide range of factors and near-infinite number of combinations involved.

SOIL FERTILITY MANAGEMENT

Farmers who practice multiple cropping and agronomists who experiment to improve these systems must make management decisions based on a complex array of factors related to soil fertility and plant nutrition. Site-specific factors include the climate, soil properties (physical, chemical, and biological), and socioeconomic considerations. These factors are relatively fixed for a given site. However, when one considers the areas where multiple cropping is practiced worldwide, variation among sites is so large as to practically defy generalizations. Additional factors that are under management control at a given site include species selection, pest and weed control practices, tillage methods and timing, spatial planting patterns, temporal planting patterns, and soil fertility management. The subject of this section is soil fertility management, but it must be recognized that decisions that influence the capacity of the soil to provide plant nutrients are not limited to fertility inputs. Every factor mentioned above—from climate to socioeconomic constraints to weed populations—has a bearing on the fertility status of the soil and management decisions. Conversely, fertilizers, green manure, rotation with legumes, and other nutrient inputs have a great influence on pest and weed incidence, stand density and composition, and profitability (Oelsligle et al., 1976). An excellent review of fertility relationships in multiple cropping systems was presented by Sanchez (1976).

Three interrelated concepts or approaches to determining multiple cropping fertility schemes are prevalent in the literature. These are based on (1) the selection of species combinations, (2) a particular species' fertility response in polyculture relative to monoculture, and (3) the objectives of the multiple cropping system. These decision-making considerations overlap, interact, and may converge to help define the most appropriate way to manage fertility inputs in a given situation.

Selection of Species Combinations

Several combinations of species may be grown depending on climate, local preferences, and other site-specific factors noted above. General categories include cereal/pulse, cereal/root crop, pulse/root crop, cereal/cereal, pulse/pulse, root crop/root crop, and forage/grass systems. More than 2 species, in fact, more than 10 species in some instances (Okigbo and Greenland, 1976) may be grown, including two or more crop types. In other situations several cultivars of a single species may also be added to the diversity of a given field (Barker, 1984b). Farmers combine species with different aboveground and belowground morphologies, varying maturation dates, and varying nutrient uptake patterns in order to more completely utilize the plant growth environment (Oelsligle et al., 1976; Willey, 1979a, 1979b). In other words, spatial and temporal differences of two or more combined species utilize available resources more efficiently than either species alone. These morphological or physiological differences that are the basis for benefits from multiple cropping also pose difficulties in fertility management. The basic question asked in fertility management is: For which species should one try to optimize fertility inputs, source, placement and timing? In cereal/root crop systems, e.g., maize/sweet potato, high nitrogen rates for optimum grain yield would stimulate sweet potato vine growth at the expense of root yield. In pulse/root crop systems in the humid tropics, crops may compete for phosphorus (Mason et al., 1985; Barker, 1984a). Thus timing and placement of phosphorus may favor rapid-maturing cowpeas over cassava, which has a much later peak phosphorus-uptake period. Thus the *objectives* of the multiple cropping system will generally take precedence in fertility management of systems with very dissimilar characteristics such as cereal/root crop, cereal/pulse, root crop/pulse, and forage/grass systems.

The problem is less acute in systems where two similar species are grown, e.g., cereal/cereal systems. Both species generally require the same nutrients at compatible levels, and decisions on timing and rates could be based on uptake curves over time. For example, in a wheat/sorghum double-crop system in temperate areas, nitrogen management of the fall-seeded wheat would be adjusted to allow optimum wheat yield, and residual nitrogen would affect the subsequent sorghum nitrogen requirement (Langdale et al., 1984). Any residual nitrogen left from the sorghum crop would affect the next wheat crop, and so on (Sanford, 1982; Sanford and Hairston, 1984; Hargrove et al., 1983). Unlike the pulse/root crop systems, the cereal/cereal system is not likely to have conflicting nutritional needs, but fertility management is more a question of refining nutrient additions with respect to uptake patterns and demands.

Performance in Monoculture vs. Polyculture

Perhaps the most common approach to multiple cropping fertility management in the past has been to ask, for a given species, How is the fertility response of a species grown as a monocrop likely to change when grown as a component of a multiple cropping system? Each species will have some baseline nutrient requirement, below which the plants' performance will suffer. Critical levels for most crop species are well known. There is evidence that, in most cases, basic nutrient uptake patterns are similar for either monocropped or multiple-cropped situations. For example, Ahmed and Rao (1982) found that the nitrogen response curve shape was the same for monocropped maize vs. maize intercropped with soybeans. On the other hand, reports often suggest that total uptake of nutrients and yield is decreased for each individual crop in a multiple cropping system (Ahmed and Rao, 1982; Faris et al., 1983; Wahua, 1983) although the total combined uptake may be greater than in monoculture (Mason et al., 1985). This suggests competition for nutrients among multiple cropped component species. Oelsligle et al. (1976) pointed out that the extent of plant interactions, and therefore nutrient competition, increases with intensity of the multiple cropping system. For example, a wheat/soybean double crop contains less species interaction than strip-cropped corn/soybeans. The most intense multiple cropping situation where competition would be most acute is in mixed intercropping where root systems and leaf canopies of unlike species are in immediate proximity on all sides, such as in maize/climbing bean systems. Recalling the question posed earlier, fertility management of a given crop may change for monocropped vs. multiplecrop situations, the more intensive the plant interactions, the more apt the fertility requirement is to change or be influenced by the companion crop. Sanchez (1976) reviews the potential reasons for differences in nutrient use by intercrops.

In most situations it appears that monocrop fertility requirements are a good basis for initial studies on multiple crop fertility management needs. Response curves and critical levels tend to be similar in polyculture and monoculture, but as the intensity of crop interactions in multiple cropping systems increases, the overall uptake may change, and optimum management techniques must be developed with care based on component species characteristics and the objectives of the multiple cropping system.

Objectives of Multiple Cropping Systems

Motivations for multiple cropping vary from the need among subsistence farmers to minimize risk, to a need for total cash grain yield maximization. Fertility management may influence the balance between crops, overall biomass production, total grain yields, total energy production, and many other parameters (Oelsligle et al., 1976). Most reports have emphasized total marketable yield or economic yield. Others, such as that of Singh et al. (1971), used total protein yield as one evaluation of multiple cropping efficiency. Thus, one must ask, How will increased nutrient levels influence objectives such as maximum protein, energy, grain, or economic yield?

Dalal (1974) found that both maize and pigeon peas produced greater yield in monoculture than when intercropped. Thus, if maximum cash return per unit land were the primary objective, a farmer would opt for the more economical monocrop and fertilize accordingly. However, Dalal (1974) also found that maize and pigeon pea intercropped in alternate rows produced more crude protein per week per hectare. Crude protein production was furthermore correlated to inorganic nitrogen levels associated with pigeon peas. Thus a fertilizer management scheme aimed at optimizing biological nitrogen fixation in pigeon peas when intercropped with maize might be the most economical means to produce maximum crude protein for subsistence farmers' diets.

Another common objective for multiple cropping is to maximize the yield of a primary crop and harvest the secondary crop(s) as a "bonus." In this case the combined yield of grain, protein, or carbohydrate is of less importance than optimizing the primary crop yield. Barker (1984a) found the maximum yield for the primary crop, sweet potato, was coincidentally the maximum point for the secondary crop, cowpeas, as well. Both crops were limited by available phosphorus and produced maximum yield at highest applied-phosphorus rates. In other situations, such as the nitrogen nutrition of legume/cereal multiple cropping systems, nitrogen fertilizer applications may seriously decrease legume yield and be less economical in the long term than fertilizing the secondary crop (e.g., grain legume) to provide maximum return of biologically fixed nitrogen via residues to subsequent primary crops (Langdale et al., 1984; Waghmare and Singh, 1984a, 1984b; Yadav, 1981).

In summary, the three elements of multiple crop fertility management (species selection, crop performance in monoculture vs. polyculture, and the objectives of multiple cropping) considered together provide a basis for a logical approach to determining fertility management schemes. A number of problems still exist in regard to multiple crop nutrition. Given the large number of options and site-specific factors encountered in multiple cropping situations, it is expensive, cumbersome, time-consuming, and perhaps physically impossible to do field trials for every possible set of conditions, or even every region. With the spectre of worldwide food shortages, particularly in areas where multiple cropping is most prevalent in lesser-developed countries, there is obvious need to concentrate research efforts on multiple cropping systems in cost- and time-effective approaches. Thus some speculations regarding future research approaches are offered.

Research in Multiple-Crop Nutrition

Multiple-Crop Fertility Response Estimation Examples of fertility response equations developed specifically for multiple cropping situations are scarce. Having to consider more than one crop and more than one nutrient simultaneously presents problems. Waghmare and Singh (1984a) and Barker (1984a) used multiple regression response equations for predicting yield response in multiple crops. Such multiple regression equations could be extended to prediction and optimization of total energy production, protein production, economic yield, or whatever the multiple cropping objective may be. Such an empirical approach is fine for specific areas and data sets, but leaves much to be desired in terms of broader application, conceptual guidelines, and time or cost effectiveness.

A more robust approach was developed by Wahua (1983) who proposed a "nutrient supplementation index" (NSI).

$$\mathrm{NSI}_A = 100\,\frac{N_a + N_b - 1}{N_A}\%$$

= nutrient supplementation index for species A

= the percent of the usual uptake for a given nutrient by sole-crop A which should be added to the multiple crop to meet the combined requirements of species $A + B$

where N_A = nutrient uptake by sole-crop A per unit land area (e.g., kg N/ha)

N_a = nutrient uptake of mixture A for the same land area as sole-crop A

N_b = nutrient uptake of B in mixture for the same land area as sole-crop A

The NSI attempts to adjust total fertility inputs into the multiple cropping system based on the relative uptake of each crop component species in monoculture. This index could be used for more than two species as well. Further development of similar indices based on the desired yield goal of more than one multiple crop component species would appear to be a promising avenue for future research. Treating a multiple crop essentially as a unique "species" or unit is an approach worth pursuing both in fertility management and plant breeding for multiple crops.

Computer Simulation When an array of potential species is selected, objectives for multiple cropping are defined, and nutrient uptake or yield response curves for those species are available, it may be possible to simulate nutrient demands of the various crop combinations. By contrasting numerous possible planting patterns, species combinations, and nutrient inputs, much of the guesswork in multiple crop fertility management could be eliminated in simulation prior to field trials. For example, suppose that maximum energy production per unit land area per week were the known objective of the cropping system. In the area of interest two root crops and three pulses were selected as socially and economically acceptable. If generalized equations for the uptake of nitrogen, phosphorus, and potassium, and predicted energy yield for each species over their respective growing seasons were known, much could be learned about the optimum timing of nutrient additions and species combinations prior to establishing trials simply by contrasting growth and response curves. As more information became available, for instance moisture release curves, biological values of the protein complements of alternative species and cultivars of each species, the simulation of multiple crop nutrient management would gain accuracy and utility. Such an approach should be considered as a time- and cost-effective supplement to expensive field trials given the large number of multiple crop combinations available to farmers and researchers.

Complementation vs. Competition Advantages of multiple cropping vs. monocropping are generally attributed to *complementation*, the greater utilization of resources in time and space. Yet less-frequently reported examples of multiple crop inefficiency and failure also result from greater exploitation of resources available for plant growth, i.e., *competition*. Multiple cropping research has not often separated the processes that constitute complementation vs. competition. The literature is full of seemingly conflicting reports due to site-dependent factors. For example, Remison (1978) found no competition for nitrogen and phosphorus in a maize/cowpea intercrop in which performance was attributed to nutrient complementation. The assumption would be that the maize and cowpea root systems exploited different soil volumes to some extent, and that the timing of nutrient demands were compatible for both crops to perform to maximum potential. Since nitrogen is mobile in the soil, there must have been ample soil nitrogen to supply both crops. Phosphorus is quite immobile, hence the feeding zones of each crop may have indeed been complementary, and roots must have also come in contact with an adequate supply of phosphorus. In another situation, Wahua (1983) found that maize and cowpeas competed for nitrogen and to a lesser extent phosphorus. Part of the reduced nitrogen uptake by cowpeas was attributed to shading of the cowpeas by maize plants resulting in reduced photosynthesis and subsequent mobilization of nitrogen products to the pods. Available nitrogen and phosphorus were apparently insufficient in these soils, unlike those in the area of Remison's study.

Such apparent contradictions are not surprising. The fact is that site-specific factors influence results, and the underlying processes for success or failure of multiple cropping systems relative to monocultures are often not discernible, or are attributable to more than one cause. For example, Wahua (1983) found that reduced nitrogen uptake by cowpeas was due to shading of the cowpeas by maize plants. Other studies (Waghmare and Singh, 1984a; Searle et al., 1981; Nambiar et al., 1983) suggest that nitrogen application and/or shading inhibits nodulation of legumes, although these effects are generally inseparable.

Oelsligle et al. (1976) suggested the need to document conditions under which multiple cropping leads to increased nutrient utilization efficiency. This admonition is still needed. In multiple cropping studies where nutrient competition is anticipated, or nutrient complementation is hypothesized to provide benefits of multiple cropping over monoculture, careful documentation of soil fertility status, leaf tissue status, and meteorological data is needed. Additional measurements may be necessary to explain why competition or complementation occurs. In the case of grass/legume or cereal/legume studies where the legume crop is expected to transfer nitrogen during the life cycle of the grass species (Whitney et al., 1967; Waghmare and Singh, 1984a; Wahua and Miller, 1978) or to subsequent cereal crops (Jones, 1974; Ahmed and Rao, 1982; Narwal et al., 1983; Yadav, 1981; Giri and De, 1980), careful documentation of nitrogen fixation potential should be made. A good example of such documentation is found in the paper by Nambiar et al. (1983), who reported that intercropping cereals significantly reduced the number of nodules, nodule weight, and nitrogenase activity in peanuts. The inhibitory effect of the cereal intercrop increased with increasing nitrogen application, but the role of applied nitrogen per se vs. shading could not be clearly separated. Thus the need for measuring all factors related to complementation/competition processes cannot be overemphasized. As suggested by Ahmed and Rao (1982) it is not enough to simply document multiple cropping yield advantage or gross nutrient uptake advantages. One must know why such advantages occur in order to understand and improve multiple cropping systems nutrient management.

CONCLUSIONS

A recurrent theme in this chapter is the complexity, large number of possible combinations, and variations within and among multiple cropping systems factors. Consider the list of factors involved in multiple cropping systems worldwide that has been presented:

Site-specific factors: climate, socioeconomic constraints, soil properties, labor and time constraints

Field operations factors: land preparation, tillage, cultivation, weed and pest control

Planting factors: dates, patterns, densities, species, cultivars

Fertility factors: species combinations, performance in monoculture vs. polyculture, objectives of multiple cropping systems

A large army of agronomic and social scientists would be required to address all the possible permutations of the factors listed above and evolve optimum multiple cropping systems for each area of the world. Such cropping systems research is indeed desperately needed, not only for the poor small farmer of lesser developed countries, but also for economically pressured farmers in the United States and other developed countries. Personnel and funding for research is limited to a relatively small number of scientists; however, there is growing support for multiple cropping research and a cooperative constituency of farmer end users. Thus a three-dimensional approach to the problem of multiple cropping systems development, from an agronomic viewpoint, is proposed:

1 *Make use of available component technology.* When one reviews the literature on multiple cropping, it becomes evident that work tends to be repetitive. We wonder how thoroughly some researchers review literature prior to establishing field trials. Granted, part of the problem is the multiplicity of possible growing conditions, objectives, and species. Rather than duplicate past efforts, there is a need to apply available information and existing technology for given crops and multiple cropping systems to new, innovative systems design and improvement. A critical corollary to applying component technology is the careful documentation of underlying processes rather than simple comparisons. By explaining underlying trends and processes, such as those reported by Nambiar et al. (1983), the study of multiple cropping systems can be advanced rather than merely substantiated or recorded.
2 *Explore the possibilities of using computer simulation models.* Given the huge number of processes involved in multiple cropping decision making, and the fact that most of these processes have been, or could be, described mathematically, the advances in simulation of crop growth factors should be applied in multiple crop models. Modeling, even in simple systems, is a complex process and not a cure-all, but eventually it could be a time-saving and money-saving tool to allow agronomists to rapidly screen combinations, sequences, and inputs. Better yet, it could provide a means for predicting optimum cropping systems inputs which could then be verified by farmers on their own fields.
3 *Develop on-farm research methodology and communication.* While the number of research scientists and stations is relatively small, the number of potential collaborators—farmers themselves—is large and represents the ultimate possible number of test sites. Communication is a critical link in on-farm research which has not been fully explored. While traditional research and extension channels have been successful in many regions of the world, they tend to be slow, confined to geographic and/or political boundaries, and often cater only to a select constituency. There is promise that more rapid and site-specific systems for development may be facilitated by communication within regional and international organizations. Examples would be the international institute multinational research networks, for example, International Rice Research Institute (IRRI), Centro Internacional de Mejoramiento de Maiz y Trigo (CIMMYT), the Center for Rural Affairs network in Nebraska, and the Regenerative Agriculture Association (RAA). Within such research net-

works, means of coordinating and communicating farmer-implemented cropping systems trials should be developed. This would allow the results of a large number of simple trials to be channeled through scientific processes to accomplish multiple cropping systems development more rapidly, accurately, and appropriately than otherwise possible.

Specific research avenues have been suggested in each section. By reviewing the literature cited, the reader can grasp the essential experimental methods and contacts needed to develop well planned, meaningful multiple cropping trials. We hope that the opportunities to use available component technologies, crop modeling, and on-farm research will develop in time as channels, focal points, and support linkages for individual efforts to improve the agronomy of multiple cropping systems.

REFERENCES

Ahmed, S., and Rao, M. R. 1982. Performance of maize-soybean intercrop combination in the tropics: Results of a multilocation study, *Field Crops Res.* 5:147–161.

Aiyer, A. K. Y. N. 1949. Mixed cropping in India, *Indian J. Agr. Sci.* 19:439–543.

Barker, T. C. 1983. Growing camote: Ikalahan style, *PAFID Fieldnotes* 2:6–7 (Philippine Association for Intercultural Development, Quezon City, Philippines).

———. 1984a. Shifting cultivation production constraints in the upland Philippines, Ph.D. dissertation, University of Missouri-Columbia, Missouri.

———. 1984b. *Shifting Cultivation among the Ikalahans,* UPLB-PESAM Working Series 1, Program on Environmental Science and Management, University of the Philippines at Los Baños, College, Laguna, Philippines.

Beets, W. C. 1982. *Multiple Cropping and Tropical Farming Systems,* Westview Press, Boulder, Colorado.

Bradfield, R. 1966. *Toward More and Better Food for the Filipino People and More Income for Her Farmers,* Agricultural Development Council, New York.

Brown, K. W., and Rosenberg, N. J. 1975. Annual windbreaks. *Crops Soils,* April-May, pp. 8–11.

Carter, D. C., Francis, C. A., Youngquist, W. C., and Pavlish, L. A. 1985. Evaluation of a constant soybean:sorghum density ratio for intercropping "replacement series" experiments, (unpublished).

Dalal, R. C. 1974. Effects of intercropping maize with pigeon peas on grain yield and nutrient uptake, *Exp. Agric.* 10:219–224.

Faris, M. A., Burity, H. A., Dos Reis, O. V., and Mafra, R. C. 1983. Intercropping of sorghum or maize with cowpeas or common beans under two fertility regimes in northeastern Brazil, *Exp. Agric.* 19:251–261.

Francis, C. A. 1981. Rationality of farming systems practiced by small farmers, in: *Symposium on Small Farms in a Changing World: Prospects for the Eighties,* Kansas State University, Manhattan, Kansas.

Francis, C. A., Prager, M., and Tejada, G. 1982. Effects of relative planting dates in bean (*Phaseolus vulgaris* L.) and maize (*Zea mays* L.) intercropping patterns, *Field Crops Res.* 5:45–54.

Francis, C. A., and Sanders, J. H. 1978. Economic analysis of bean and maize systems: Monoculture versus associated cropping, *Field Crops Res.* 1:319–335.

Gallaher, R. N. 1980. Potentials and projection for increasing agronomic production with multiple cropping and/or minimum tillage systems in Florida, University of Florida, Multicropping Minimum Tillage Circular 479.

Giri, G., and De, R. 1980. Effect of preceding grain legumes on growth and nitrogen uptake of dryland pearl millet, *Plant Soil.* 56:459–464.

Gomez, A. A., and Gomez, K. A. 1983. *Multiple Cropping in the Humid Tropics of Asia.* IDRC, Ottawa, Ontario, Canada, IDRC-176e.

Gomm, F. B., Sneva, F. A., and Lornz, R. J. 1976. Multiple cropping in the western United States, in: *Multiple Cropping,* (R. I. Papendick, P. A. Sanchez, and G. B. Triplett, eds.), Amer. Soc. Agron. Spec. Publ. 27, pp. 41–50.

Hargrove, W. L., Touchton, J. T., and Johnson, J. W. 1983. Previous crop influence on fertilizer nitrogen requirements for double-cropped wheat, *Agron. J.* 75:855–859.

ICRISAT (International Crops Research Institute for the Semi-Arid Tropics). 1981. *Proc. Int. Workshop on Intercropping,* 10–13 January 1979, ICRISAT, Hyderabad, India.

IDRC (International Development Research Corporation). 1976. *Proc. Symp. Intercropping in Semi-Arid Areas,* 10–12 May, 1976, Morogoro, Tanzania, IDRC-076e.

———1980. *Proc. Second Symp. Intercropping in Semi-Arid Areas,* 4–7 August 1980, Morogoro, Tanzania, IDRC-186e.

IRRI (International Rice Research Institute). 1977. *Proc. Symposium on Cropping Systems Research and Development for the Asian Rice Farmer,* 21–24 September, 1976, IRRI, Los Baños, Philippines.

Jones, M. J. 1974. Effects of previous crop on yield and nitrogen response of maize at Samaru, Nigeria, *Exp. Agric.* 10:273–279.

Langdale, G. W., Hargrove, W. L., and Gibbens, J. 1984. Residue management in double-crop conservation tillage systems, *Agron. J.* 76:689–694.

Lewis, W. M., and Phillips, J. A. 1976. Double cropping in the eastern United States, in: *Multiple Cropping,* (R. I. Papendick, P. A. Sanchez, and G. B. Triplett, eds.), Amer. Soc. Agron. Spec. Publ. 27, pp. 41–50.

Mason, S. C., Leihner, D. E., and Vorst, J. J. 1986. Cassava-cowpea and cassava-peanut intercropping. III. Nutrient concentrations and removal, *Agron. J.* vol. 78.

McIntosh, J. L. 1984. Institutionalizing FSR/E: The Indonesian Experience, in: *Proc. Farming Systems Research and Extension Workshop,* October, 1984, Kansas State University, Manhattan, Kansas.

Misbahulmunir, M. Y., Sammons, D. J., and Weil, R. R. 1984. Maize-groundnut intercropping: effects of relative planting times on associated intercrops, (unpublished).

Nambiar, P. T. C., Rao, M. R., Reddy, M. S., Floyd, C. N., Dart, P. J., and Willey, R. W. 1983. Effect of intercropping on nodulation and N_2-fixation by groundnut, *Exp. Agric.* 19:79–86.

Narwal, S. S., Malik, D. S., and Malik, R. S. 1983. Studies in multiple cropping. II. Effects of preceding grain legumes on the nitrogen requirement of wheat, *Exp. Agric.* 19:143–151.

Oelsligle, D. D., McCollum, R. E., and Kang, B. T. 1976. Soil fertility management in tropical multiple cropping, in: *Multiple Cropping* (R. I. Papendick, P. A. Sanchez, and G. B. Triplett, eds.), Amer. Soc. Agron. Spec. Publ. 27, pp. 275–292.

Okigbo, B. N., and Greenland, D. J. 1976. Intercropping systems in tropical Africa, in: *Multiple Cropping* (R. I. Papendick, P. A. Sanchez, and G. B. Triplett, eds.), Amer. Soc. Agron. Spec. Publ. 27, pp. 63–101.

Osiru, D. S. O., and Willey, R. W. 1972. Studies on mixtures of dwarf sorghum and beans (*Phaseolus vulgaris*) with particular reference to plant population, *J. Agric. Sci. Camb.* 79:531–540.

Papendick, R. I., Sanchez, P. A., and Triplett, G. B. 1976. *Multiple Cropping,* Amer. Soc. Agron. Spec. Publ. 27.

RAA. 1985. *Profitable Farming Now,* (The New Farm, eds.), Rodale Press, Emmaus, Pennsylvania.

Remison, S. U. 1978. Neighbour effects between maize and cowpea at various levels of N and P, *Exp. Agric.* 14:205–212.

Rockwood, W. G., and Lal, R. 1974. Mulch tillage: A technique for soil and water conservation in the tropics, *Span.* 17:77–79.

Sanchez, P. A. 1976. *Properties and Management of Soils in the Tropics,* Wiley and Sons, Inc., New York.

Sanford, J. O. 1982. Straw and tillage management practices in soybean-wheat double-cropping, *Agron. J.* 74:1032–1035.

Sanford, J. O., and Hairston, J. E. 1984. Effects of N fertilization on yield, growth, and extraction of water by wheat following soybeans and grain sorghum, *Agron. J.* 76:623–627.

Searle, P. G. E., Comudom, Y., Shedden, D. C., and Nance, R. A. 1981. Effect of maize + legume intercropping systems and fertilizer nitrogen on crop yields and residual nitrogen, *Field Crops Res.* 4:133–145.

Singh, S. D., Misra, D. K., Vyas, D. L., and Daulay, H. S. 1971. Fodder production of sorghum in association with different legumes under different levels of nitrogen, *Indian J. Agric. Sci.* 41:172–176.

Unger, P. W., and Stewart, B. A. 1976. Land preparation and seedling establishment practices in multiple cropping systems, in: *Multiple Cropping* (R. I. Papendick, P. A. Sanchez, and G. B. Triplett, eds.), Amer. Soc. Agron. Spec. Publ. 27 pp. 255–273.

Waghmare, A. B., and Singh, S. P. 1984a. Sorghum-legume intercropping and the effects of nitrogen fertilization. I. Yield and nitrogen uptake by crops, *Exp. Agric.* 20:251–259.

———. 1984b. Sorghum-legume intercropping and the effects of nitrogen fertilization. II. Residual effect on wheat, *Exp. Agric.* 20:261–265.

Wahua, T. A. T. 1983. Nutrient uptake by intercropped maize and cowpeas and a concept of nutrient supplementation index (NSI), *Exp. Agric.* 19:263–275.

Wahua, T. A. T., and Miller, D. A. 1978. Effects of intercropping on soybean N_2-fixation and plant composition on associated sorghum and soybeans, *Agron. J.* 70:292–295.

Whitney, A. S., Kanehiro, Y., and Sherman, G. D. 1967. Nitrogen relationships of three tropical forage legumes in pure stands and in grass mixtures, *Agron. J.* 59:47–50.

Willey, R. W. 1979a. Intercropping—Its importance and research needs. Part 1. Competition and yield advantages, *Field Crop Abstr.* 32:1–10.

———. 1979b. Intercropping—Its importance and research needs. Part 2. Agronomy and research approaches, *Field Crop Abstr.* 32:73–85.

Willey, R. W., and Osiru, D. S. O. 1972. Studies on mixtures of maize and beans (*Phaseolus vulgaris*) with particular reference to plant population, *J. Agr. Sci. Camb.* 79:517–529.

Yadav, R. L. 1981. Intercropping pigeon pea to conserve fertilizer nitrogen in maize and produce residual effects on sugarcane, *Exp. Agric.* 17:311–315.

Chapter 9

Insect, Weed, and Plant Disease Management in Multiple Cropping Systems

Miguel A. Altieri
Matt Liebman

Multiple cropping systems constitute agricultural systems diversified in time and space. Much evidence suggests that this vegetational diversity often results in significant reduction of insect pest problems (Altieri and Letourneau, 1982; Cromartie, 1981; Perrin, 1980). A large body of literature cites specific crop mixtures that affect particular insect pests (Andow, 1983; Litsinger and Moody, 1976; Perrin, 1977; Perrin and Phillips, 1978), while other papers explore the ecological mechanisms involved in pest regulation (Root, 1973; Bach, 1980; Risch, 1981). Clearly, much knowledge has accumulated, and this acquired information is slowly providing a basis for designing crop systems so that pest problems and the need for active control measures are minimized (Murdoch, 1975).

Research on the effects of multiple cropping on weeds, pathogens, and nematodes has started to emerge, and studies indicate that their populations change in response to diversification of cropping systems (Bantilan et al., 1974; Sumner et al., 1981; Egunjobi, 1984). The effects of intensive systems on pests and weeds can neither be generalized nor predicted because of the enormous variety of systems utilized throughout the world. As the temporal and spatial dimensions of vegetational diversity change, so does the magnitude of the effects on pest populations (Perrin, 1980). For example, strip cropping systems can preferentially act as trap crops or as sources of natural enemies which move from one strip to another. In intercropping systems where crops are more closely intermingled, other mechanisms (i.e., repellency, masking, natural enemy enhancement, physical barriers) may affect insect pests. Moreover, a particular crop mix might be of value in controlling one pest in one area [i.e., *Heliothis virescens* in corn (*Zea mays*)/cotton (*Gossypium* sp.) strip cropping in Peru], while increasing the same pest in other areas (i.e., *H. virescens* in Tanzania) (Smith and Reynolds, 1972).

Insect herbivore species were found to be less abundant in multiple crops than in monocultures (Risch et al., 1983). Predictive trends for weed populations are harder to establish because of the relative lack of quantitative studies. Plant pathogens, in turn, seem to be buffered in multispecies crop associations, especially in systems of high genetic diversity and with high populations of antagonists in the soil (Browning, 1975).

There are many reviews examining the effects of diversifying agroecosystems on insect pest abundance; therefore we emphasize recent findings and questions that are central to crop manipulation and agroecosystem design. Our review concentrates on the dynamics of insect, pathogen, and weed communities in intercropping systems. For discussion of pest relations in other systems (i.e., cover cropping in orchards and vineyards, agroforestry, living mulches, strip cropping) there are several available literature sources (see Cromartie, 1981; Altieri and D. K. Letourneau, 1982; Altieri and D. L. Letourneau, 1984; and references cited therein). The scarce information on the effects of multiple cropping on weed abundance and disease incidence is also assembled; however, it seems that much work is needed before a general theory of the effects of these patterns on weeds and diseases can be developed.

PATTERNS OF INSECT ABUNDANCE

In recent years, ecologists have conducted experiments in multiple cropping systems to test the theory that increased plant diversity fosters stability of insect populations (Pimentel, 1961; Root, 1973; van Emden and Williams, 1974). A recent examination of all available studies on the effects of these patterns on insect pest populations tends to support the theory, although confusion may arise depending on how diversity and stability are defined (Risch et al., 1983). In multiple cropping, structural and species vegetational diversity (a measure of the biotic, structural, and microclimatic complexity arising from the mixing of different plants) results from the addition of crop plants in time and space. Stability can refer to low pest population densities over time. Of 198 herbivore species examined, 53 percent exhibited lower abundance in multicrops than in monocultures, 18 percent were more abundant in multicrops, 9 percent showed no difference, and 20 percent showed a variable response (Andow, 1983; Risch et al., 1983).

Examples of specific crop mixtures that result in reduced pest incidence can be found in Litsinger and Moody (1976), Altieri and D. K. Letourneau (1982), and Andow (1983). Results where no differences were observed or where higher pest incidence occurred in multicrops usually are not found. In Nigeria, populations of flower thrips (*Megalurothrips sjostedti*) were reduced by 42 percent on cowpea (*Vigna unguiculata*)/maize polycultures. However, cropping pattern had no effect on infestations of *Maruca testulalis,* pod-sucking bugs, and meloid beetles (Matteson et al., 1984).

In Nigeria, early infestations of *Maruca* were no different in monocrops and polycultures of maize and cowpea, but 12 weeks after planting infestations were significantly higher in the monocrops. Similar shifts were observed with *Laspeyresia* and thrips (Matteson et al., 1984). In India, larval populations of *Heliothis armigera* were higher in sorghum (*Sorghum bicolor*)/pigeon pea (*Cajanus cajan*) intercropping systems than in sole pigeon pea plots, which led to high grain losses in polycrops (Bhatnagar and Davies, 1981). In home garden plots of beans (*Phaseolus vulgaris*) bordered by marigolds (*Tagetes* spp.), Latheef and Irwin (1980) reported that their designs did not favor control of *Heliothis zea* and *Epilachna varivestris*. In Georgia, Nordlund et al. (1984) did not find significant reductions of *Heliothis zea* damage in maize ears, bean pods, or tomato (*Lycopersicon esculentum*) fruits in polycultures of maize, bean, and tomato. In the Philippines, Hasse and Litsinger (1981) found that intercropping maize with legumes did not reduce the numbers of egg masses laid by corn borers (*Ostrinia furnacalis*).

Ecological Hypotheses to Explain Insect Trends

A reduced insect pest incidence in multicrops may be the result of increased parasitoid and predator populations, availability of alternate food for natural enemies, decreased colonization and reproduction of pests, chemical repellency, masking and/or feeding inhibition from nonhost plants, prevention of pest movement and/or emigration and optimum synchrony between pests and natural enemies (Matteson et al., 1984). Perrin and Phillips (1978) described the stages in pest population development and dynamics that may be affected by mixed cropping. At the crop colonization stage they postulate that disruption of olfactory and visual responses, physical barriers, and diversion to other hosts are important mechanisms regulating herbivores in multiple cropping systems. Once the pests become established in the field, their populations may be regulated by limitation of dispersal, feeding disruption, reproduction inhibition, and mortality imposed by biotic agents.

Hasse and Litsinger (1981) described several mechanisms that have been put forward to explain pest reduction in intercropping systems. A list of the proposed mechanisms is given in Table 9.1. All available evidence suggests that the biotic, structural, and microclimatic complexity of multicrop systems work

Table 9.1 Possible Effects of Intercropping on Insect Pest Populations

Factor	Explanation	Example
	Interference with host-seeking behavior	
Camouflage	A host plant may be protected from insect pests by the physical presence of other overlapping plants	Camouflage of bean seedlings by standing rice stubble for beanfly
Crop background	Certain pests prefer a crop background of a particular color and/or texture	Aphids, flea beetle, and *Pieris rapae* are more attracted to cole crops with a background of bare soil than to ones with a weedy background
Masking or dilution of attractant stimuli	Presence of nonhost plants can mask or dilute the attractant stimuli of host plants leading to a breakdown of orientation, feeding, and reproduction processes	*Phyllotreta cruciferae* in collards
Repellent chemical stimuli	Aromatic odors of certain plants can disrupt host finding behavior	Grass borders repel leafhoppers in beans, populations of *Plutella xylostella* are repelled from cabbage/tomato intercrops).
	Interference with population development and survival	
Mechanical barriers	All companion crops may block the dispersal of herbivores across the polyculture. Restricted dispersal may also result from mixing resistant and susceptible cultivars of one crop by settling on nonhost components.	
Lack of arrestant stimuli	The presence of different host and nonhost plants in a field may affect colonization of herbivores. If a herbivore descends on a nonhost it may leave the plot quicker than if it descends on a host plant.	
Microclimatic influences	In an intercropping system favorable aspects of microclimate conditions are highly fractioned, therefore insects may experience difficulty in locating and remaining in suitable microhabitats. Shade derived from denser canopies may affect feeding of certain insects and/or increase relative humidity which may favor entomophagous fungi.	
Biotic influences	Crop mixtures may enhance natural enemy complexes. (See natural enemy hypothesis in text.)	

Source: Data from Hasse and Litsinger, 1981.

synergistically to produce an "associational resistance" (Tahvanainen and Root, 1972). Root (1973) hypothesized two ways that associational resistance may be achieved: (1) natural enemy and (2) resource concentration.

Natural Enemy Hypothesis This proposition predicts that there will be a greater abundance and diversity of natural enemies of pest insects in polycultures than in monocultures. Predators tend to be polyphagous and have broad habitat requirements, so they would be expected to encounter a greater array of alternative prey and microhabitats in complex environments. As a result, relatively stable populations of generalized predators can persist in these habitats because they can exploit the wide variety of herbivores that become available at different times or in different microhabitats. Specialized predator populations are less likely to fluctuate widely because the refuge provided by a complex environment enables their prey to escape widespread annihilation (Root, 1973). Moreover, diverse habitats offer many important requisites for adult predators and parasites, such as nectar and pollen sources, which are not available in a monoculture, reducing the probability that they will leave or become locally extinct (Risch, 1981).

Resource Concentration Hypothesis Insect herbivore populations can be influenced directly by the concentration or spatial dispersion of their food plants. Many herbivores, particularly chrysomelid beetles, are more likely to find and remain on hosts that are growing in dense or nearly pure stands (Root, 1973), and which are thus providing concentrated resources and homogeneous physical conditions. Bach (1980) found that diobroticite beetles spent a longer time on host rather than on nonhost plants. Thus, in habitats with higher concentrations of host plants, the beetles were more numerous on a per-plant basis.

In diverse systems, the visual and chemical stimuli from host and nonhost plants affect both the rate of colonization of herbivores and their behavior. A herbivorous insect approaching a habitat usually will have greater difficulty in locating a host plant when the relative resource concentration is lower. Theoretical questions relating to this hypothesis have been discussed in detail by Bach (1980) and Risch (1981).

Experimental Case Studies to Test the Hypotheses

Andow et al. (1983) contend that herbivore movement patterns are more important than the activities of natural enemies in explaining the reduction of monophagous pest populations in diverse annual crop systems. There are two classic studies that support this view. The first study (Risch, 1981) looked at the population dynamics of six chrysomelid beetles in monocultures and polycultures of maize/bean/squash (*Cucurbita pepo*). In polycultures containing at least one nonhost plant (maize), the number of beetles per unit was significantly lower relative to the numbers of beetles on host plants in monocultures. Measurement of beetle movements in the field showed that beetles tended to emigrate more from polycultures than from host monocultures. Apparently this was due

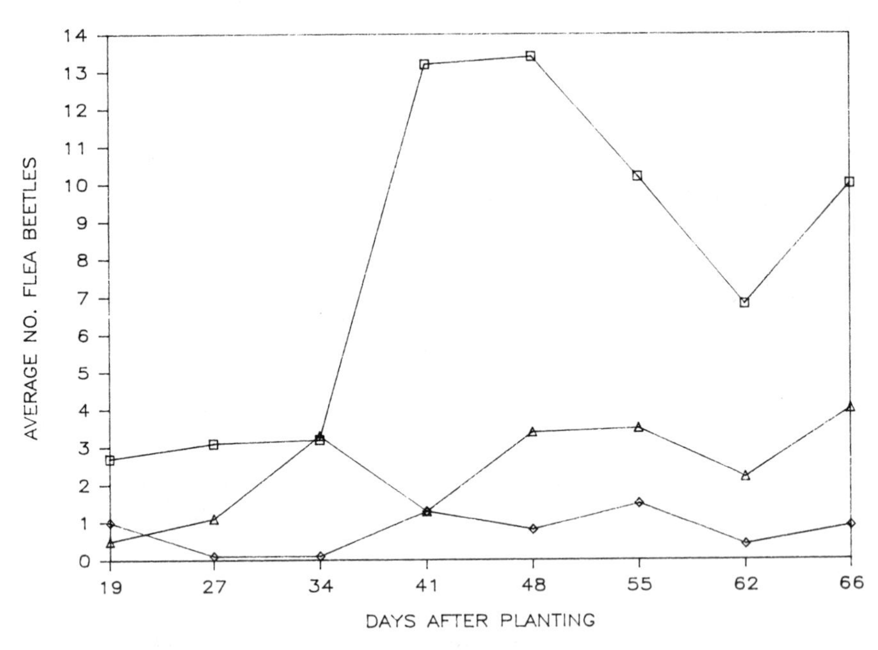

Figure 9.1 Population densities of flea beetles (*Phyllotetra cruciferae*) in collard monocultures (□), and collards associated with a nonhost crop, barley (△), and a host crop, wild mustard (◇), in Albany, California.

to several factors: (1) beetles avoided host plants shaded by maize, (2) maize stalks interfered with flight movements of beetles, and (3) as beetles moved through polycultures they remained on nonhost plants for a significantly shorter time than on host plants. There were no differences in rates of parasitism or predation of beetles between systems.

The second study examined the effects of plant diversity on the cucumber beetle, *Acalymma vittatum* (Bach, 1980). Population densities were significantly greater in cucumber (*Cucumis sativus*) monocultures than in polycultures containing cucumber and two nonhost species. Bach also found greater tenure time of beetles in monocultures than in polycultures. She also determined that these differences were caused by plant diversity *per se,* and not by differences in host plant density or host plant size. Many studies do not control differences in host plant density or size, thus they do not reveal if differences in numbers of herbivores between monocultures and polycultures are due to diversity or rather to the interrelated and confounding effects of plant diversity, plant density, and host plant patch size.

In northern California, densities of cabbage aphids (*Brevicoryne brassicae*) and flea beetles (*Phyllotreta cruciferae*) were significantly lower on cauliflower (*Brassica oleracea botrytis* var. Early Snowball) plants grown in association with weeds or vetch (*Vicia* sp.) than in clean cultivated monocultures (Altieri, unpublished data). The depression of crop growth and biomass in the diverse plots added a confounding effect in that it was not clear if herbivore reduction

Table 9.2 Effect of Flower Removal on the Abundance of Flea Beetles, *Phyllotreta cruciferae*, in Collard/Wild Mustard Intercrops

	Flea beetle densities per collard plant[a]	
System	**Early July**	**Late July**
Collard monoculture	3.0	5.0
Collard/wild mustard polyculture with flowers	0.5	1.5
Collard/wild mustard without flowers	0.7[b]	8.0

[a]Means derived from 10 plants per plot (6 × 6 m, each replicated 3 times).
[b]Before flowers were removed.
Source: Altieri, unpublished data.

resulted from poorer plant quality, which made cauliflowers less attractive to the herbivores.

In another study, flea beetle numbers were significantly lower in weedy collards (*Brassica oleracea acephala* var. Georgia), dominated by wild mustard (*Brassica campestris*), than in weed-free monocultures (Altieri and Gliessman, 1983). Flea beetles preferred this plant over collards, thus flea beetles were diverted from collards resulting in diluted feeding on the collards. The authors argue that wild mustards have higher concentrations of allylisothiocyanate (a powerful attractant to adults of *P. cruciferae*) than collards, and therefore the preference of flea beetles for wild mustard simply reflected different degrees of attraction to the foliage levels of this particular glucosinolate in the weeds and collards. Figure 9.1 illustrates this preference in the field by showing that population densities of flea beetles on collard plants grown as monocultures are greater than on collards intercropped with wild mustards and with nonhost barley (*Hordeum vulgare*) (Altieri and Schmidt, unpublished data). Although the barley effect might support the resource concentration hypothesis, the trap cropping effect of wild mustards exerts a stronger influence on beetle abundance in this case. A recent study also showed that removal of flowers of wild mustards results in a substantial reduction of the attractant effect (Table 9.2). Consequently collard plants in flowerless intercrops experienced greater flea beetle loads than the intercrop with flowers and even monocultures.

Risch et al. (1983) collected additional data that does not support the enemies hypothesis. They found that predation rates on egg masses of the European corn borer (*Ostrinia nubilalis*) by a predaceous beetle (*Coleomegilla maculata*) was significantly higher in maize monocultures than in the more densely planted maize/bean/squash polyculture. They argue that in polycultures, the beetles apparently spent more time foraging on plants (beans and squash) that contained no food, thus decreasing their foraging effectiveness. Even if prey densities per maize plant were the same in the two culture types, beetles might forage less efficiently in the polyculture due to unrewarded time spent foraging on bean and squash plants. This lower reward rate leads to faster emigration of beetles from polycultures (Wetzler and Risch, 1984).

Wrubel (1984) contends that visual camouflage from nonhost plants may have resulted in more Mexican bean beetles colonizing soybean (*Glycine max*) monoculture than maize/soybean intercropped plots. Conversely, the higher concentration of food resources in clover/soybean (two legumes) than soybean monoculture plots may explain the slightly higher abundances of polyphagous acridids that Wrubel found in the clover/soybean plots. Differences in the structure of the crop canopy in tall maize/soybean and short maize/soybean plots appeared to affect the behavior of several groups of herbivores, with lower abundance of Japanese beetles due to shading of the soybean canopy by the taller maize plants.

There are studies that support the enemies hypothesis. In tropical corn/bean/squash systems, Letourneau (1983) studied the importance of parasitic wasps in mediating the difference in pest abundance between simple and complex crop arrangements. A squash-feeding caterpillar, *Diaphania hyalinata* (Lepidoptera: Pyralidae), occurred at low densities on intercropped squash in tropical Mexico. Part of the effect of the associated maize and bean plants may have been to render the squash plants less apparent to ovipositing moths. Polyculture fields also harbored greater numbers of parasitic wasps than did squash monocultures. Malaise trap captures of parasitic wasps in monoculture consisted of one-half the number of individuals caught in mixed culture. The parasitoids of the target caterpillars were also represented by higher numbers in polycultures throughout the season. Not only were parasitoids more common in the vegetationally diverse, traditional system, but the parasitization rates of *D. hyalinata* eggs and larvae on squash were higher in polycultures. Approximately 33 percent of the eggs in polyculture samples over the season were parasitized and only 11 percent of eggs in monocultures. Larval samples from polycultures showed an incidence of 59 percent parasitization for *D. hyalinata* larvae whereas samples from monoculture larval specimens were 29 percent parasitized.

Another study conducted in Davis, California, tested whether predator colonization rates can be manipulated through vegetational diversity (Letourneau and Altieri, 1983). We compared the densities of *Orius tristicolor* and its preferred prey, *Frankliniella occidentalis,* between squash monocultures and polycultures of squash, corn, and cowpea. The patterns of predator colonization rates and pest densities in these two cropping systems paralleled those documented for predator/prey interactions in the mite grapevine systems of Flaherty (1969). In both studies, the colonization rate of predators was increased in diverse habitats, and the prey (pest) populations in each case reached lower maximum levels. In Flaherty's study, the causative factor was the close proximity of the source of colonizing predators. The great variation between levels of Willamette mite (*Eotetranychus willamettei*) infestation on individual grape vines was caused by their variable proximity to clumps of Johnson grass (*Sorghum halepense*). The grass supported an alternate host for a predatory mite, *Metaseiulus occidentalis*. The predators then colonized contiguous vines sufficiently early to suppress the pest mite populations. In our study, the sources of colonizers were presumably at similar distances to randomly assigned plots of monoculture and polyculture. We suggest that the determining factor for differential colonization by *Orius* sp. in monocultures and polycultures of squash was attraction to the early season crop habitat during the host location process.

Our results showed that mean density of thrips on squash leaves was initially much greater in monoculture than in polyculture and remained at significantly higher levels until 65 days after sowing. The *Orius* sp. density, however, was significantly higher on squash early in the season (days 30 and 42) in polyculture. A decrease in prey density accompanied an increase in adult *Orius* sp. colonization in both treatments until thrips were at low densities.

Predator manipulation experiments conducted in field cages, in which *Orius* sp. populations were either included or excluded, showed that the density of thrips was influenced by predation by *Orius* sp. (Letourneau and Altieri, 1983). On uncaged control plants, the mean density of thrips per leaf declined steadily from day 50, as it had in the general field samples. Inside the exclusion-inclusion cages, thrips density more than tripled the first week after *Orius* and *Erigone* spp. spiders were eliminated. When predators, equal in number to those that were eliminated, were added to cage 1, the thrips density fell in this cage.

Altieri (1984) recently found that brussel sprouts (*Brassica oleracea gemmifera* var. Jade Cross) grown in polycultures with fava beans or wild mustard supported more species of natural enemies (six species of predators, and eight species of parasites) than monocultures (three species of predators and three species of parasites). Apparently, the presence of flowers, extrafloral nectaries, and alternate prey/hosts associated with the companion plants allowed this enhancement.

Another aspect that has barely been considered in the enemies hypotheses is direct plant/natural enemy interactions. Price et al. (1980) have argued that the third trophic level is especially relevant in multispecies crop associations, not only because the herbivore/enemy interaction on one plant species can be influenced by the presence of associated plants, but also because the herbivore/enemy interactions on one plant species can be influenced by the presence of herbivores on associated plant species.

To our knowledge there are a few studies that have documented direct relations between plants and natural enemies. These studies show that some entomophagous insects are attracted to particular plants, even in the absence of host or prey, or by chemicals released by the herbivore's host plant or other associated plants. A few of these studies tested the attractance of plants to parasitoids and showed that certain parasitoids prefer particular plants over others. Others showed that parasitization of a pest was higher on a certain crop than on others. Of significant practical interest are the findings of Altieri et al. (1981), which showed that parasitization rates of *H. zea* eggs by *Trichogramma* sp. were greater when the eggs were placed on soybeans next to corn, *Desmodium* sp., *Cassia* sp., or *Croton* sp., than on soybeans grown alone. Although the same numbers of eggs were placed on the associated plants, few of these eggs were parasitized; results suggested that these plants were not actively searched by *Trichogramma* sp., but in some way the plants enhanced the efficiency of parasitization on the associated soybean plants. A possibility is that they emited volatiles with kairomonal action. Further tests showed that application of water extracts of some of these associated plants to soybeans enhanced parasitization of *H. zea* eggs by *Trichogramma* spp. wasps. The authors stated that a better attraction and retention in the extract-treated plots may have caused the higher parasitization levels. The possibility that vegetationally complex plots are more chemically diverse than monocultures and therefore more acceptable and arousing to parasitic wasps opens many new dimensions for biological control through habitat management and behavior modification.

Entomological Studies Conducted in Traditional Farmers' Fields

For centuries small farmers have developed and/or inherited complex cropping systems. These systems are based on countless distinct crop varieties and crop mixtures that were selected by farmers over thousands of years for resistance to pests and for other characteristics. These varieties and mixtures are grown using agricultural practices that usually enhance cultural and biological pest control. Thus, many traditional cropping systems have built-in pest-control mechanisms (Matteson et al., 1984). Unfortunately, there are few references on pest dynamics in these systems, and on the control methods commonly used in traditional agriculture. The scattered information is of an anthropological nature, collected by nonentomologists, so that the identities of pests are only conjecture. Brown and Marten (1984) recently reviewed the literature on traditional pest management practices with the intention of providing information on which elements to retain in the course of agricultural modernization. Pest-control features of some traditional systems are described in Matteson et al. (1984) and descriptions of pest-control practices utilized by Southeast Asian peasants can be found in Litsinger and Moody (1976).

During 1982 and 1983, studies were conducted in Tlaxcala, Mexico, on the dynamics of insect communities associated with maize managed with traditional technologies (Trujillo and Altieri, unpublished data). In this state, maize is grown in a variety of situations: as monoculture, intercropped with fava beans (*Vicia faba*), strip-cropped with alfalfa (*Medicago sativa*), weedy, and in association with apple (*Malus pumila*) trees and forest trees. The results indicate that populations of the pestiferous scarab beetle, *Macrodactylus* sp., and a number of predaceous species (e.g., Coccinellidae, *Collops* sp., *Orius* sp.) vary significantly, depending on the location and size of the field, the associated plants, the surrounding vegetation and the type of cultural management. For example, in some areas, predaceous beetles (e.g., Coccinellidae and *Collops* sp.) were more abundant in maize rows adjacent to well-established alfalfa strips than in the center rows of the same maize field. In other areas, where the alfalfa was recently cut, this gradient was no longer apparent. In the area of Coaximulco, *collops* beetles were more abundant in maize fields intercropped with apples than in corresponding monocultures. Coccinellid beetles, however, exhibited the opposite trends.

A survey of insect communities associated with maize was conducted near Lembang, Indonesia. Two types of cropping systems are common in the area: (1) *Kebun,* a mixture of annual crops (e.g., maize/cauliflower, maize/sweet potatoes (*Ipomoea batatas*), maize/bean), and (2) *Kebun-campuran,* a mixture of annuals and perennial shrubs and trees [e.g., maize, cauliflower, and cassava (*Manihot esculenta*) grown under a canopy of citrus (*Citrus* sp.), coffee (*Coffea arabica*), banana (*Musa paradisiaca*), and clove (*Syzgium aromaticum*)]. Four agroecosystems in which maize was a dominant crop were surveyed in August 1983 (Altieri and Marten, unpublished data). Each system was mapped and measurements were taken on weed diversity and percentage of cover, insect abundance, species diversity, and level of pest damage. Observations were complemented by interviews with farmers about crop management practices and socioeconomic aspects. Table 9.3 summarizes the profile of predator abundance and pest damage in the four cropping systems. Although our survey revealed variations in insect pest incidence and natural enemy abundance between fields, it is not clear whether these differences were related to the vegetational structure of the systems, or were merely a consequence of differential management, location, or chance. It would seem worthwhile to examine the elements of natural pest control built in these cropping systems, so that these elements can be retained in the course of agricultural modernization.

Methodologies in the Study of Insect Dynamics

Many approaches have been tried in plots of monocultures and polycultures to explain the ecological mechanisms underlying the effects of diversity. Risch

Table 9.3 Profile of Pest Damage and Relative Abundance of Predators in Maize Agroecosystems in Lembang, West Java[a]

	Results[b]			
Parameter	System 1	System 2	System 3	System 4
Maize plants infested with aphids (%)[c]	38	52	56	100
Maize plants with *Heliothis* damage (%)[c]	0	10	2	13
Maize plants with *Spodoptera* damage (%)	63	46	18	68
No. of coccinellid adults per 50 maize plants	37	4	19	8
No. of coccinellid larvae per 50 maize plants	24	11	47	62
No. of syrphid larvae per 50 maize plants	7	4	13	18
No. of spiders per 50 maize plants	93	130	78	187

[a]Means derived from 50 surveyed maize plants in each system, 70 days after planting.
[b]System 1: Maize field surrounded by a belt of cassava, banana, citrus, coffee, and other trees.
System 2: Maize/sweet potato intercropping system.
System 3: Terraced cauliflower field with maize rows in the edge of the terrace.
System 4: Maize/cauliflower intercropping system.
[c]Mostly small aphid colonies and light leaf damage, respectively.

(1981) studied beetle movement behavior to see if this could account for lower numbers of beetles in maize/bean intercrops. He placed directional Malaise insect traps on each side of every plot. When beetles flew out of the plot, some of them landed on the vertical trap walls and were caught in the collecting jars. By counting these trapped beetles and estimating the total number of beetles in the plot at that time from direct counts, he calculated a ratio of the two groups and called the ratio "tendency to emigrate," which measures the beetle's relative tendency to leave a plot once it has arrived. After 60 to 65 days, there was a much greater tendency to emigrate from the maize monoculture and the polycultures than from the bean monoculture. This corresponds with the observation that there were far fewer beetles on beans planted with maize in the polycultures than in the bean monocultures, and that this large difference became apparent approximately 65 days after planting. Maize has some kind of inhibitory effect on the presence of this insect species.

How does maize exert its inhibitory effect? Beans grown with maize are shaded more than beans in monocultures. One possibility is that the beetles avoid feeding in shaded areas, preferring to feed on plants that are not shaded. This was tested directly by constructing two large shade screens and suspending them 80 cm above the ground. One screen provided little shade, allowing 65 percent light transmission, and the other provided much more shade, allowing only 25 percent light transmission. Squash and bean plants were grown in the greenhouse and placed under these screens. Then the numbers of beetles on the plants were counted over a series of days. The results showed that there were always significantly more beetles under the light shade screen than the dark.

Yet shade might not be the only way that the presence of maize interferes with beetle flight behavior. To determine if a vertical obstruction, like a maize stalk, could discourage beetle colonization in other ways, dry maize stalks were staked among potted bean plants and a light screen was erected over the plants. Potted beans without maize stalks were also placed in a nearby area with a darker screen over them, so that the total amount of light reaching the plants in both areas was identical. Risch consistently found many more beetles in the beans without maize stalks, indicating that maize physically inhibited beetle colonization in ways other than by just increasing the overall shade of a microhabitat.

Although the above experiments provided an indication of the underlying causes of the reduction in beetle numbers in maize/bean polycultures, they did not help in predicting numbers of beetles in different variations of the entire maize/bean/squash system. Risch and Andow (unpublished data) also studied the influence of size of the plot and relative proportions of maize, beans, and squash on the number of beetles in the field. They observed and modeled the movement of one beetle, *Acalymma vittatum,* a squash specialist that is much more abundant in monocultures of squash than in maize/bean/squash polycultures in Ithaca, New York. The variables Risch and Andow thought might be important in ultimately determining the rate at which a beetle leaves a maize/bean/squash polyculture versus a squash monoculture are the following: the time a beetle spends on a maize, bean, or squash plant; the probability of moving to a maize, bean, or squash plant; the distance a beetle flies when it leaves a maize, bean, or squash plant and flies over an intercrop or monoculture; and its orientation behavior at the edge of a plot.

Wetzler and Risch (1984) examined the behavior of a coccinellid beetle in the field in four diffusion experiments. Each involved the release of beetle populations in a matched pair of agricultural plots (10 × 10 m each) planted to various combinations of maize, beans, and squash. The day preceding each release, all *Coleomegilla* individuals were aspirated from every plant, ensuring "clean" fields for each experiment.

In the first experiment 314 beetles were released in a high-density maize monoculture and an equal number in a maize/bean/squash polyculture 34 days after planting. Maize density was 3 times higher in the monoculture than polyculture. Releases in the two fields were completed within 20 minutes of each other by gently pouring dormant (chilled) beetles onto basal leaves. All beetles were marked in the first release with Testor's model paint applied to an eletryum. Sight-counting census was done at 1, 3, 6, 12, and 24 hours after release along the east-west, northeast-southwest, and northwest-southeast radii. This sampling arrangement established concentric annuli spaced 0.5 m apart.

In the second experiment, 175 beetles were released in each of the same two plots at 40 days after planting. In the third experiment, 65 days after planting, maize plants were selectively removed from the dense maize monoculture so that the total maize density was the same in the maize monoculture and maize/bean/squash polyculture. Thus, total plant density was 3 times higher in the polyculture. In the last experiment, the fourth release, 71 days after planting, was preceded by removal of all squash and bean plants from the polyculture to yield two sparse maize monocultures. The influence of food resources on mobility was tested by pruning all anthers, which contained the bulk of aphids and pollen fed upon by *Coleomegilla,* from each maize plant in one field and intertwining them among the intact anthers of the other monoculture.

Wetzler and Risch (1984) conducted another experiment to determine whether differences in diffusion rates from monocultures and polycultures might be caused in part by differences in the average time a beetle spent on maize, bean, and squash plants (i.e., tenure time per plant). Maize, squash, and bean plants were first grown in pots until all the plants were in flower. Approximately half the maize plants had large numbers of the corn aphid *Rhopalosiphum maidis*. In the first trial, 50 *C. maculata* were placed on five aphid-infested maize plants, 50 beetles were placed on five bean plants, and 50 beetles were placed on five squash plants (10 beetles per plant). The beetles were cooled to approximately 6°C before being placed on plants. The number of beetles remaining on the plants was counted approximately every 10 minutes for a period of 100 minutes.

Sight counting is an effective means of population censusing since the beetles are highly visible, thus avoiding problems associated with trapping. Careful collection of individuals for release enabled uniform, almost equivalent releases of adults. Since each experiment was run for only 24 hours and was preceded by a minimum of handling of beetles, mortality was extremely low

(0.5 percent) and complications due to beetle reproduction were nonexistent. The timing of the 1-, 3-, and 6-hour censuses was arranged to correspond with maximum periods of diurnal activity to ensure that the most conservative diffusion estimates would arise during the final censuses. Since all experiments were conducted within a 5-week interval, seasonal variability (i.e., migratory movements) of *Coleomegilla* activity was restricted.

Bach (1980) focused on the response of one specialist herbivore, the striped cucumber beetle (*Acalymma vittata*) to cucumber monocultures versus cucumber/broccoli/maize polycultures. By controlling total plant density, host-plant density, and plant diversity, Bach was able to distinguish the effects of these three confounding variables. Applying a three-way analysis of variance to censuses of beetles per cucumber plant, Bach reported a significant effect of both plant density and diversity on *Acalymma* abundance, but the results only partially support the resource-concentration hypothesis. Although an increase in stand purity yielded the expected increase in beetles per cucumber plant, an increase in cucumber density reduced the number of beetles per plant (Bach, 1980). Bach also found that cucumber plants were on average smaller in polycultures than in monocultures and that beetle density was positively correlated with plant size. There were two polyculture plots with cucumbers equal in size to monoculture cucumbers; in these two thriving polycultures, beetles were still significantly fewer per plant than in monocultures. Thus, it is clear that the reduced beetle numbers in polycultures cannot simply be attributed to smaller cucumber plants. In a later study with *Acalymma vittata,* Bach (1980) provided evidence for a surprising reduction in foliage palatability associated with cucumber/tomato polycultures. Beetles given a laboratory choice between cucumber leaves grown in monoculture and cucumber leaves grown in a tomato/cucumber mixture significantly preferred monoculture leaves. This illustrates the subtle links that are possible between plant diversity and plant quality, quite apart from the conventional ideas concerning the influence of resource concentration on herbivores (Kareiva, 1983).

In their studies of corn/cowpea/squash polycultures, Letourneau and Altieri (1983) found visual inspection sampling of thrips and *Orius* to produce a more representative measure of density than did sticky traps, pan traps, or Malaise traps, each of which showed very low catches. Ten hills of squash (each hill consisting of two plants) were randomly selected, and the plant most southwest in the hill was sampled by gently turning each leaf and recording the numbers of *Orius* adults and nymphs (as well as any other common arthropods). Thrips were counted on one medium-sized leaf of each plant. During the season, *Orius* densities increased and plants grew so large that the number of leaves sampled was reduced to five per plant: the growing shoot, two young, and two old leaves. Biomass estimates were made at 2-week intervals by measuring leaf widths on all the leaves of 10 plants per plot. The leaf width of squash plants was highly correlated with leaf biomass, determined as dry weight of the leaf blade ($r = 0.93$). To standardize for possible leaf size differences between treatments (and thus, searching-area differences), predator numbers per plant were converted to numbers per 5 g of leaf biomass. Individual leaves sampled for thrips were also measured to allow for conversion of thrips per leaf to thrips per 5 g of leaf biomass.

To determine whether predators were concentrated within treatments on plants with higher thrips densities, an index of aggregation was calculated on day 30. Ratios of mean thrips density on plants with *Orius* to those without *Orius* present would be significantly greater than 1 if *Orius* were showing such a preference within a treatment. These examples illustrate the range of methodologies that have been employed.

Management Considerations

Multiple crop management is basically the design of spatial and temporal combinations of crops in an area. There are many possible crop arrangements and each can have different effects on insect populations. For insects, the attractiveness of crop habitats in terms of size of field, nature of surrounding vegetation, plant densities, height, background color and texture, crop diversity, and weediness is subject to manipulation.

In intercrop systems, the choice of a tall or short, early- or late-maturing, flowering or nonflowering companion crop can magnify or decrease the effects on particular pests (Altieri and Letourneau, 1982). The inclusion of a crop that bears flowers during most of the growing season can condition the buildup of parasitoids, thus improving biological control. Similarly, the inclusion of legumes or other plants supporting populations of aphids and other soft-bodied insects that serve as alternate prey/hosts can improve survival and reproduction of beneficial insects in agroecosystems. The presence of a tall associated crop such as maize or sorghum may serve as a physical barrier or trap to pests invading from outside the field. The inclusion of strongly aromatic plants such as onion (*Allium cepa*), garlic (*Allium sativum*), or tomato (*Lycopersicon esculentum*) can disturb mechanisms of orienation to host plants by several pests.

The date of planting of component crops in relation to each other can also affect insect interactions in these systems. An associated crop can be planted so that it is at its most attractive growth stage at the time of pest immigration or dispersal, diverting pests from other more susceptible or valuable crops in the mixture. Planting of okra (*Hibiscus esculentus*) to divert flea beetles (*Podagria* spp.) from cotton in Nigeria is a good example (Perrin, 1980). Maize planted 30 and 20 days earlier than beans reduced leafhopper population on beans by 66 percent compared with simultaneous planting. Fall armyworm damage on maize was reduced by 88 percent when beans were planted 20 to 40 days earlier than maize when compared to simultaneous planting (Altieri et al., 1978).

We still understand little of how spatial arrangements (e.g., row spacings) of crops affect pest abundance in intercrops. For example, there is greater reduction in damage to cowpea flowers by *Maruca testulalis* in intrarow rather than interrow mixtures of maize and cowpea. Selection of proper crop varieties can also magnify insect suppression effects. In Colombia, lower whorl damage by *Spodoptera frugiperda* was observed in maize associated with bush beans than in maize mixed with climbing beans. In the same trials, maize hybrid H-207 seemed to exhibit lower *Spodoptera* damage than hybrid H-210, when intercropped with beans (Altieri et al., 1978). Clearly, much further work is needed before appropriate crop mixtures and row spacings are to be achieved.

The manipulation of weed abundance and composition in intercrops can also have major implications on insect dynamics (Altieri et al., 1977). When weed and crop species grow together, each plant species hosts an assemblage of herbivores and their natural enemies; thus trophic interactions become very complex. Many weeds offer important requisites for natural enemies such as alternate prey/hosts, pollen, or nectar as well as microhabitats that are not available in weed-free monocultures (van Emden, 1965). The beneficial insects associated with many weed species have been surveyed (Altieri and Whitcomb, 1979). Relevant weeds that support rich natural enemy faunas include the perennial stinging nettle (*Urtica dioica*), Mexican tea (*Chenopodium ambrosioides*), camphorweed (*Heterotheca subaxillaris*), and goldenrod (*Solidago altissima*). In the last 20 years, research has shown that outbreaks of certain types of crop pests are more likely to occur in weed-free fields than in weed-diversified crop systems (Altieri et al., 1977). Crop fields with a dense weed cover and high diversity usually have more predaceous arthropods than in weed-free fields. Ground beetles, syrphids, and lady beetles (Coccinellidae) are abundant in weed-diversified systems. Relevant examples of cropping systems in which the presence of specific weeds has enhanced the biological control of particular pests can be found in Altieri and Whitcomb (1974) and Altieri and D. K. Letourneau (1982). These observations suggest that selective weed control may change the mortality of insect pests caused by natural enemies. The ecological basis for obtaining crop/weed mixtures that enhance insect biological suppression awaits further development.

EFFECTS OF MULTIPLE CROPPING ON PLANT DISEASES

In the wild, the dispersion of host plants among other plant species restricts the spread of pathogens (Browning, 1975). The situation is little different in traditional agroecosystems in which many crop species are grown in species-rich or genetically diverse mixtures (Thresh, 1982). Such diversity gives a measure of stability in that the failure of some species or genotypes due to diseases may be compensated by the improved performance of others if adequate time is available for compensatory growth. The presence of immune or resistant plants in the mixtures impedes pathogen spread and increases the separation between susceptible plants. Based on this rationale, Larios and Moreno (1977) have convincingly argued that the most suitable agroecosystems to avoid disease damage in tropical areas are multiple cropping systems that resemble the local natural system.

Larios and Moreno (1977) documented evidence of disease buffering in various tropical intercropping systems. For example, *Ascochyta phaseolerum* was less prevalent in cowpea interplanted with maize than in cowpeas growing alone. The total number of diseased plants as well as the speed of dissemination of the pathogen was less in the polyculture. The maize plants apparently act as a natural barrier to the free spread of the fungus propagules. In this same crop association, cowpea virus diseases were less prevalent. The total number of infected plants with cowpea mosaic virus (CPMV) and chlorotic cowpea mosaic virus (CCMV) was less in polycultures than in monocultures, apparently because fewer numbers of vector chrysomelid beetles (e.g., *Diabrotica cerotoma*) were present in the mixed stands. A similar situation occurred in Malawi, where beans trapped aphids, thus decreasing the spread of rosette disease of groundnut in mixed stands (Thresh, 1982). Radish mosaic in Japan decreased when radishes (*Raphanus sativus*) were sown between rows of rice (*Oryza sativa*) or trefoil, and pigeon pea in Haiti was completely protected from virus diseases when grown between rows of tall sorghum (Palti, 1981).

These data indicate that the use of nonhost crops in interplantings can significantly reduce the rate of virus spread in the field. A buffer crop such as maize, when grown between the source of peanut mottle and a susceptible soybean crop, can reduce the amount of separation required to prevent disease spread (Zitter and Simons, 1980). Mosaic virus of alfalfa is more prevalent in monocultures than in mixtures with cocksfoot grass (Thresh, 1982). The only example in which multiple crops have been adopted is for protecting sugar beet (*Beta saccharifera*) seed crops from aphid-borne viruses by intersowing with barley or other crops. Growing of mustard or barley which grows to heights of 66 and 41 cm, respectively, together with sugar beet stecklings lowers the incidence of beet mild yellowing virus in the stecklings (Palti, 1981).

Another example of disease reduction occurs when maize is grown in association with cassava. Cassava scab (*Sphaceloma sp.*) is notably reduced in cassava/maize polycultures. By doubling over the maize stalks before harvesting, there is an increase in the disease, thus showing the protective effect of the maize (Larios and Moreno, 1977). Similarly, dieback in cassava caused by *Glomerella cingulata* was more intense in the monocultures. On the other hand, the severity of angular leaf spot of beans caused by *Isariopsis griseola* was highest in bean polycultures that included maize and lowest in systems where beans were intercropped with sweet potato or cassava (Moreno, 1977).

Some crop associations, due to the microclimate modifications imposed by the dominant crop, result in increases in relative humidity and shade which may favor the incidence of diseases, such as angular leaf spot and wilt (*Thanatephorus cucumeris*) of the common bean. In such cases it is necessary to modify the spatial arrangement of the associations to alter the microclimate and thus minimize the negative impact of these diseases. In general, the shielding effect of the companion crops against airborne pathogens should more than offset the microclimatic advantage pathogens may derive from the dense foliage of mixed crops (Palti, 1981).

In summary, mixtures of different crop species buffer against disease losses by delaying the onset of the disease, reducing spore dissemination, or modifying microenvironmental conditions such as humidity, light, temperature, and air movement. Certain associated plants can function as repellants, antifeedants, growth disrupters, or toxicants. In the case of soilborne pathogens, some plant combinations may enhance soil fungistasis and antibiosis through indirect effects on soil organic matter content; however, research in this area is largely lacking (Sumner et al., 1981). Provided the pathogen/host/environment relationship is understood in a particular cropping system, the use of mixed cropping and therefore a diversity of genotypes shows great possibilities for disease reduction. A parallel approach involves the use of multilines in cereal crops to achieve high genetic diversity (Browning, 1975).

IMPACT OF MULTIPLE CROPPING SYSTEMS ON NEMATODE POPULATIONS

The concept of crop diversification for the management of nematode populations has been applied mainly in the form of decoy and trap crops. Decoy crops are nonhost crops which are planted to make nematodes waste their infection potential. This is effected by activating larvae of nematodes in the absence of hosts that would enable them to continue their development. A list of nematodes that can be decoyed in this way is presented in Table 9.4.

Trap crops are host crops sown to attract nematodes but destined to be harvested or destroyed before the nematodes manage to hatch. This has been advocated for cyst nematodes, sowing crucifers to be plowed in before the nematodes of beets can develop fully. The same objective is achieved in pineapple (*Ananas comosus*) plantations by planting tomatoes and destroying them before root-knot nematodes can produce eggs (Palti, 1981).

There is also evidence that some plants adversely affect nematode populations through toxic action. Oostenbrink et al. (1957) showed that several varieties of *Tagetes erecta* and *T. patula* reduced the population of certain root-infecting nematode species such as *Pratylenchus, Tylenchorchynchus, Paraty-*

Table 9.4 Decoy Crops Used for the Reduction of Nematode Populations

Crop	Nematode species	Decoy crops
Eggplant	*Meloidogyne incognita, M. javanica*	*Tagetes patula, Sesamum orientale*
Tomato	*M. incognita, Pratylenchus alleni*	*T. patula*, castor bean, chrysanthemum
Tomato	*M. incognita*	*T. patula*, groundnuts
Narcissus, tomato, okra	*Meloidogyne* sp.	*T. patula*
Soybean	*Rotylenchulus* sp., *Pratylenchus* sp.	*T. minuta, Crotalaria spectabilis*
Various	*Pratylenchus penetrans*	*T. patula*, hybrids of *Gaillardia* and *Hellenium*
Various	*Pratylenchus neglectus*	Oil-radish (*Raphanus oleiferus*)
Oats	*Heterodera avenae*	Maize
Various	*Trichodorus* sp.	Asparagus

Source: Data from Palti, 1981.

lenchus, and *Rotylenchus*. The effect of marigolds on *Pratylenchus* eelworms appears to be due to the nematicidal action of the growing plant roots which exude alpha-terthienyl. In subsequent studies, Visser and Vythilingum (1959) also reported that these two marigold species considerably decreased *Pratylenchus coffeae* and *Meloidogyne javanica* populations in tea (*Camellia sinensis*) soil. The cultivation of marigolds reduced nematodes more quickly and more effectively than keeping the tea soil fallow. There are other plants whose root extracts show nematicidal action. For example, *Ambrosia* spp. and *Iva xanthiifolia* reduce populations of *Pratylenchus penetrans* (Hijink and Suatmadji, 1967).

Little work has been conducted on nematode suppression in intercropping systems. The nematode *Anguina tritici*, which enters wheat seedlings from the soil and infests the ears, has been partially controlled in India by growing *Polygonum hydropiper* with wheat (*Triticum* sp.). Also in India, the plant *Sesamum orientale* has been found to produce root exudates that are nematicidal to root-knot nematodes and to decrease root-knot infestation in *Abelmoschus esculentus* growing alongside (Atwal and Mangar, 1969).

Egunjobi (1984) recently studied the ecology of *Pratylenchus brachyurus* in traditional maize cropping systems of Nigeria. Nitrogen/phosphorus/potassium fertilizer applications increased the numbers of the nematode more in soil under monoculture maize than in plots with maize intercropped with cowpea, groundnut, or green gram.

ECOLOGY AND MANAGEMENT OF WEEDS IN INTERCROP SYSTEMS

Weed management in intercrops has been the subject of very little research but is an important topic for four reasons. First, intercropping is a cultural method used by the majority of farmers in Latin America (Pinchinat et al., 1976; Francis et al., 1976), Africa (Okigbo and Greenland, 1976), and Asia (Harwood and Price, 1976). While intercropping is commonly practiced on small farms where capital inputs are limited, intercropping yield advantages are by no means restricted to such situations (Osiru and Willey, 1972; Bantilan et al., 1974; Andrews and Kassam, 1976; Sanchez, 1976). Second, in spite of their many beneficial uses (Van de Goor, 1954; Mishra, 1969; Kapoor and Ramakrishnan, 1975; Altieri et al., 1977; Bye, 1981; Chacon and Gliessman, 1982; Weil, 1982), weeds can seriously limit food production in both monoculture and intercropping systems (Muzik, 1970; Holm, 1971; Moody, 1977). Weed control is often the operation with the highest labor demand during the cropping cycle (Moody, 1977; Compton, 1982). Third, weed management is a central point of coordination for many farm operations. Effective weed management involves integration of soil fertility and water management, tillage practices, choice of crop sequences, crop density, crop species and cultivars, insect management, labor, available traction power, and cash inputs (Bantilan et al., 1974). Finally, interactions between three or more plant species can be complex and might not be predictable from knowledge of monoculture and binary mixture results (Haizel and Harper, 1973). Crop cultivars, proportions, densities, spatial arrangements, and management practices that give high intercrop yields under weed-free conditions may not give acceptable yield and weed suppression under weedy conditions (Plucknett et al., 1977). Solutions to this problem will require detailed studies of weeds in intercrop systems, although knowledge about how crops and weeds grow alone and in binary mixtures will be very useful.

Weed management for any cropping system can involve the use of many kinds of biological, physical, and chemical techniques to promote crop dominance over weeds. There is no question that herbicides, tillage, and water management can be important components of weed management programs for intercrop systems (Bantilan et al., 1974; Plucknett et al., 1977; Akobundu, 1980a; Moody and Shetty, 1981). However, this review concentrates on biological factors affecting intercrop/weed balance. There are several reasons for this.

There is growing recognition among weed scientists that farmers in both industrialized and developing nations cannot rely entirely on herbicides for weed control (Akobundu, 1980a; Walker and Buchanan, 1982). Many weed species are evolving herbicide resistance, making chemical control less effective and more costly (Plucknett et al., 1977; Lebaron and Gressel, 1982). Many farmers lack the cash or credit necessary for the purchase of herbicides and the appropriate application devices, and in many areas extension education is inadequate for selection of appropriate chemicals for specific weed problems (Akobundu, 1980a). If they are affordable, available herbicides may be undesirable because of safety and environmental considerations (Pimentel et al., 1980; Revkin, 1983). Because herbicides are often crop specific it has been difficult to find compounds that will control a broad spectrum of weeds without causing damage to the component crops of the intercrop (Moody and Shetty, 1981).

Physical manipulations of the intercrop environment for weed control very closely resemble those used for sole crops. Technical problems of tillage operations for intercrops remain to be worked out for many crop combinations, but solutions appear possible with adaptations of existing technologies (Anderson, 1981).

In contrast to chemical and physical means of weed control in intercrop systems, the biological factors that promote intercrop dominance over weeds are complex and poorly understood. These factors are a farmer's first line of defense in preventing crop losses to weed interference and have critical importance in the design of effective weed management strategies for intercrop systems.

Intercrop weed management combines two qualitatively different aspects of plant/plant interactions. To increase intercrop yields, complementarity in patterns of resource use by the component crops must be emphasized. The goal is to minimize the degree of overlap in resource use by intersown crop species such that more resources are exploited and more yield can be harvested per unit of ground area. In contrast, to achieve weed control, the similarity of requirements of crop and weed species, the consequent competition for limited resources, and the suppression of growth and yield of the associated species are emphasized. Weed scientists and farmers work to create an environment that is detrimental to weeds and favorable to crops. Intercropping has potential as a means of weed control because it offers the possibility of a mixture of crops capturing a greater share of available resources than in monocropping, preempting their use by weeds.

Intercropping can be practiced in two distinct ways. A farmer can desire a certain amount of yield from each component of an intercrop and use a cropping system that balances interference between crops to give a desirable result. Alternatively, a farmer can be primarily interested only in the yield of one main crop and intersow other species for insurance against crop failure, minor economic uses, erosion control, soil fertility maintenance, and/or weed control. In the first case it is the summed yield of the two or more components viewed on some sort of relative basis (e.g., land equivalent ratio, LER) that is important. In the latter case, the intercrop is meant to give full yield of the main crop, with the intersown minor species providing some additional benefit.

Provided that interference between crop components is weaker than that between crops and weeds, both types of intercropping can suppress the growth of weeds more than sole cropping (Yih, 1982). It should also be noted that intercropping may suppress weeds no more than sole-cropping but still provide more yield because of exploitation of otherwise unoccupied niches by one or more of the added crop components. This is shown (Table 9.5) by the results of two experiments conducted with barley and pea (*Pisum sativum*) (Liebman, unpublished data). Weed growth was similar in the intercrop and monoculture

Table 9.5 Barley, Pea, and Weed Yields from Two Experiments Conducted near Ithaca, New York[a]

Experiment	Yield (g/m²)	
	Warren Farm, 1982	Turkey Farm, 1983
Total seed crop:		
Weedy monoculture—barley	175.0 ± 4.1	153.3 ± 10.6
Weedy intercrop—barley	154.4 ± 13.6	139.9 ± 15.1
(Partial LER)	(0.88)	(0.92)
Weedy monoculture—pea	214.7 ± 8.1	63.4 ± 8.5
Weedy intercrop—pea	207.3 ± 13.1	62.6 ± 10.7
(Partial LER)	(0.96)	(0.99)
(Total LER)	(1.84)	(1.91)
Total aboveground weed biomass:		
"Weeds alone" control	523.9 ± 9.9	302.5 ± 24.3
Weedy pea monocultures	446.7 ± 50.6	266.6 ± 26.2
Weedy barley monocultures	183.2 ± 12.6	165.4 ± 11.4
Weedy intercrop	174.2 ± 5.6	166.1 ± 15.3

[a]Intercrops were additive mixtures of the two component monocultures. Weeds (mostly *Brassica* spp.) were intentionally sown at high densities, and no weed control was practiced during the course of the experiments. Means and their standard errors are shown.

Source: Liebman, unpublished data.

barley, however intercrop yield advantages were substantial because of the added yield of pea.

Factors Affecting Intercrop/Weed Balance

Although intercropping appears to offer considerable potential as a means of increasing crop dominance over weeds, the effectiveness of weed control by intercrop systems differs not only among intercrop combinations but also among replicated trials of a single combination (Moody, 1980). In a review of the intercrop/weed literature, Moody and Shetty (1981) noted that the growth of weeds in an intercrop may be either severely depressed or hardly affected, relative to crop-free control treatments. Moreover, weed growth in intercrops may be lower than in all component monocultures, lower than in one of the monocultures, or equal in the intercrop and monocultures. Intercrop yield data are similarly variable. Much research remains to be done before the reasons for these different results become clear, but the following factors have been suggested to influence intercrop/weed relationships. Their effects are reviewed by Moody (1980) and Moody and Shetty (1981). Zimdahl (1980) and Walker and Buchanan (1982) provide relevant reviews of the role of these factors for weed control in monocultures.

Crop Density Crop density is one of the most easily manipulated factors affecting crop production, and it is well known that increased seeding rates can promote crop dominance over weeds in monoculture cropping systems (Godel, 1935; Staniforth and Weber, 1956). Highest yields for many intercrop combinations grown under weed-free conditions are obtained with increased crop population densities (Osiru and Willey, 1972; Willey and Osiru, 1972; Willey, 1979). Similarly, maximum intercrop yields and weed suppression are obtained under weedy conditions with total crop densities significantly higher than those used for monocultures (i.e., the superposition of normal or even higher density monocultures of the components to form "additive" crop mixtures) (Moody and Shetty, 1981). Data from Shetty and Rao (1981) illustrate this principle very clearly (Table 9.6). Highest combined crop yields and the greatest degree of weed suppression were obtained from a sorghum/pigeon pea mixture with a normal density of pigeon pea sown with a twice normal population of sorghum.

Arny et al. (1929) found that weed growth in a flax/wheat intercrop was inversely proportional to crop density. As the seeding rate for either or both components of the mixture increased, weed weights decreased. Weeds were particularly sensitive to the presence of wheat, and the authors concluded that intercropping the two species made it possible to grow flax on land that was too weedy for flax alone. Economic returns from the intercrop were generally as good or better than those from the highest earning sole crop.

"Smother" intercrops and "live mulch" intercrops are high-density, additive crop mixtures that appear to offer great promise as means of weed control. In these situations low-growing weed-suppressive species are sown between rows

Table 9.6 Influence of Crop Density of Sorghum/Pigeon Pea Intercrops on Crop Yields and Weed Growth[a]

Treatment[b]	Yield (kg/ha)		LER	Weed count per meter of sorghum row	Weed dry matter at harvest (g/m²)	
	Sorghum	Pigeon pea			Sorghum	Pigeon pea
N-sorghum	4043	—	—	22	30	—
N-pigeon pea	—	1704	—	—	—	142
0.5N-sorghum + 0.5N pigeon pea	2108	809	1.0	21	36	118
0.5N-sorghum + N-pigeon pea	2438	970	1.2	15	32	95
0.5N-sorghum + 2N-pigeon pea	2540	1002	1.2	17	25	43
N-sorghum + 0.5N pigeon pea	2895	804	1.2	21	23	52
N-sorghum + N-pigeon pea	2615	1062	1.2	17	15	51
N-sorghum + 2N-pigeon pea	2913	1375	1.5	12	18	46
2N-sorghum + 0.5N pigeon pea	2675	661	1.0	15	10	45
2N-sorghum + N-pigeon pea	3168	1295	1.6	16	10	26
2N-sorghum + 2N-pigeon pea	3118	1071	1.4	12	9	31
LSD (0.05)[c]	902	517	—	—	39	15

[a]One initial hand weeding was given to all treatments 3 weeks after planting
[b]N-sorghum = "normal" sorghum population of 180,000 plants per hectare.
N-pigeon pea = "normal" pigeon pea population of 40,000 plants per hectare.
0.5N = one-half the normal; 2N = twice the normal.
[c]LSD, Least Significant Difference
Source: Data from Shetty and Rao, 1981.

of main crop species. Akobundu (1980a) reported that melon (*Citrullus lanatus*) and sweet potato could replace three hand weedings when intersown into sole-cropped yam or yam intercropped with maize and cassava. The vining smother crops not only served as a labor-saving means of weed control, but also provided erosion control through increased soil coverage. Two legume species, *Centrosema pubescens* and *Psophocarpus palustris*, gave excellent control of weeds when intersown between maize rows (Akobundu, 1980b). Maize yields were significantly higher in the live mulch plots that received no fertilizer than in unfertilized conventionally tilled and no-till plots. When fertilizer was applied to the treatments, maize yields in the live mulch plots were equal to or better than the conventionally tilled and no-tillage plots. Akobundu (1980b) concluded that the intersown legume species contributed nitrogen to the maize and that this production system offers the opportunity for improving soil fertility, crop yield, and weed control on otherwise impoverished soils of the humid tropics.

The legume species, *Desmodium heterophyllum* and *Phaseolus vulgaris*, sown between rows of cassava, gave similar weed control to that achieved with continuous manual weeding (CIAT, 1980). The seed cost of the legumes was offset by the value of the bean crop and the long-lasting cover, erosion control, nitrogen fixation, and forage material from *D. heterophyllum*. Williams (1972) found that undersowing barley or fava bean with ryegrass (*Lolium* sp.) or red clover (*Trifolium pratense*) greatly decreased the growth and survivorship of weed seedlings. Although yields were not assessed, the barley crop appeared to be high yielding.

Shetty and Rao (1981) reported that interplanting cowpea and mung bean (*Vigna radiata*) into sorghum or pigeon pea minimized weed growth after one hand weeding. The weed suppression due to the smother crops was about the same as that obtained with two hand weedings. The smother crops had no significant effect on either sorghum or pigeon pea and provided additional grain yield themselves. However, the smother crops were ineffective for weed control in sorghum/pigeon pea intercrops and lowered yield of both intercrop components. Thus the results of the multicomponent mixture could not be predicted from knowledge of growth of the species in simpler combinations.

Robinson and Dunham (1954) found that soybean sown in narrow rows with winter wheat or winter rye as companion crops yielded as much or more than soybean without companion crops whether in noncultivated narrow rows or in cultivated wide rows. Weed control with companion crops was about equal to that achieved by cultivation. Intersowing wheat or rye into soybean was a relatively inexpensive method of weed control that could reduce soil erosion and organic matter losses associated with normal monoculture production methods.

It should be noted that intersown live mulch crops can greatly reduce the yields of main crop species if competition for water and/or nutrients is strong. Kurtz et al. (1952) reported that when maize was sown between strips of previously established legume and grass sods, maize yield was severely depressed. Decreases in maize yield from intercrop plots could be partially reversed with either added nitrogen or water and more fully reversed with both nitrogen and water, although maize yields from sole crop plots usually exceeded those from fertilized and irrigated intercrop plots by about 15 percent. The ability of live mulch crops to compete with main crop species may be limited by the use of growth retardants (Akobundu, 1980b) and low doses of herbicides (Vrabel et al., 1980b), use of less aggressive species and cultivars (Vrabel et al., 1980a; Nicholson, 1983), simultaneous planting dates for main crop and live mulch species (Robinson and Dunham, 1954; Vrabel et al., 1980a), and mowing the mulch species (Vrabel et al., 1980b).

Spatial Arrangement Crop and weed plants are sessile, and the capture of growth resources is a very local phenomenon. Because of this, the intensity of interference between neighboring plants tends to increase as the distance between them decreases (Harper, 1977). More equidistant spacing for a single crop species grown under weedy conditions should thus decrease the strength of interference between crop plants and increase interference between the crop and associated weeds. Walker and Buchanan (1982) note that at equal seeding rates sole crops are generally more weed-suppressive when planted in narrower (i.e., more equidistant) rather than wider (i.e., less equidistant) rows.

The effects of intercrop spatial arrangement on associated weeds have not yet received much attention from researchers. Because interference between crop components can have indirect effects on crop/weed relationships (Yih, 1982), the effects of intercrop spatial arrangement are more complex than for sole crops. Bantilan et al. (1974) found that with equal maize populations, mung bean reduced weed growth more when it was intercropped between wide maize rows than when it was sown between narrow maize rows. The researchers noted a similar but less pronounced effect for intercropped peanut.

Relative Proportions of Component Crops Shetty and Rao (1981) reported that sole-cropped pearl millet (*Pennisetum americanum*) suppressed the growth of weeds much more strongly than sole-cropped groundnut, and that weed growth in mixtures of the two crops closely reflected varying proportions of the components. Weed yields increased as the proportion of millet in the mixture decreased. A 1:3 pearl millet/groundnut row arrangement gave the highest combined crop yield (LER = 1.15), although it gave a small amount of weed suppression relative to that obtained from mixtures containing a higher proportion of millet.

Crop Species and Cultivar In intercrop/weed experiments in which crop components are varied, large differences in weed-suppression ability have been noted among species (Robinson and Dunham, 1954; Williams, 1972; Bantilan et al., 1974; Shetty and Rao, 1981). These reflect differences in the timing and nature of resource capture and are manifest in differences in growth form between species. For example, in maize-based intercrop systems, Bantilan et al. (1974) found that mung bean was more weed-suppressive than peanut and ascribed this to its more rapid early growth and more uniform canopy structure. Experiments with sole crops have shown that within species large differences in weed-suppression ability exist among genotypes (Kawano et al., 1974; McWhorter and Hartwig, 1972; Yip et al., 1974; Stilwell and Sweet, 1974). Few researchers have evaluated the effects of crop cultivars on weed suppression in intercrop systems. Bantilan et al. (1974) reported that when two cultivars of mung bean were intercropped with maize the more prostrate cultivar was more weed-suppressive. However, the cultivars were not compared in the same experiment. Rao and Shetty (1976) observed differences in weed suppression between two pigeon pea cultivars when they were sole cropped, but no such difference was apparent when they were intercropped with sorghum. Maize cultivars did show differences in weed suppression when intercropped with sweet potato (Moody and Shetty 1981).

Fertility Differential response to applied fertilizer, especially nitrogen, can greatly alter the nature of interference between associated plant species (Blaser and Brady, 1950; Stern and Donald, 1962; de Wit et al., 1966; Appleby et al., 1976). The possibility of interactions between cropping systems and fertility levels must be considered carefully. Bantilan et al. (1974) reported that as the amount of applied nitrogen increased, maize/peanut (*Arachis hypogaea*) and maize/sweet potato intercrops were more weed-suppressive, whereas a maize/mung bean intercrop became less weed-suppressive. Crop yields (measured by LER) were thus determined by interactions between cropping systems, soil fertility level, and method of weed control.

Other Factors Other factors, such as soil moisture, weed density, weed community composition, herbivores, and pathogens undoubtedly have important influences on intercrop/weed relationships, but clarificiation of their roles awaits further investigation.

Shifts in Weed Species Composition in Intercrops

It has long been recognized that different species of weeds are commonly associated with different crops (Plucknett et al., 1977; Muenscher, 1980). This is the result not only of differences in weed control techniques normally associated with specific crops (e.g., tillage, herbicides), but also of differences in the nature of crop/weed interference (Buchanan et al., 1975). These species-specific types of interactions result in shifts in weed species dominance between crops (IRRI, 1975), and for this reason, crop rotation can be used as an effective means of preventing population increases of any single weed species (Walker and Buchanan, 1982).

In contrast to rotational sequences of crop monocultures which combine the weed-suppressive effects of different crops over an extended period of time, intercropping combines the weed-suppressive effects of different crops within a single season. It is therefore of interest to compare the composition of the weed flora associated with intercrops to the weed flora of each component monoculture.

Shetty and Rao (1981) found that the species composition of weeds associated with groundnut/pearl millet intercrops was greatly influenced by varying proportions of the intercrop components. In sole-cropped pearl millet, the weed flora was a mixture of many species, whereas in sole-cropped groundnut the predominant weeds (80 percent of total weed biomass) were species of only *Celosia, Digitaria,* and *Cyperus*. As more rows of groundnut were introduced in place of pearl millet rows, there was a striking increase in both numbers and biomass of the tall and competitive *Celosia*. The buildup of *Celosia* was found only in groundnut rows, for it occurred in negligible numbers in and around pearl millet rows. These results strongly suggest that with regard to weed community composition, crop/weed interactions are extremely localized and that shifts in weed floristics due to intercropping might be predicted from knowledge of weed community composition in sole crops of the components.

Mohler and Liebman (unpublished data) grew weedy barley and pea monocultures and intercrops and found that the influence of cropping treatment on the relative composition of the weeds paralleled effects on total weed productivity. As weed yields decreased due to increased interference from crops, the relative importance of the dominant weed species (*Amaranthus retroflexus* or *Brassica kaber*) decreased. These results are similar to those of Shetty and Rao (1981) who noted more mixed assemblages of weeds in more competitive crop systems. The extent that intercropping can increase interference from crops to weeds may thus be closely related to the usefulness of intercropping in preventing shifts in weed community composition toward dominance by a few, highly competitive species.

Mechanisms of Weed Suppression in Intercrops

Much of the recent research devoted to weed-free intercrop systems has stressed ecophysiological mechanisms by which overyielding takes place (Trenbath, 1975, 1981; Willey and Roberts, 1976; Wahua and Miller, 1978a, 1978b; Natarajan and Willey, 1981; Reddy and Willey, 1981; Martin and Snaydon, 1982). This research has placed crop growth and development in the context of resource capture and conversion and has shown very clearly how niche differences between crop species can lead to increased biological efficiency and yield advantages over sole cropping (Willey, 1979; Vandermeer, 1981). This situation is in marked contrast to that of intercrop/weed relationships, where hypotheses of niche preemption, competitive exclusion, and allelopathic interference cannot be adequately evaluated for lack of sufficient ecophysiological data. Collection of such data should prove exceptionally fruitful for the design and improvement of weed-suppressive intercrop systems.

Bantilan et al. (1974) provided evidence that light interception by maize/mung bean, maize/peanut, and maize/sweet potato intercrops was greater than that by the component monocultures. This effect was evident within 30 days after sowing the crops. The authors concluded that the intercrops were better at suppressing weeds because of increased preemptive use of light effected by earlier canopy closure.

The importance of belowground resource use in influencing intercrop/weed interactions should not be ignored. Mohler and Liebman (unpublished data) measured predawn water potential of the weed *Amaranthus retroflexus* growing in barley and pea monocultures, a replacement series barley/pea intercrop, and unplanted control plots. Large differences were detected among treatments (Table 9.7). In particular, water deficits for *Amaranthus* were greater in plots containing barley (i.e., the barley monoculture and the intercrop) than in plots without barley (i.e., the pea monoculture and the unplanted control). Growth of *Amaranthus* in association with pea was much depressed relative to the crop-free control treatment (Table 9.7), indicating that factors other than late-season soil moisture were also important in influencing crop/weed balance.

Table 9.7 Predawn Water Potentials and Final Aboveground Biomass of the Weed *Amaranthus retroflexus* Grown in Crop-free Control Plots, Barley and Pea Monocultures, and a Replacement Series Intercrop

Treatment	Predawn water potential (bars)[a]	Biomass at harvest (g/m^2)
Crop-free control	−24.8 ± 2.7	516.7 ± 39.5
Pea monoculture	−23.2 ± 3.5	249.1 ± 53.3
Pea/barley intercrop	−33.9 ± 3.5	112.6 ± 44.6
Barley monoculture	−37.3 ± 3.5	73.7 ± 53.3

[a]Water potential (measured 58 days after planting) data reflect effects of long drought period. Results given as means and their standard errors.

Source: Mohler and Liebman, unpublished data.

The role of allelochemical interference between intercrop components and weeds has scarcely been explored, although this type of weed control has been shown to be potentially useful in monoculture cropping systems (Putnam and Duke, 1974; Fay and Duke, 1977; Lockerman and Putnam, 1979). The question of allelochemical control of weeds in intercrop systems is complicated by the possibility of interference between crop species. For allelopathy to be effective for weed control in intercrops there must be selectivity in the effects of toxins released by the crops: weed species must be more susceptible than crop components. Recent work by Gliessman (1983) shows that this selectivity may indeed be possible.

Gliessman (1983) evaluated the effect of squash leaf extract on radicle elongation of maize, cowpea, and cabbage (*Brassica oleracea capitata*). The extract was found to have a stronger inhibitory effect on cabbage than on the other two species. Gliessman suggested that the interplanting of squash into maize/cowpea intercrops by farmers in southeastern Mexico is an effective means of weed control (Gliessman et al., 1981; Letourneau, 1983) not only because of the shade cast by the squash leaves, but also because of selective allelochemical inhibition. Gliessman (1983) also found that extracts of certain "nonweed" species intentionally maintained by local farmers in maize/cowpea fields (Chacon and Gliessman, 1982) had only weak inhibitory effects against maize and cowpea, but strong inhibitory effects against cabbage. He proposed that noncrop species offer possibilities for selective allelochemical weed control.

Recommendations for Further Reseach

It is clear from the preceding discussion that manipulations of the biological characteristics of intercrop systems can strongly affect crop/weed relationships. However, understanding of the biology of intercrops and associated weeds has not yet developed to the point where meaningful predictions of crop and weed responses to altered environmental conditions can be made.

Knowledge of patterns of resource use by intercrops and weeds would clearly be helpful in generating general theories of intercrop/weed interactions. Such ecophysiological studies should provide invaluable information about the altered environments that intercropping presents to weeds as well as the responses of intercrops and weeds to resource deficits resulting from biotic interactions. The role of allelopathy in intercrop/weed dynamics may be important and warrants investigation. In both resource preemption and allelochemical approaches to intercrop weed control emphasis should be placed on understanding and exploiting the differential response of crop and weed species to altered environmental conditions.

Current knowledge of genotype resource environment, and cropping system interactions is very poorly developed. Morphological and physiological characteristics of species and cultivars exhibiting different yield and interference abilities should be investigated in intercrop/weed communities. Better understanding of the ecophysiological factors responsible for shifts in weed species composition might allow manipulation of intercrop systems to prevent population increases of particularly noxious weeds.

CONCLUSIONS

A considerable amount of work has emerged in the last decade, showing that diversification of crop habitats frequently results in reduced pest incidence. The studies have disproportionately focused on insect dynamics, with little attention given to the effects of multiple cropping systems on disease epidemiology and especially weed ecology. Research projects that integrate the simultaneous effects of polycultures on all biotic components of the agroecosystems are sorely lacking. There is every reason to expect an increase of multitrophic level interactions as the crop systems become richer in plant, insect, and microorganism species diversity. Unraveling these complex relationships can lead to pest management systems that integrate cropping practices, weed control measures, and soil management to provide effective and harmonious means of disease, weed, and insect control.

It is clear that these complex systems affect insect populations by either interference with herbivore movement and colonization or by increased herbivore mortality caused by natural enemies. Whatever the underlying mechanisms accounting for pest reduction, data are of some predictive value. However, generalizations and recommendations are difficult for yet untried systems. It is here where studies of traditional polycultures may be of value in guiding the design of pest resistant cropping systems. Evidence suggests that in many areas, peasants have kept pest damage within acceptable bounds by employing a wide variety of traditional management practices centered around the use of polycultures. Some mixtures, like the maize/bean/squash of Central America and Mexico or the genetically rich potato fields of the Andes, have persisted for centuries, exhibiting an array of stabilizing properties (Wilken, 1977).

It has been suggested that multiple cropping potentials are restricted to less developed countries where low-input agriculture is practiced, because they are capital-restricted, labor demanding, and management-intensive production systems. It is also implied that these systems cannot be efficiently mechanized, limiting their adoption in developed countries. One of the main reasons why the cotton/alfalfa strip cropping system, which efficiently reduced *Lygus* bugs in California, was not adopted was because of added costs in the alfalfa cutting operations and different water needs of both crops, thus upsetting irrigation schedules. Some agronomists argue that mixed agriculture cannot be implemented within the actual structure of U.S. agriculture (large farms with capital-intensive operations). In an era of increasing costs of chemical-based agriculture and accelerating concern about the contamination of an environment, these multiple species systems provide an alternative on farms of all sizes. Further research is needed to explore the applications of multiple cropping systems as one component of a management-intensive approach to insect, pathogen, and weed control.

ACKNOWLEDGMENTS

We thank Linda Schmidt for her helpful editorial comments, and the Jessie Smith-Noyes Foundation, New York, for supporting much of our multiple cropping systems research.

REFERENCES

Akobundu, I. O. 1980a. Weed control strategies for multiple cropping systems of the humid and subhumid tropics, in: *Weeds and Their Control in the Humid and Subhumid Tropics,* (I. O. Akobundu, ed.), International Institute for Tropical Agriculture (IITA), Ibadan, Nigeria, pp. 80–100.

———. 1980b. Live mulch: A new approach to weed control and crop production in the tropics, *Proc. 15th British Weed Control Conf.*, pp. 377–382.

Altieri, M. A. 1984. Patterns of insect diversity in monocultures and polycultures of brussels sprouts, *Prot. Ecol.* 6:227–232.

Altieri, M. A., Francis, C. A., van Schoonhoven, A., and Doll, J. D. 1978. A review of insect prevalence in maize (*Zea mays* L.) and bean (*Phaseolus vulgaris* L.) polycultural systems, *Field Crops Res.* 1:33–50.

Altieri, M. A., and Gliessman, S. R. 1983. Effects of plant diversity on the density and herbivory of the flea beetle, *Phyllotreta cruciferae* Goeze, in California collard (*Brassica oleracea*) cropping systems, *Crop Prot.* 2:497–501.

Altieri, M. A., and Letourneau, D. K. 1982. Vegetation management and biological control in agroecosystems, *Crop Prot.* 1:405–430.

———. 1984. Vegetation diversity and insect pest outbreaks, *CRC Crit. Rev. Plant Sci.* 2:131–169.

Altieri, M. A., Lewis, W. J., Nordlund, D. A., Gueldner, R. C., and Todd, J. W. 1981. Chemical interactions between plants and *Trichogramma* wasps in Georgia soybean fields, *Prot. Ecol.* 3:259–263.

Altieri, M. A., van Schoonhoven, A., and Doll, J. 1977. The ecological role of weeds in insect pest management systems: a review illustrated by bean (*Phaseolus vulgaris*) cropping systems, *PANS* 23:195–205.

Altieri, M. A. and Whitcomb, W. H. 1974. The potential use of weeds in the manipulation of beneficial insects, *Hort. Sci.* 14:12–18.

Anderson, D. T. 1981. Seeding and interculture mechanization requirements related to intercropping in India, in *Proc. Int. Workshop on Intercropping,* International Crops Research Institute for the Semi-Arid Tropics (ICRISAT), Patancheru, India, pp. 328–336.

Andow, D. A. 1983. Plant diversity and insect populations: Interactions among beans, weeds and insects, Ph.D. dissertation, Cornell University, Ithaca, New York, 201 pp.

Andrews, D. J., and Kassam, A. H. 1976. The importance of multiple cropping in increasing world food supplies, in: *Multiple Cropping,* (R. I. Papendick, P. A. Sanchez, and G. B. Triplett, eds.), Amer. Soc. Agron. Spec. Publ. 27, Madison, Wisconsin, pp. 1–10.

Appleby, A. P., Olsen, P. O., and Colbert, D. R. 1976. Winter wheat yield reduction from interference by Italian ryegrass, *Agron. J.* 68:463–466.

Arny, A. C., Stoa, T. E., McKee, C., and Dillman, A. C. 1929. *Flax cropping mixture with wheat, oats, and barley* U.S. Dep. Agric. Tech. Bull. 133, Washington, D. C.

Atwal, A. S., and Mangar, A. 1969. Repellent action of root exudates of *Sesamum orientale* against the root-knot nematode *Meloidogyne incognita* (Heteroderidae: Nematoda), *Indian J. Entomol.* 31:286–289.

Bach, C. E., 1980. Effects of plant density and diversity in the population dynamics of a specialist herbivore, the striped cucumber beetle, *Acalymma vittata* (Fab.), *Ecology* 61:1515–1530.

Bantilan, R. T., Palada, M. C., and Harwood, R. K. 1974. Integrated weed management: I. Key factors affecting crop-weed balance, *Phil. Weed Sci. Bull.* 1(2):14–36.

Bhatnagar, V. S., and Davies, J. C. 1981. Pest management in intercrop subsistance farming, in: *Proc. Int. Workshop on Intercropping, ICRISAT.,* Patancheru, India, pp. 249–257.

Blaser, R. E., and Brady, N. C. 1950. Nutrient competition in plant associations, *Agron. J.* 42:128–135.

Brown, B. J., and Marten, C. G. 1984. The ecology of traditional pest management in Southeast Asia, Working Paper, East-West Center, Honolulu, Hawaii.

Browning, J. A. 1975. Relevance of knowledge about natural ecosystems to development of pest management programs for agroecosystems, *Proc. Amer. Phytopathol. Soc.* 1:191–194.

Buchanan, G. A., Hoveland, C. S., Brown, V. L., and Wade, R. H. 1975. Weed population shifts influenced by crop rotations and weed control programs, *Proc. South. Weed Sci. Soc.* 28:60–71.

Bye, R. A., Jr. 1981. Quelites—ethnoecology of edible greens—past, present, and future, *J. Ethnobiol.* 1(1):109–123.

Chacon, J. C., and Gliessman, S. R. 1982. use of the "non-weed" concept in traditional tropical agroecosystems of south-eastern Mexico, *Agro-ecosystems* 8:1–11.

CIAT (Centro Internacional de Agricultura Tropical). 1980. *Cassava Program 1979 Annual Report,* CIAT, Cali, Colombia.

Compton, J. A. F. 1982. *Small Farm Weed Control,* Intermediate Technical Development Group, Ltd. and International Plant Protection Center, London, U.K.

Cromartie, W. J. 1981. The environmental control of insects using crop diversity, in: *CRC Handbook of Pest Management in Agriculture,* vol. 1, (D. Pimentel, ed.), CRC Press, Boca Raton, Florida, pp. 223–251.

de Wit, C. T., Tow, P. G., and Ennik, G. L. 1966. Competition between legumes and grasses. *Vers. Landbouwkd. Onderz.* 687:1–30.

Egunjobi, O. A. 1984. Effects of intercropping maize with grain legumes and fertilizer treatment on populations of *Pratylenchus brachyurus* (Nematoda) and on the yield of maize (*Zea mays*), *Prot. Ecol.* 6:153–167.

Fay, P. K., and Duke, W. B. 1977. An assessment of allelopathic potential in *Avena* germ plasm, *Weed Sci.* 25:224–228.

Flaherty, D. 1969. Ecosystem trophic complexity and the willamete mite, *Eotetranychus willametei* (Acarina: Tetranychidae) densities, *Ecology* 50:911–916.

Francis, C. A., Flor, C. A., and Temple, S. R. 1976. Adapting varieties for intercropped systems in the tropics, in: *Multiple Cropping,* (R. I. Papendick, P. A. Sanchez, and G. B. Triplett, eds.), Amer. Soc. Agron. Spec. Publ. 27, Madison, Wisconsin, pp. 235–254.

Gliessman, S. R. 1983. Allelopathic interactions in crop/weed mixtures: Applications for weed management, *J. Chem. Ecol.* 9(8):991–999.

Gliessman, S. R., Garcia E. R., and Amador A. M. 1981. The ecological basis for application of traditional agricultural technology in the management of tropical agroecosystems, *Agro-ecosystems* 7:173–185.

Godel, G. L. 1935. Relation between rate of seeding and yield of cereal crops in competition with weeds, *Sci. Agric.* 16:165–168.

Haizel, K. A., and Harper, J. L. 1973. The effects of density and the timing of removal on interference between barley, white mustard, and wild oats, *J. Appl. Ecol.* 10(1):23–31.

Harper, J. L. 1977. *Population Biology of Plants,* Academic Press, New York.

Harwood, R. R., and Price, E. C. 1976. Multiple cropping in tropical Asia, in: *Multiple Cropping,* (R. I. Papendick, P. A. Sanchez, and G. B. Triplett, eds.), Amer. Soc. Agron. Spec. Publ. 27, Madison, Wisconsin, pp. 11–40.

Hasse, V., and Litsinger, J. A. 1981. The influence of vegetational diversity on host finding and larval survivorship of the Asian corn borer, *Ostrinia furnacalis,* IRRI Saturday Seminar, Entomology Dep., International Rice Research Institute (IRRI), Philippines.

Hijink, M. J., and Suatmadji, W. R. 1967. Influence of different compositae on population density of *Pratylenchus penetrans* and some other root-infesting nematodes, *Neth. J. Plant Pathol.* 73:71–82.

Holm, L. R. 1971. The role of weeds in human affairs, *Weed Sci.* 19:485–490.

IRRI (International Rice Research Institute). 1975. Cropping systems program report for 1975. IRRI, Los Baños, Philippines.

Kapoor, P., and Ramakrishnan, P. S. 1975. Studies on crop-legume behavior in pure and mixed stands, *Agro-ecosystems* 2:61–74.

Kareiva, P. 1983. Influence of vegetation structure on herbivore population: Resource concentration and herbivore movement, in: *Variable Plants and Herbivores in Natural and Managed Systems,* (R. Denno, ed.), Academic Press, New York.

Kawano, K., Gonzalez, H., and Lucena, M. 1974. Intraspecific competition, competition with weeds, and spacing response in rice, *Crop. Sci.* 14:841–845.

Larios, J. F., and Moreno, R. A. 1977. Epidemiologia de algunas enfermedades foliares de la yuca en diferentes sistemas de cultivo. II. Roya y muerte descendente, *Turrialba* 27:151–156.

Latheef, M. A., and Irwin, R. D. 1980. Effects of companionate planting on snap bean insects, *Epilachna varivestis* and *Heliothis zea, Environ. Entomol.* 9:195–198.

Lebaron, H. M., and Gressel, J. 1982. *Herbicide Resistance in Plants,* John Wiley & Sons, New York.

Letourneau, D. K. 1983. The effects of vegetational diversity on herbivorous insects and associated natural enemies: examples from tropical and temperate agroecosystems, Ph.D. dissertation, University of California, Berkeley, California.

Letourneau, D. K., and Altieri, M. A. 1983. Abundance patterns of a predatar, *Orius tristicolor* (Hemiptera: Anthocoridae), and its prey, *Frankliniella occidentalis* (Thysanoptera: Thripidae): Habitat attraction in polycultures versus monocultures, *Environ. Entomol.* 12:1464–1469.

Litsinger, J. A. and Moody, K. 1976. Integrated pest management in multiple cropping systems, in: *Multiple Cropping,* (R. I. Papendick, P. A. Sanchez, and G. B. Triplett, eds.), Amer. Soc. Agron. Spec. Publ. 27, Madison, WI, pp. 293–316.

Lockerman, R. H., and Putnam, A. R. 1979. Field evaluation of allelopathic cucumbers as an aid to weed control, *Weed Sci.* 27:54–57.

Matteson, P. C., Altieri, M. A., and Gagne, W. C. 1984. Modification of small farmer practices for better management, *Annu. Rev. Entomol.* 29:383–402.

McWhorter, C. G., and Hartwig, E. E. 1972. Competition of johnsongrass and cocklebur with six soybean varieties, *Weed Sci.* 20:56–59.

Mishra, M. N. 1969. Economic utilization of weeds in India, *Labdev. J. Sci. Tech.* 7(3):195–199.

Moody, K. 1977. Weed control in multiple cropping, in: *Int. Rice Research Inst. Proc. Symposium Cropping Systems Research and Development for the Asian Rice Farmer,* 21–24 September 1976. IRRI, Los Baños, Philippines, pp. 281–294.

———. 1980. Weed control in intercropping in tropical Asia, in: *Weeds and their Control in the Humid and Subhumid Tropics,* (I. O. Akobundu, ed.), IITA, Ibadan, Nigeria, pp. 101–108.

Moody, K., and Shetty, S. V. R. 1981. Weed management in intercropping systems, in: *Proc. Int. Workshop on Intercropping,* ICRISAT, Patancheru, India, pp. 229–237.

Moreno, R. A. 1977. Efecto de diferentes sistemas de cultivo sobre la severidad de la mancha angular del frijol (*Phaseolus vulgaris*) causada por *Isariopsis griseola, Agron. Cost.* 1:39–42.

Muenscher, W. C. 1980. *Weeds (2d ed).* Comstock Publ. Assoc., Cornell University Press, London, U.K.

Murdoch, W. W. 1975. Diversity, complexity, stability and pest control, *J. Appl. Ecol.* 12:795–807.

Muzik, T. J. 1970. *Weed Biology and Control,* McGraw-Hill, New York.

Natarajan, M., and Willey, R. W. 1981. Growth studies in sorghum/pigeon pea intercropping with particular emphasis on canopy development and light interception, in: *Proc. Int. Workshop on Intercropping,* ICRISAT, Patancheru, India, pp. 180–187.

Nicholson, A. G. 1983. Screening turfgrasses and legumes for use as living mulches in vegetable production. M. S. thesis, Cornell University, Ithaca, New York.

Nordlund, D. A., Chalfont, R. B., and Lewis, W. J. 1984. Arthropod populations, yield and damage in monocultures and polycultures of corn, leaves and tomatoes, *Agriculture, Ecosystems, and Environment,* 11(4):353–367.

Okigbo, B. N., and Greenland, D. J. 1976. Intercropping systems in tropical Africa, in: *Multiple Cropping* (R. I. Papendick, P. A. Sanchez, and G. B. Triplett, eds.), Amer. Soc. Agron. Spec. Publ. 27, Madison, Wisconsin, pp. 63–101.

Oostenbrink, M., Kuiper, K., and S'Jacob, J. J. 1957. *Tagetes als feindpflanzen von Pratylenchus* Arten, *Nematologica* 2 (Suppl.):424–433.

Osiru, D. S. O., and Willey, R. W. 1972. Studies on mixtures of dwarf sorghum and beans (*Phaseolus vulgaris*) with particular reference to plant population, *J. Agric. Sci. Camb.* 79:531–540.

Palti, J. 1981. *Cultural Practices and Infectious Crop Diseases,* Springer Verlag, Berlin.

Perrin, R. M. 1977. Pest management in multiple cropping systems, *Agro-ecosystems* 3:83–118.

———. 1980. The role of environmental diversity in crop protection, *Prot. Ecol.* 2:77–114.

Perrin, R. M., and Phillips, M. L. 1978. Some effects of mixed cropping on the population dynamics of insect pests, *Ent. Exp. Appl.* 24:385–393.

Pimentel, D. 1961. Species diversity and insect population outbreaks, *Ann. Entomol. Soc. Amer.* 54:76–86.

Pimentel, D., Andow, D., Dyson-Hudson, R., Gallahan, D., Jacobson, S., Irish, M., Kroop, S., Moss, A., Schreiner, I., Shepard, M., Thompson, T., and Vinzant, B. 1980. Environmental and social costs of pesticides: A preliminary assessment, *Oikos* 34:126–140.

Pinchinat, A. M., Soria, J., and Bazan, R. 1976. Multiple cropping in tropical America, in: *Multiple Cropping,* (R. I. Papendick, P. A. Sanchez, and G. B. Triplett, eds.), Amer. Soc. Agron. Spec. Publ. 27, Madison, Wisconsin, pp. 51–61.

Plucknett, D. L., Rice, E. J., Burrill, L. C., and Fisher, H. H. 1977. Approaches to weed control in cropping systems, in: *Proc. Symp. Cropping Systems Research and Development for the Asian Rice Farmer,* 21–24 September, 1976, IRRI, Los Baños, Philippines, pp. 295–308.

Price, P. W., Bouton, C. E., Gross, P., McPherson, B. A., Thompson, Z. N. and Weide, A. E. 1980. Interactions among three trophic levels: Influence of plants on interactions between insect herbivores and natural enemies, *Annu. Rev. Ecol. Syst.* 11:41–65.

Putnam, A. R., and Duke, W. B. 1974. Biological suppression of weeds: Evidence for allelopathy in accessions of cucumbers, *Science* 185:370–372.

Rao, M. R., and Shetty, S. V. R. 1976. Some biological aspects of intercropping systems on crop-weed balance, *Indian J. Weed Sci.* 8(1):32–43.

Reddy, M. S. and Willey, R. W. 1981. Growth and resource use studies in an intercrop of pearl millet/groundnut, *Field Crops Res.* 4:13–24.

Revkin, A. C. 1983. Paraquat: A potent weed/killer is killing people, *Sci. Dig.* June 1983, p. 36.

Risch, S. J. 1981. Insect herbivore abundance in tropical monocultures and polycultures: An experimental test of two hypotheses, *Ecology* 62:1325–1340.

Risch, S. J., Adow, D., and Altieri, M. A. 1983. Agroecosystem diversity and pest control: Data, tentative conclusions and new research directions, *Environ. Entomol.* 12:625–629.

Robinson, R. G., and Dunham, R. S. 1954. Companion crops for weed control in soybeans, *Agron. J.* 46:278–281.

Root, R. B. 1973. Organization of a plant-arthropod association in simple and diverse habitats: The fauna of collards (*Brassica oleracea*), *Ecol. Monogr.* 43:95–124.

Sanchez, P. A. 1976. *Properties and Management of Soils in the Tropics,* John Wiley & Sons, New York.

Shetty, S. V. R., and Rao, M. R. 1981. Weed-management studies in sorghum/pigeonpea and pearl millet/groundnut intercrop systems—some observations, in: *Proc. Int. Workshop on Intercropping, ICRISAT,* Patancheru, India, pp. 238–248.

Smith, R. F., and Reynolds, H. T. 1972. Effects of manipulation on cotton agroecosystems in insect pest populations, in: *The Careless Technology* (T. Favar, ed.), Natural History Press, New York.

Staniforth, D. W., and Weber, C. R., 1956. Effects of annual weeds on the growth and yield of soybeans, *Agron. J.* 48:467–471.

Stern, W. R., and Donald, C. M. 1962. Light relationships in grass-clover swards, *Aust. J. Agric. Res.* 13:599–614.

Stilwell, E. K., and Sweet, R. D. 1974. Competition of squash cultivars with weeds, *Proc. Northeast Weed Control Conf.* 28:229–233.

Sumner, D. R., Doupnik, B., and Boosalis, M. G. 1981. Effects of reduced tillage and multiple cropping on plant disease, *Annu. Rev. Phytopathol.* 19:167–187.

Tahvanainen, J. C., and Root, R. B. 1972. The influence of vegetational diversity on the population ecology of a specialized herbivore *Phyllotreta cruciferae* (Coleoptera: Chrysomelidae), *Oecologia* 10:321–346.

Thresh, J. M. 1982. Cropping practices and virus spread, *Annu. Rev. Phytopathol.* 20:193–218.

Trenbath, B. R. 1975. Neighbor effects in the genus *Avena*. III. A diallel approach, *J. Appl. Ecol.* 12:189–200.

———. 1981. Light-use efficiency of crops and potential for improvement through intercropping, in: *Proc. Int. Workshop on Intercropping,* ICRISAT, Patancheru, India, pp. 141–153.

Van de Goor, G. A. W. 1954. The value of some leguminous plants as green manures in comparison with *Crotalaria juncea*. *J. Neth. Agric. Sci.* 2:37–43.

Vandermeer, J. 1981. The interference production principle: an ecological theory for agriculture, *BioScience* 31:361–364.

van Emden, H. F. 1965. The role of uncultivated land in the biology of crop pests and beneficial insects, *Sci. Hort.* 17:121–136.

van Emden, H. F., and Williams, G. F. 1974. Insect stability and diversity in agro-ecosystems, *Annu. Rev. Entomol.* 19:455–475.

Visser, T., and Vythilingam, M. K. 1959. The effect of marigolds and some other crops on the *Pratylenchus* and *Meloidogyne* populations in tea soil, *Tea Quart.* 30:30–38.

Vrabel, T. E., Minotti, P. L., and Sweet, R. D. 1980a. Seeded legumes as living mulches in sweet corn, paper 769, Dep. Veg. Crops, Cornell University, Ithaca, New York.

———. 1980b. Legume sods as living mulches in sweet corn, paper 776, Dep. Veg. Crops, Cornell University, Ithaca, New York.

Wahua, T. A. T. 1978. Leaf water potentials and light transmission in intercropped sorghum and soyabeans, *Exp. Agric.* 14:373–380.

Walker, R. H., and Buchanan, G. A. 1982. Crop manipulation in integrated weed management systems, *Weed Sci. 30 (Suppl.):*17–24.

Weil, R. R. 1982. Maize-weed competition and soil erosion in unweeded maize, *Trop. Agric.* 59(3):207–213.

Wetzler, R. E., and Risch, S. J. 1984. Experimental studies of beetle diffusion in simple and complex crop habitats, *J. Anim. Ecol.* 53:1–19.

Wilken, G. C. 1977. Integrating forest and small-scale farm systems in middle America, *Agro-ecosystems* 3:291–302.

Willey, R. W. 1979. Intercropping—its importance and research needs, Parts I and II. *Field Crops Abstr.* 32:1–10, 73–83.

Willey, R. W., and Osiru, D. S. O. 1972. Studies of mixtures of maize and beans (*Phaseolus vulgaris*) with particular reference to plant population, *J. Agr. Sci. Camb.* 79:517–529.

Willey, R. W., and Roberts, E. H. 1976. Mixed cropping, in: *Solar Energy in Agriculture, Proc. Joint Inter. Solar Energy Soc. Conf.*, University of Reading, U.K., pp. 44–47.

Williams, E. D. 1972. Growth of *Agropyron repens* seedlings in cereals and field beans, in: *Proc. 11th British Weed Cont. Conf.*, pp. 32–37.

Wrubel, R. P. 1984. The effect of intercropping on the population dynamics of the arthropod community associated with soybean (*Glycine max*), M. S. thesis, University of Virginia, Charlottesville, Virginia, 77 pp.

Yih, W. K. 1982. Weeds, intercropping, and mulch in the temperate zones and the tropics—some ecological implications for low-technology agriculture, Ph.D. dissertation, University of Michigan, Ann Arbor, Michigan.

Yip, C. P., Sweet, R. D., and Sieczka, J. B. 1974. Competitive ability of potato cultivars with major weed species, *Proc. Northwest Weed Control Conf.* 28:271–281.

Zimdahl, R. L. 1980. *Weed-Crop Competition—A Review,* International Plant Protection Center, Oregon State University, Corvallis, Oregon.

Zitter, T. A., and Simms, J. N. 1980. Management of viruses by attraction of vector efficiency and by cultural practices, *Annu. Rev. Phytopathol.* 18:289–310.

Chapter 10

Breeding for Multiple Cropping Systems

Margaret E. Smith
Charles A. Francis

Combining experience with the established techniques of plant breeding, agronomists have developed procedures that can determine whether breeding is a useful approach for addressing the production constraints of a monoculture cropping system. However, how can one evaluate the potential of a plant breeding approach for improving multiple cropping systems? In complex cropping patterns, the number of possible nongenetic factors that can be manipulated to increase production is much larger than in monoculture. These manipulations include changes in spatial arrangements, relative planting dates, relative crop densities, and timing and placement of fertilizer, to name a few. Still more complex is the question of whether several uniquely focused breeding programs should be carried out if a crop is grown in several different cropping systems. This approach multiplies the cost in personnel, time, space, and operating budget that the total program would require. As these questions point out, the decision to start a breeding program to select varieties for a particular multiple cropping system is not easily made.

For monoculture systems, it is fairly simple to define the objectives for varietal selection, as they generally relate directly to biological or climatic production constraints. With a multiple cropping system, however, the identifiable constraints to production are often complex and may include a limited resource environment, competition for resources, or multiple objectives of the farmer. The genetic traits that need to be altered to alleviate these types of constraints are not easily identifiable. The potential effect of changes brought about by varietal selection in one crop on the other crop (or crops) in the system must also be considered. An improvement in one crop may be compensated by new constraints it imposes on another crop, thus negating any progress.

Finally, once breeding objectives are defined for a multiple cropping system, an appropriate breeding methodology must be chosen. Here again, the choice is not as simple as for monoculture, since methodologies for multiple cropping systems have not been tested extensively. The appropriateness of monoculture selection and the point in the breeding process when other crops that are part of the system should be incorporated into testing are difficult issues. The environment that is adequate for screening is also hard to define. Many multiple cropping systems are primarily grown on marginal land with less than optimal levels of inputs. Should screening be done under these conditions, or under highly controlled and optimal conditions for the crop? What will on-farm conditions be, and what systems will the farmer be using 5 to 10 years from now when the new varieties are released?

Whether an effective breeding program can be carried out for a specific multiple cropping system depends on the genetic diversity available within a crop for performance in that system. What the objectives for that breeding program should be depend on the genetic traits that affect the crop's performance in the system. Measuring that rather nebulous quality—performance—and identifying schemes for improving it must be considered in choosing a breeding methodology. Three topics are explored in detail in this chapter: (1) genetic diversity in relation to multiple cropping systems, (2) genetic traits of importance in multiple cropping systems, and (3) appropriate breeding methodologies.

GENETIC DIVERSITY

Plant breeders involved with traditional research, who deal with a single crop in a monoculture cropping system, are accustomed to thinking of genetic diversity *within* crop species. In multiple cropping systems, however, genetic diversity between species also plays an important role. Multiple cropping may include everything from a simple multiline cereal variety (Borlaug, 1959; Jensen, 1952), to a two-species forage/grass mixture for pasture, to the highly complex cropping patterns practiced in the tropical lowlands of Asia (Harwood and Price, 1976) and Africa (Okigbo and Greenland, 1976). At the other extreme in terms of diversity are the monoculture systems such as wheat (*Triticum aestivum*) in the Great Plains or maize (*Zea mays*) in the corn belt of the United States. The latter, when it involves a single cross maize hybrid, is a system with minimal diversity even within the species: all plants are as near to being genetically identical as possible.

This total range of cropping systems, from monocultures to complex mixtures, represents a spectrum of genetic diversity, both within and among species. Francis (1981) compares this continuum of diversity to that found in natural ecosystems (Fig. 10.1). These vary from extremely diverse tropical rain forests where more than 100 different species may be found on a single hectare, to coastal salt marshes where often only a few species of *Spartina* grass may be present (Long and Woolhouse, 1977).

It frequently has been suggested in the ecological literature that diverse ecosystems are more stable than simple ones. Goodman (1975), however, after an extensive review concluded that "observational confirmation of the diversity-stability hypothesis [has] never materialized." It may be reasoned that increased diversity conveys increased stability only if food chains are connected "in parallel," where each species has several food sources. If food chains are connected primarily "in series," so that each species depends on a single species from the trophic level below, then even a diverse ecosystem may be drastically affected by a decline in just one species, and thus will not be very stable (C. A. S. Hall, personal communication).

Monocultural cropping systems can be considered as a series system, with all crop production dependent on a single species. A multiple cropping system

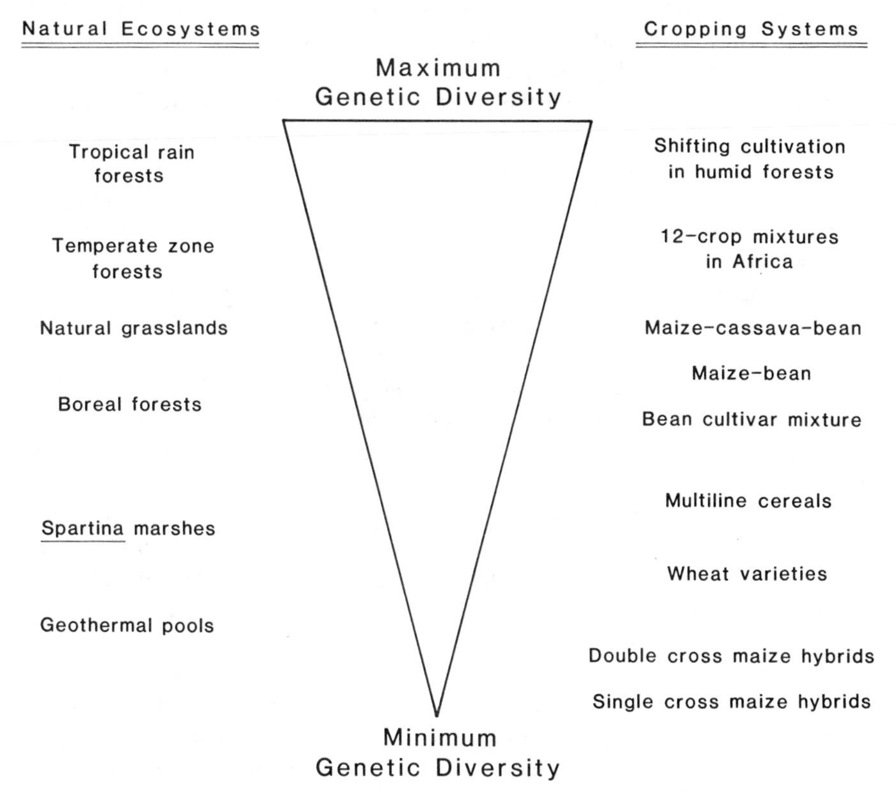

Figure 10.1 Spectrum of genetic diversity in natural ecosystems and in cropping systems. (Adapted from Francis, 1981.)

on the other hand would introduce a parallel component, with dependency then being spread over several crops, and thus providing greater food and/or income stability.

It is interesting to note that nothing on the natural ecosystem side of the diversity continuum parallels the situation of a monoculture hybrid maize crop. Apparently, no natural system remains stable over time with such a low level of diversity. The risks of genetic uniformity in cropping systems have been pointed out by various authors (Adams et al., 1971; Browning and Frey, 1969; Trenbath, 1975). The outbreak of late blight (*Phytophthora infestans*) on potatoes (*Solanum* sp.) in Ireland and the epidemic of southern corn leaf blight (*Helminthosporium maydis*, race T) in the United States in the early 1970s are classic examples of the risks associated with widespread use of genetically uniform cropping systems (Adams et al., 1971). Yet most agronomic and breeding research to date has been aimed at these types of systems, and trends in agricultural development are largely in this direction as well. Clearly there is need for a change in the emphasis of research, and in some cases in the farmer mentality that demands absolute uniformity in the field.

Plant breeding methodology that has been developed for monocultures may not be readily applicable to more complex systems. The development of methodology is much more difficult for systems such as the highly complex cropping patterns of the tropical lowlands. One such complex system used in Africa employs a series of mounds reaching 2 m or more in height (Fig. 10.2). Twelve edible species are distributed over the better-drained tops and sides of the mounds, and rice is grown in the moister areas between mounds (Okigbo and Greenland, 1976). Varietal improvement for such a system is complicated by the number of species involved, the diversity of microenvironments that any one species encounters, and the many possible variations in agronomic management of the system. The number of possible combinations of densities, planting dates, and spatial arrangements of the component crops, and of interactions between crops and their insect and disease complexes is difficult to imagine. For this reason, research that *has* dealt with multiple cropping systems to date has concentrated on the simpler two- or three-species systems. Such work should provide initial information on the potential as well as the mechanisms for crop improvement in complex systems.

In choosing varieties for multiple cropping systems, variation both between crop species and within species is important (Francis, 1981). If improvement of a cropping system requires major changes in plant maturity, plant architecture, leaf area, distribution of roots in the soil, or physical location of storage organs, variation among crop species should be considered first, as this is much greater than the variation available within species. Ideally, one could choose from the whole range of crop species adapted to a given ecological zone in designing alternative cropping patterns. Descriptions of underexploited species with potential economic value have been compiled, apparently with this "supermarket" approach to designing cropping systems in mind (NAS, 1975). Realistically, however, it is extremely difficult to change diet, culture, and tradition. Thus,

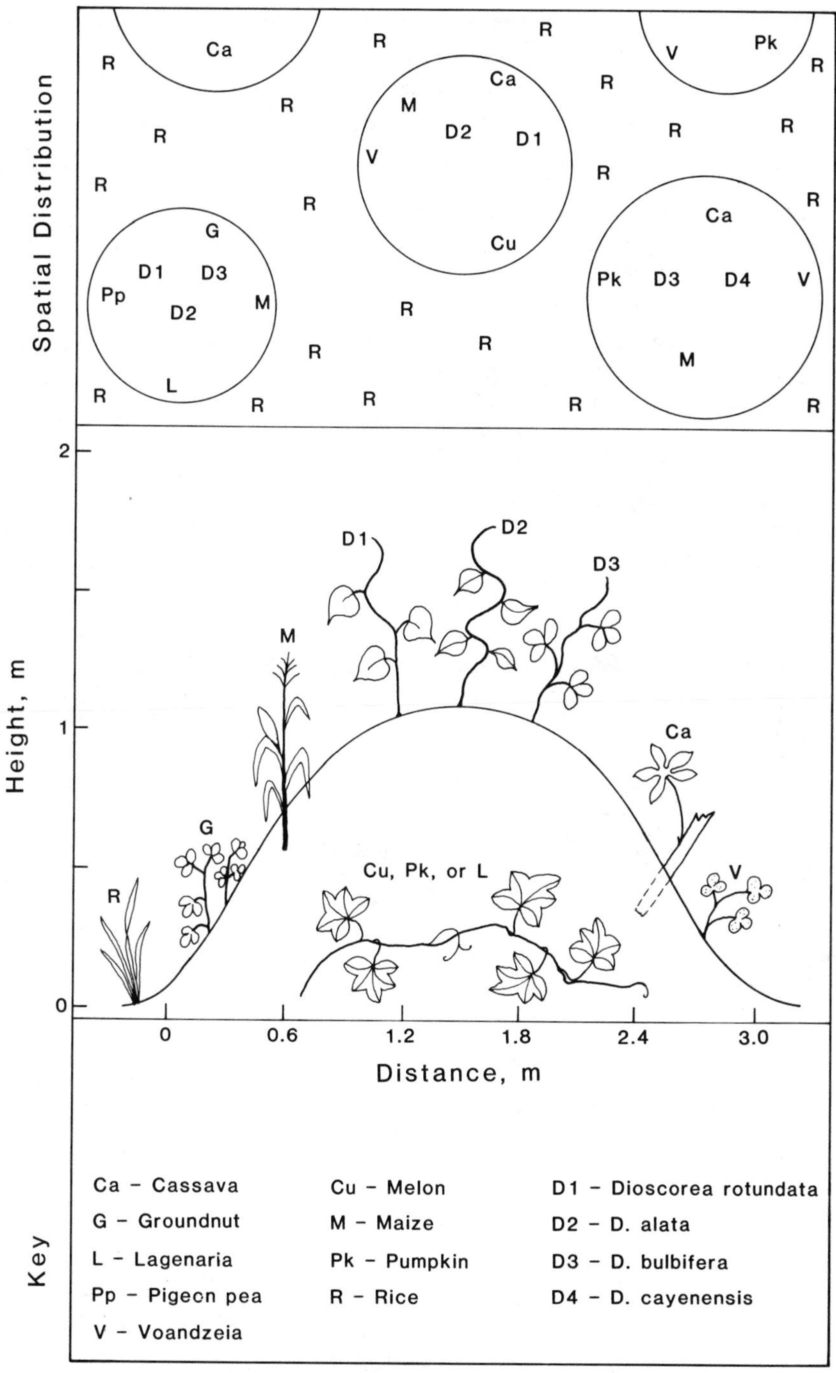

Figure 10.2 Spatial distribution of thirteen crop species on and between raised mounds in Nigeria. (Reproduced from Okigbo and Greenland, 1976.)

use of the total range of among-species diversity is not feasible in most cases, and varietal selection is constrained primarily to selecting within species that are already grown successfully and consumed in an area.

Potential for genetic improvement of cropping systems is thus a function of the diversity available within the most important crop species. Existing germ plasm resources of these species have been collected through the network of International Agricultural Research Centers and key national research programs. Each center has assumed the global responsibility for one or more major crop species (IARC collections listed in Table 10.1), concentrating on the collection, preservation, description, and use of a broad range of the available germ plasm for that species. Most improvement work to date has concentrated on screening collections for traits of immediate interest for intensive monoculture cropping patterns. Characters such as plant type, maturity class, seed color and type, protein content, and adaptation to stress conditions have been evaluated. As an example of the diversity available within one of these collections, the range of variation encountered for several characters within the collection of *Phaseolus vulgaris* at CIAT has been summarized in Table 10.2. Similar ranges of variation are found in the collections of other species. This reservoir of variability and the results of its evaluation are available to the researcher oriented toward crop improvement for multiple cropping systems.

Experience has shown that plant breeders currently use a narrow range of the total existing variability in most cultivated species. It is easier to work with the most productive available cultivars in a given agroclimatic zone, since introduction of genes from nonadapted representatives of the same species may cause an initial reduction in yield potential. These nonadapted types are used as

Table 10.1 Location of Germ Plasm Collections for the Major Food Crop Species

International Agricultural Research Center[a]	Crop species for which collections are maintained	Reference
AVRDC	Temperate and tropical vegetables	AVRDC (1983)
CIAT	Cassava, *Phaseolus* spp.	CIAT (1983)
CIMMYT	Maize, wheat	CIMMYT (1983)
CIP	Potato	CIP (1983)
ICRISAT	Sorghum, millets, groundnut, chickpea, pigeon pea	ICRISAT (1983)
IITA	Cowpea, sweet potato, yam	IITA (1983)
INTSOY	Soybean	INTSOY (1983)
IRRI	Rice	IRRI (1983)

[a]AVRDC (Asian Vegetable Research and Development Center), Tainan, Taiwan; CIAT (Centro Internacional de Agricultura Tropical), Cali, Colombia; CIMMYT, Centro Internacional de Mejoramiento de Maiz y Trigo, Mexico City, Mexico; CIP (Centro Internacional de la Papa), Lima, Peru; ICRISAT (International Center for Research in the Semi-Arid Tropics), Hyderabad, India; IITA (International Institute of Tropical Agriculture), Ibadan, Nigeria; INTSOY (International Soybean Program), Urbana, Illinois, USA; IRRI (International Rice Research Institute), Los Baños, Philippines.

Table 10.2 Range of Variation Observed in the *Phaseolus vulgaris* Collection at CIAT for Several Plant Characters

Character	Range of variation observed
Growth habit	Determinate bush to indeterminate vigorous climber
Seed color	White, cream, yellow, pink, red, maroon, purple, black, mottled
Seed shape	Round, elongated, kidney
Days to flowering	27 to 70
Duration of flowering	3 to 66 days
Number of nodes at flowering	6 to 32
Number of nodes at maturity	6 to 40
Number of pods per plant	1 to 99
Number of seeds per pod	2 to 9

Source: Data from CIAT, 1980.

sources of genes for resistance to insects or pathogens, for tolerance to a given stress condition (such as low temperature or acid soils), or for a specific seed color or grain texture preferred by the consumer but not present in high yielding germ plasm sources. It would appear, then, that there is ample genetic variability available for further improvement of most crop species. Almost invariably, when plant breeders have looked for variation in a trait not previously studied, the search has been successful. By inference then, there should exist variation in the germ plasm banks of the major crop species for adaptation or performance in multiple cropping systems. Once specific traits that lead to improved performance in these systems are identified, the genetic resources of these collections can be used in the improvement of multiple cropping systems.

GENETIC TRAITS OF IMPORTANCE

Specific genetic traits that are needed in varieties for multiple cropping systems have been reviewed in several articles (Francis, 1981; Francis et al., 1976; Lantican, 1977; Villareal and Lai, 1977). Many of these traits are also logical breeding objectives for monoculture cropping systems, such as adaptation to the target environment (temperature, rainfall, and photoperiod regimes), tolerance or resistance to prevailing insect or disease problems, responsiveness to favorable environmental changes (such as additional fertilizer or better water availability), and reasonably high and stable agronomic yield potential. In the case of multiple cropping, various other traits might be needed for an improved variety as well, such as greater or less competitive ability, tolerance to shading, or modifications in plant architecture. As the number of traits included in a selection scheme increases, the rate of improvement for each trait decreases. Thus the challenge to plant breeders working with multiple cropping systems is to sort through these many potential selection objectives and focus on a limited number of clearly defined traits that will improve total system productivity or stability.

Defining objectives is both more difficult and more critical in breeding for multiple cropping systems, due to the great number of interactions between and among crop species and to the effects of variation in agronomic practices. The phenotype of an intercropped variety is the product of many generations of natural selection (Finlay, 1976), not of short-term selection for one or a few specific traits. Wien and Smithson (1981) suggest screening a wide range of pure lines in the multiple cropping system to identify traits that are important to adaptation to the system. They caution, however, that one must also determine the extent to which these same characters are expressed in monoculture. Clements and Morris (1982) provide evidence suggesting that cropping systems *do* affect phenotypic expression of soybeans (*Glycine max*) grown before or after rice (*Oryza sativa*). Thus one might identify a desirable phenotype for the multiple cropping system, but have to select for something rather different if selections were done in monoculture. Finally, phenotypic traits that improve the agronomic performance of the cropping system are not the only traits that must be considered. The farmers' needs and objectives also must be taken into account. These may include traits such as crop quality and durability in storage, yield stability, potential markets, or other factors that do not directly bear on the biological productivity of the system.

In selecting objectives for a breeding program, the researcher must decide which of the desired modifications in the crop or the whole system are most appropriately addressed through plant breeding, and which would be more readily dealt with through changes in the cropping pattern or cultural practices. Breeding new varieties may be the most rapid and cost-effective method to achieve resistance to an insect or a pathogen for which other methods of control are ineffective. Finding new selections that are adapted to colder temperatures at planting may be much more difficult than changing planting date to a more favorable time for that crop. Breeding a maize variety with adequate tolerance to drought stress may be less realistic than changing from maize to sorghum (*Sorghum bicolor*) in some regions. (See Chap. 13.)

The breeder must also speculate on the future of farmers' cropping systems, and anticipate the changes that will occur between the time a decision is made on breeding objectives and the time a variety selected based on those objectives will be available to farmers. Typically, the lead time for variety development is 5 to 10 years, depending on the pollination habits of the crop, the complexity of inheritance of the trait(s) involved, and the intensity of the program or number of generations per year that can be accomplished. Francis (1980) attempted to predict the changes which would occur in monoculture corn and sorghum cropping systems in the United States, and the implications of those changes for corn and sorghum breeding programs. Few researchers have attempted to do the same for multiple cropping systems, yet the breeder must make such predictions when choosing breeding objectives.

In the following sections, some of the traits that should be considered in choosing objectives for a multiple crop breeding program are discussed. The list of traits is certainly not exhaustive, but is meant to highlight those of generalized importance.

Crop Maturity

Desirable maturity for crops in a multiple cropping system is defined not only by the limitations imposed by the environment, as is the case for monocultures, but also by the relation of each crop to others in the system. Early maturity has often been cited as important in most multiple cropping systems (Moseman, 1966; Swaminathan, 1970). It allows intensification of cropping over time, as in the case of upland crops planted before or after rice, or several consecutive rice crops grown in one year (Gomez and Gomez, 1983; Lantican, 1980). Research on cereal/pigeon pea (*Cajanus cajan*) intercrops has shown that earlier maturing cereals resulted in better system productivity, primarily due to the larger pigeon pea yields obtained with earlier cereal varieties (Rao and Willey, 1980, 1983). Earlier maturing varieties are also less sensitive to environmental variation (Saeed and Francis, 1984; Saeed et al., 1984). The possibility of avoiding environmental stresses is increased because their planting dates can be adjusted.

In certain cropping systems, however, late maturity is necessary. For rice interplanted with maize in Central America, tall long-cycle rice varieties are desirable (Hart, 1977). The maize is doubled over after physiological maturity, allowing light to penetrate to the lower story rice crop. The rice flowers and matures amidst the maize which is drying down in the field. Intercropped maize and sorghum in Central America functions in a similar fashion. Sorghum varieties that mature in 8 or 9 months combine with an early maize crop to produce a system that takes excellent advantage of the bimodal rainfall distribution in the area (Fig. 10.3). This system also provides a high level of harvest security in areas with erratic rainfall distribution. In the event of a maize crop failure, the sorghum, which is inherently more drought tolerant and produces an extensive root system while the maize is growing, can develop and produce a crop in even the worst of years (Hawkins, 1983). Studies of mung beans (*Vigna radiata*) in Asia showed that late maturing cultivars performed better than early ones under rainfed conditions with low light intensity and high temperatures (Poehlman, 1978).

Several of the studies mentioned above suggest that the combination of two or more crops with different maturities is advantageous. The difference in maturity between crops is probably more important than the absolute maturity of any one crop in a multiple cropping system. Tables 10.3 and 10.4 summarize the yield and economic advantages attributable to maturity differences in two-crop systems. Rao and Willey (1983) conclude that cereal/pigeon pea mixtures will yield more pigeon pea as the maturity difference between the two crops increases. Results from studies of cereal mixtures are similar. In sorghum/millet (*Pennisetum americanum*) intercrops, the combinations with the earliest millet and the latest sorghum and with the earliest sorghum and the latest millet were the most productive (ICRISAT, 1977). Long-cycle cotton (*Gossypium* spp.) varieties can be grown with shorter-cycle maize, millet, or sorghum to give greater economic returns than any of the possible monocultures (Baker, 1978).

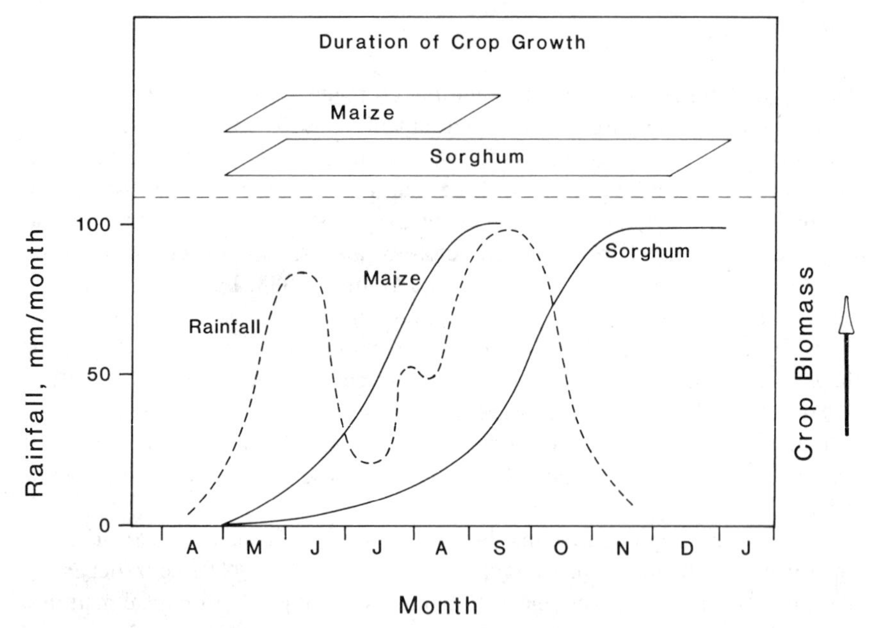

Figure 10.3 Rainfall distribution and crop growth curves for a maize and sorghum intercrop in Central America. (From Hawkins, 1983.)

Table 10.3 Yield Advantage in Two-Crop Patterns Attributed to Temporal Differences in Development of Component Species

Early species and days to harvest		Late species and days to harvest		Yield advantage	Reference
Pearl millet	85	Sorghum	150	80%	Andrews (1972b)
Several intercrops	80–100	Pigeonpea	180	≤73	Krantz et al. (1976)
Maize	85	Groundnut	120	20–60	IRRI (1974)
Maize	90	Rice	160	30–40	IRRI (1975)
Beans	85	Sorghum	120	55	Osiru and Willey (1972)
Beans	85	Maize	120	38	Willey and Osiru (1972)
Sorghum	106	Pigeon pea	173	20–65	Rao and Willey (1983)
Millet	86	Pigeon pea	173	50–90	Rao and Willey (1983)
Sorghum	106	Sorghum	160	20	Baker (1979)
Setaria	80	Pigeon pea	185	105	Rao and Willey (1980)
Pearl millet		Pigeon pea	185	90	Rao and Willey (1980)
Bean type I[a]	70	Maize	150	44	Francis et al. (1982)
Bean type II	75	Maize	150	41	Francis et al. (1982)
Bean type III	80	Maize	150	34	Francis et al. (1982)
Bean type IV	90	Maize	150	20	Francis et al. (1982)

[a]Bean types: I = determinate, II = semideterminate, III = indeterminate, nonclimbing, IV = indeterminate climbing.

Table 10.4 Economic Advantage in Two-Crop Patterns Attributed to Temporal Differences in Development of Component Species

Early species and days to harvest		Late species and days to harvest		Economic advantage	Reference
Several intercrops	90–100	Maize	130	24–76%	Anand Reddy et al. (1980)
Several intercrops	90–100	Maize	130	4–56	Anand Reddy et al. (1980)
Several intercrops	90–100	Sorghum	130	15–75	Anand Reddy et al. (1980)
Millet	75	Groundnut	128	52	Baker (1978)
Maize	110	Groundnut	128	18	Baker (1978)
Groundnut	128	Sorghum	160	43	Baker (1978)

Andrews (1972a, 1972b) suggests that the advantages of systems combining early and late maturing crops are derived from their increased capacity to use the entire available supply of growth factors, due to the temporal differences in demand between the two crops. Relative crop maturity, then, is an important consideration in selecting varieties for intensive multiple cropping systems.

Photoperiod Sensitivity

Development of photoperiod insensitive varieties has often been advocated for multiple cropping systems, as it allows total flexibility of planting dates while maintaining the possibility of predicting crop maturity, based primarily on available moisture and on temperatures (Moseman, 1966; Swaminathan, 1970). Photoperiod insensitivity has been an important and useful breeding objective in the rice program at IRRI (Coffman, 1977), permitting the harvest of two or three sequential crops per year. It has also been a priority in legume breeding (i.e., Tiwari, 1978), allowing much greater flexibility in the incorporation of legumes into cereal-based cropping systems.

In certain cases, however, photoperiod sensitivity is desirable (Gomez and Gomez, 1983). Figure 10.4 illustrates the effect of photoperiod sensitivity on flowering in sorghum. This effect permits the close adjustment of crop cycles to rainfall distribution in the maize/sorghum system (Fig. 10.3). Traditional rice varieties in Asia flower and fill grain during the shorter days of November and December so that maturity coincides with the end of the rainy season and the start of the dry season. At this time, the crop still has enough residual moisture to mature but can be safely harvested and dried in the sun without risk of heavy rains. Photoperiod sensitivity can be used to ensure that a crop will have an extended life cycle and complement an associated short-season crop, or to ensure that flowering and maturity occur at favorable times during the year.

Plant Morphology

There has been more speculation about aboveground plant architecture and complementarity of different plant types than about any other aspect of species choice

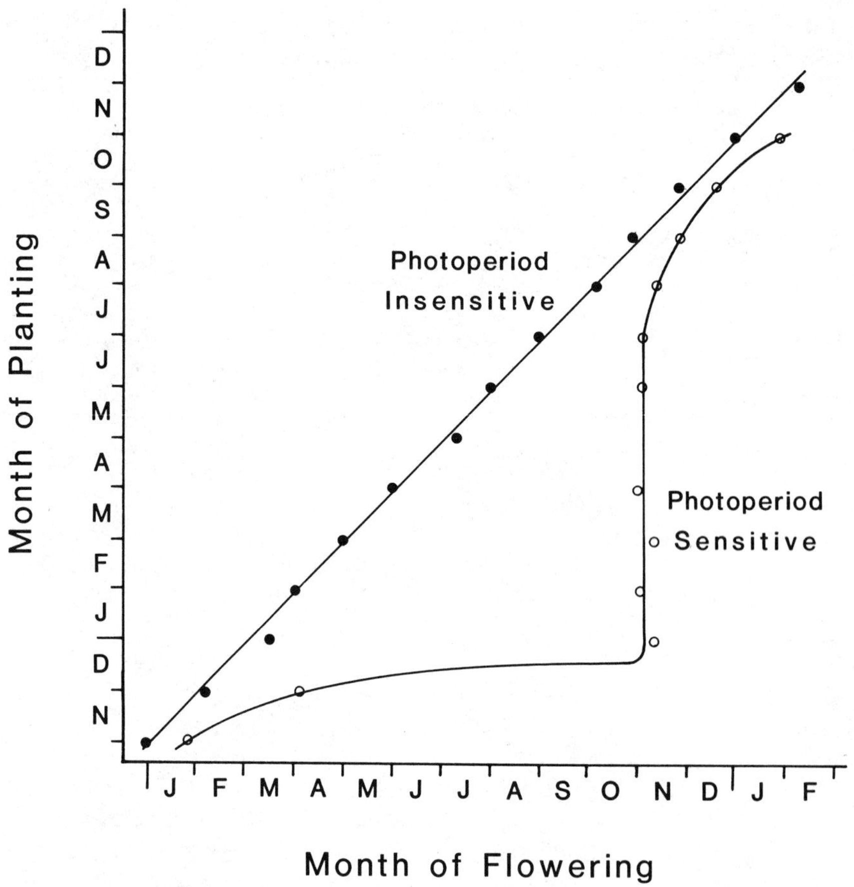

Figure 10.4 Planting date and flowering date for photoperiod sensitive and insensitive sorghum varieties in southern Honduras. (From Meckenstock, 1984.)

and selection for multiple cropping systems. Often these conclusions or conjectures are based on limited evidence and a single genotype of each plant type. Despite these limitations, some guidelines on plant types for multiple cropping are beginning to emerge.

Plant height is a morphological trait for which considerable within-species variability exists, and which one intuitively expects to affect crop performance in mixtures. Consequently it has been extensively studied. Reducing the height and thus the shading or competitive effect of a dominant crop generally allows increased productivity of lower story crops (Andrews, 1972a; Davis and Garcia, 1983; Davis et al., 1984). In intercrops of other cereals with sorghum, the taller cereals reduced sorghum yields more than shorter cereals (Rao and Willey, 1980). Soybeans yielded more in association with short maize than with tall maize, even though the plant height differences achieved with the maize varieties were minimal (Thompson et al., 1976). Plant height of cassava (*Manihot esculenta*) has also been shown to influence the yield of the second of two cowpea (*Vigna unguiculata*) intercrops (Rodriguez et al., 1982). Shorter, more erect-leaved varieties of various crops have resulted in better yields in monocultures (i.e., Jennings and Herrera, 1968). These types may prove useful in multiple cropping systems as well, for minimizing shading and thus light competition to associated crop species.

Plant type of understory crops also influences the productivity of crop mixtures. In some cases, determinate growth and medium or short plant height appear to be desirable for intercropped legumes (Catedral and Lantican, 1978; Gomez, 1976, 1977), while in other cases cultivars with indeterminate or climbing types gave better results (Francis et al., 1982; Hart, 1977). Determinate soybeans were more productive than indeterminate soybeans in intercrops in India (Tarhalkar and Rao, 1975), but the reverse was true for soybeans relay planted into winter wheat or spring oats (*Avena sativa*) (Chan et al., 1980). Prostrate indeterminate cowpeas were best suited to intercrop systems in Nigeria (Ojomo, 1976). The most desirable plant type for understory crops appears to depend quite heavily on the particular multiple cropping system being considered.

Because total productivity is more important than individual component yield, the effect of understory plant type on the taller crop also must be considered. Although indeterminate cowpeas appear better suited to intercrops in Nigeria than determinate ones, the latter had less effect on yields of associated maize (Ojomo, 1976; Wien and Nangju, 1976). Maize yields in Colombia were unaffected by intercropping with determinate bush beans, but associated climbing beans reduced maize yields by 31 percent (Francis et al., 1982). A later, leafier cowpea cultivar reduced maize yield in association more than an earlier cowpea with lower leaf area index (Wanki et al., 1982). In India, a pigeon pea cultivar with no basal branching was better for intercropping than traditional branching types, and castor varieties with a dwarf habit and converging branches, erect petioles, and long spikes were better for intercropping than traditional tall types (Tarhalkar and Rao, 1975). As with crop maturity, the differences in plant height and canopy structure between component crops in a mixture appear to be more important than the actual morphology of any one component (ICRISAT, 1977).

The relative morphology and yield of individual species in crop mixtures are influenced not only by the presence of other crops in the mixture, but also by densities and spatial arrangements of the component crops (Zandstra and Carangal, 1977; Chui and Shibles, 1984; Francis, 1978). Furthermore, variables such as total system yield, protein production, eonomic return, or stability of the system are more important than actual yield of any one crop component (Francis, 1981). Economic returns vary with the changing relative prices of component crops (Francis and Sanders, 1978). In this complex biological, economic, and social context, the choice and design of an "optimum" plant phenotype is not simple.

Root Systems

Compared to what has been published on aboveground plant morphology in multiple cropping systems, relatively limited information is available on root systems (Willey, 1979a, 1979b). One intuitively expects that intercropped species with distinct rooting patterns should have the potential to more completely exploit available water and nutrients than either species alone (Francis, 1981; Swaminathan, 1970). A shallow root system was suggested as a desirable trait for vegetable crops to be intercropped with longer season crops like sugarcane (*Saccharum spontaneum*) in order to reduce direct competition in the root zone (Villareal and Lai, 1977). Shallow root systems were also recommended for crop varieties to be grown after rice, in this case not to avoid root zone competition but to enhance tolerance to the structure of paddy soils (Lantican, 1977).

Research on species interaction and competition in the root zone is just beginning to be published. A technique for separating aboveground and belowground interactions in intercropping has been presented (Willey and Reddy, 1981), and root growth in a pearl millet and groundnut (*Arachis* sp.) intercrop was studied (Gregory and Reddy, 1982). The latter authors report that the distribution and quantity of roots was different when the crops were grown together compared to their respective monocultures. Cowpea is known for its vigorous belowground competition for water and nutrients (Rao and Willey, 1980), yet a study of a sorghum/cowpea intercrop indicated that the two crops were equally competitive for soil water (Shackel and Hall, 1984). Much basic research is still needed in order to begin to understand the interactions between crop root systems in intercrops. Species appear to vary in their temporal needs for water (Teare et al., 1973) and for nutrients (Willey, 1979b), and these variations may turn out to be more important than variations in root system morphology.

Stress Tolerance

Crop growth rates depend largely on the temperature regime and availability of water during the growing season. Temperature is unlikely to be altered to any significant degree by the presence of several species; however, moisture stress could alter growth rates in regions with limited rainfall. When alternatives for multiple cropping systems are designed planting dates are often altered or new crops are added at the beginning or end of the principal crop growing season. Whereas in a monoculture each species is planted when environmental conditions are optimal at a given location, manipulations to intensify and improve multiple cropping systems may result in temperature and moisture regimes that are less than optimal for some crops. Thus tolerance to extremes of temperature, to drought stress, or to waterlogging stress may be needed.

A temperate-zone relay involving a winter cereal (wheat) plus a summer legume (soybean) may require both shorter-cycle cultivars and a temperature adaptation not required for monocultures. If soybean planting were delayed until after wheat harvest, a successful soybean cultivar would need to germinate and grow at higher initial soil temperatures than a monoculture soybean cultivar. Similarly, a delayed planting of wheat in the fall could require a cultivar which would establish rapidly and achieve enough fall growth to withstand the winter dormancy period. Intensified cropping may also require changes in cultural practices, such as no-till seeding of a second crop directly into the stubble from the first. With minimum tillage, soil temperatures generally remain lower than with conventional tillage. For such systems, cultivars that can germinate and grow at low initial soil temperatures would be required.

Intensification of rice-based cropping systems in Asia has concentrated on planting upland crops before or after the main rice crop. For the prerice environment, tolerance to early season drought stress (while rains are still sporadic) and late season waterlogging stress (when the rains have become well established) is required (Lantican, 1980). For the postrice environment, the needs are reversed: waterlogging tolerance is required during early crop development, and drought tolerance later in the season, as the crop matures on residual soil water (Lantican, 1977, 1980). Variation for tolerance to drought and waterlogging has been reported for various crop species (del Rosario and Santos, 1982; Kass, 1984).

Tolerance to brief periods of temperature or water stress is important both in monoculture and multiple cropping systems. Selection for relatively insensitive cultivars, however, allows greater flexibility in choosing planting dates, a secondary benefit of use in multiple cropping systems. It may also lead to cultivars with broader adaptation (Tiwari, 1978). Selection for broad adaptation is a desirable breeding objective, yet it is important primarily to national research programs or commercial companies with an interest in selling seed. Farmers are concerned with the range of temperature and rainfall conditions expected on their own farms. In some cases this range will be less than the range used to test for broad adaptation. In others, without irrigation, year-to-year rainfall variation may be greater than the location-to-location variation used in testing for broad adaptation. Breeders must be careful to avoid either extreme—selecting for too broad a range of environments, which results in losses in yield potential, or selecting for adaptation that is too specific and greatly limits the usefulness of varieties produced.

Density Response

The importance of total population of plants in an intercrop as well as the proportions of component species that make up this total population (or plant density) has been reviewed and discussed by Willey (1979b). Multiple cropping studies of simultaneous planting of component crops have used the replacement series, where a proportion of one crop is substituted for a proportion of the other, and less frequently an additive model where monoculture densities of two or more species are planted together. The replacement series is a preferred methodology, unless the objectives include maintaining the maximum possible yield of one component, often a taller cereal species. One drawback to the replacement series technique is that farmers, in optimizing their cropping systems, are likely to vary not only the relative proportions of crops in a mixture, but also the plant density of one or more of the crops. Replacement series methodology also assumes that interspecific and intraspecific competition are equal, and calculates a species ratio (the number of plants of species *A* that provides competitive pressure equivalent to that of one plant of species *B*) in order to maintain that equivalence through the replacement series. Recent experimental evidence suggests that a single species ratio does not satisfy the equal competition constraint (Carter et al., 1985).

In multiple cropping, the identification of varieties that can respond to increases in density allows greater flexibility in the design of alternative cropping patterns (Francis, 1981; Francis et al., 1976). Individual crop densities influence the yields and yield components of each species, but recent results suggest that a wide range of combinations of densities may give similar total yields and gross returns in some patterns (Carter et al., 1983). An optimum combination of crop densities depends not only on the yield response of each crop, but on the spatial arrangements that are feasible for that pattern. Studying density response of component crops in a multiple cropping system is greatly complicated by the quantity of interactions with other factors, which must either be fixed or accounted for in the interpretation of data.

Insect and Disease Resistance

Only recently have reviews appeared that discuss insect and disease incidence and management in multiple cropping systems (Perrin, 1977; van Rheenen et al., 1981). The authors of these reviews point out the limited research done to date on pest problems associated with multiple cropping. Various studies have been done comparing pest incidence in monoculture to that in crop mixtures, and the results of some of these are summarized in Table 10.5. Altieri (1977) concluded that species diversification in crop mixtures or in mixtures of crops and weeds was an effective approach to tropical pest management. The results in Table 10.5 generally support that conclusion. Thus one may view multiple cropping as an additional tool in the farmer's array of cultural pest control measures (see also Chap. 9).

These differences in relative insect or pathogen incidence between monoculture and multiple cropping suggest that different priorities should be set by the plant breeder working for one specific system (Francis, 1981). Although the potential for fall armyworm (*Spodoptera frugiperda*) attack in maize is an ever-present threat in the Cauca Valley in Colombia (Altieri, 1977), damage can be ameliorated by intercropping with beans. Thus, genetic improvement of maize for intercropping could deemphasize selection for fall armyworm tolerance in favor of other traits of immediate importance to the mixed crop microenvironment.

Although it may be less important to select for specific resistances to insects and diseases in breeding for multiple cropping systems, it remains important to monitor the field resistance to pests of varieties being developed. Many traditional crop varieties carry generalized resistance to a wide range of pests, and new varieties that do not at least maintain that level of resistance will not be readily accepted by farmers. Litsinger (1982) advocates developing general insect-pest resistance through selecting for traits such as early maturity (which permits fewer insect generations per crop cycle), seedling vigor, uniformity of plant development (which minimizes the period when plants at a susceptible growth stage are present in the field), and plant architecture that is nonpreferred by insects (i.e., loose sorghum heads).

Table 10.5 Disease and Insect Control Effects of Some Multiple Cropping Systems

Cropping system and location	Disease and insect control and effects of system	Reference
Maize and beans, Kenya	Less halo blight, common mosaic, anthracnose, common blight, angular leafspot, and bollworm on beans	van Rheenen et al. (1981)
Maize and beans, Colombia	Less fall armyworm on maize and less leaf beetle on beans	Altieri et al. (1978)
Maize and beans, Costa Rica	Less rust and angular leafspot on beans	Mora (1978)
Beans and sweet potatoes, beans and cassava, Costa Rica	Less angular leafspot on beans	Moreno (1977)
Maize and beans, Costa Rica	More angular leafspot on beans	Moreno (1977)
Maize and cassava, Costa Rica	Less fall armyworm and *Cyrtomenus bergi* and more *Diatraea* spp. borers on maize	Carballo and Peairs (1984)

Grain Quality

If a crop is grown primarily for consumption on the farm, specific grain quality characteristics such as color, texture, cooking aroma, and taste may be critical breeding objectives. If part or all of the crop is commercialized, the grain quality that the market demands is equally important, yet it may differ from that required by the farm family. Characters such as weathering tolerance in the field, resistance to storage pests, and keeping quality of the grain may also be important for maintaining quality until the grain is harvested and used.

Considerable interest and effort has been expended by researchers in breeding for higher protein quality, especially in the cereals. Although there is no doubt that cereal grains have less than optimal nutritional value, it is questionable whether plant breeding advances in this area will be recognized and rewarded in the marketplace. Changes achieved in the protein quality of cereal grains to date are small enough that their effect on improving human nutritional status has been negligible when the diet included *any* other protein sources (legumes, dairy products, fish, or meat). It may well be better, from a nutritional standpoint, to aim for diversifying diets and providing more food through increased production rather than improving the quality of cereal proteins.

Cultivar Uniformity

Single-cross hybrids of maize, sorghum, and other species, or narrowly based pure line varieties of self-pollinated crops, such as wheat or common bean, provide an extreme of genetic uniformity to farmers. This uniformity has the advantages of synchronous flowering and harvest maturity, which simplify cultural practices and mechanical harvesting. Yet it also brings associated risks of susceptibility to a pathogen or insect that is particularly virulent on a uniform, widely grown cultivar, and of susceptibility to environmental stresses occurring at critical plant growth stages. Highly uniform varieties are especially risky under the conditions of rain-fed, limited resource farming (Francis et al., 1976).

Some level of stress tolerance could be incorporated into hybrids or varieties at the expense of the extreme uniformity found in currently available commercial cultivars. Hybrids with some spread in days of flowering could avoid total crop loss if a severe drought occurred during pollination. A mixture of phenotypically similar but genetically different hybrids could provide benefits similar to those from multiline cereal varieties for resistance to insects and pathogens (Francis, 1980; Jensen, 1952). Three-way crosses offer advantages in slightly lowered genetic uniformity, wider adaptation, and lower seed cost (due to higher yields in hybrid seed production) when compared with single crosses (Hookstra and Ross, 1982). In multiple cropping systems, there may be a need for uniformity of grain type and color for commercialization of the crop, and at times a need for uniformity of development and maturity (for example, when a crop must fill a particular time-limited niche in the system) (IRRI, 1973). If a crop is grown for use on the farm, a range in maturities may be preferable, as this extends the period of direct availability of the crop to the farm family. Thus, component crop uniformity is desirable for some traits and undesirable for others in multiple cropping systems, and brings with it certain risks as well.

Yield Stability

Crop yield stability is a complex product of genetic yield potential and tolerance to stress conditions, and is of critical importance to subsistence farmers who practice multiple cropping. Even more important than the stability of individual cultivars or component crops is the yield and income stability of the total cropping pattern during the year and over many years. Some progress has been made in the statistical evaluation of yield stability, and this is currently being translated into genetic improvement of crops which can become more stable components of cropping systems.

The theoretical and practical basis for studying yield stability was set forth in two papers published in the 1960s (Eberhart and Russell, 1966; Finlay and Wilkinson, 1963). Since then, most work has concentrated on the stability of yields and yield components in monoculture crops. A recent study by Saeed and Francis (1983) reports that sorghum yield stability is primarily a function of crop maturity differences, with earlier maturing varieties exhibiting greater yield stability. They suggest that yield stability is best evaluated in groups of cultivars with minimal differences in maturity. In another recent study of stability in grain sorghum, Heinrich et al. (1983) reported that high yield potential and stability are not mutually exclusive traits, and traits that confer stability can be identified and selected for in varieties with high yield potential.

Little work has been done on the stability of crop mixtures such as those found in multiple cropping. The underlying assumption is that when one crop in a complex pattern suffers short-term stress from drought or temperature, or from an insect or disease problem specific to that crop, then another crop will compensate by using available growth resources and yield more than expected based on its proportion in the mixture. Willey (1979a) points out that such compensation is not possible if the crops are grown separately. Therefore growing several species simultaneously in monocultures will give a certain degree of stability to farm production but not as much as could be achieved by intercropping the various species. Harwood and Price (1976) argue the contrary, pointing out that stress conditions often occur after considerable vegetative growth has taken place, and that a nonyielding component crop will still exert competitive pressure and reduce the yield of a more successful companion. Thus, the mixture would be less stable than concurrent and separate monoculture plantings.

Recently, two studies have compared the stability of monoculture and multiple cropping systems, based on compilations of data from many experiments in many locations. One was conducted in Colombia, and compared monoculture maize, monoculture climbing beans on trellises, and a maize-climbing bean intercrop (Francis and Sanders, 1978). The other study, conducted in India, compared sorghum and pigeon pea monocultures and an intercrop of the two (Rao and Willey, 1980). Results from these analyses are summarized in Table

Table 10.6 Yield Stability of Intercropping Patterns Based on Multiple Experiment Analyses

	Yield		Income		Probability of an income ≥ Y/ha	
Cropping pattern	kg/ha	CV	Total (CP/ha)	CV	Y ≥ 0	Y≥ 10,000 CP
Maize and climbing bean:[a]						
Maize monoculture	4986	23.6	1,944	242.5	0.65	0.04
Bean monoculture	2941	29.4	16,061	86.2	0.80	0.55
Maize and beans	6114	22.9	16,521	66.0	0.92	0.73
			Total (Rs/ha)	CV	Y≥ 250 Rs	Y≥ 3,250 Rs
Sorghum and pigeon pea:[b]						
Sorghum monoculture	3208	47.0	3,208	—	0.95	0.40
Pigeon pea monoculture	1446	42.7	2,892	—	0.91	0.28
Sorghum and pigeonpea	3656	39.0	4,473	—	1.00	0.65

Note: CV = Coefficient of variation; Y = Variable yield per hectare; Rs = Indian rupees; CP = Colombian pesos.

[a]Twenty-trial analysis in Colombia, 1975 to 1976 (Francis and Sanders, 1978).

[b]Ninety-four-trial analysis in India, 1972 to 1978 (Rao and Willey, 1980).

10.6. In both cases, the intercrops appeared to be more stable than the monocultures.

Similar to the situation for disease and insect resistance, multiple cropping appears to be a practice that can increase yield stability under given environmental conditions over that attainable with monocultures of the same species. Certainly an improvement in yield stability of the component species would contribute greater stability to total system yield, if this component stability were in fact developed and measured in the relevant intercrop pattern. Intercropping places a form of stress on each component crop, with competition for nutrients and moisture on all components of the pattern and an additional light stress on the understory crop (or crops). If a cultivar is developed that has tolerance to this stress, and it produces moderate to high seed yield even under some stress conditions, this cultivar should contribute both greater yield and potential stability to the intercrop pattern. The several characteristics discussed in this section as desirable traits for intercropping all should contribute to greater stability.

BREEDING METHODOLOGY

Breeding for multiple cropping systems can successfully make use of conventional crop improvement procedures for self- and cross-pollinated crops, as long as the breeding objectives are clearly defined. The specific traits needed for a successful intercrop variety must be known in order to design a crop improvement program. Many traits that might be considered were discussed in the previous section. Improving yield potential, which is probably the most universal and fundamental goal of plant breeding programs, was not discussed as it is equally important in monocultures and in multiple cropping systems. It is also a complex character, controlled by many genes and highly influenced by environment.

The environmental influence on yield potential can be considered to include the influence of the particular cropping system in which a species is grown. Evaluation of the genotype by cropping system interaction, especially for yield potential, is a critical step in designing a breeding program (Francis et al., 1976; Francis, 1981; Gomez and Gomez, 1983). Simply stated, the absence of interaction suggests that selection and evaluation in one cropping system will lead to development of new varieties that are optimal across a range of systems. If there is a strong genotype by system interaction, it may be necessary to test and choose different varieties for each system. This is a critical decision as plant breeding is a long and costly process. Combining yield stability (broad adaptation) and high yield potential, for instance, requires evaluation of potential selections in replicated trials over multiple years or locations or both. If genotype by system interaction is highly significant, it may be necessary to carry out such evaluations for each important cropping system. The remainder of this section discusses some of the techniques used for evaluating genotype by cropping system interactions, and some of the issues involved in choosing breeding methods and selection criteria.

Yield Reduction in Intercrops

A simple method of investigating genotype by cropping system interactions and of selecting promising genotypes is to measure monoculture and intercropped yields for a series of genotypes and calculate the yield reduction under intercropping. This method has the advantage of being statistically quite simple, and it also allows the breeder to evaluate yield potential in monoculture (where genotypic differences are usually expressed to the maximum extent possible) and yield potential in the intercrop system most prevalent on farms.

This technique was used to study the performance of 20 maize genotypes in three cropping systems: monoculture, association with bush beans, and association with climbing beans (Francis et al., 1983). Results showed that if a breeder had selected the four highest yielding maize genotypes in monoculture (20 percent selection pressure), only one of the best four for intercropping with bush bean and two of the best four for intercropping with climbing bean would have been included. This gives a probability of 0.25 and 0.50, respectively, of selecting the best genotypes for these intercrop patterns by testing only in monoculture. If the 10 highest yielding genotypes in monoculture had been selected (50 percent selection pressure) the success rate would have increased to 0.80 for each intercrop pattern in this example. Using yield reduction from monoculture to intercrop as a selection criterion rather than monoculture yield, 8 of the 10 best genotypes for intercropping with bush beans and 7 of the 10 best for intercropping with climbing beans would have been selected.

In selecting for systems such as these, the effects of maize genotypes on bean yields must also be taken into account. This can be done in terms of total grain yield (maize plus beans) or of gross income from the two intercrop systems. Using either of these criteria to measure intercrop productivity, 7 to 9 of the 10 best genotypes for intercrops would have been chosen by selecting the 10 best yielding maize genotypes in monoculture (50 percent selection pressure). It is clear that a relaxed selection criterion or multiple system testing needs to be done to avoid eliminating some of the most promising genotypes for intercrops.

This technique is a simple one, requiring no statistics other than the mean yield of each genotype in each cropping system. It is a rather empirical approach to genotype selection, but probably reflects the same types of criteria that have been used in the gradual selection of farmers' varieties for intercropping.

Correlation Analysis

Francis (1981) and Gomez and Gomez (1983) have discussed the rationale behind using simple linear correlation coefficients between genotype yields in two cropping systems to evaluate genotype by environment interaction. This method is attractive in that it allows evaluation of data from published sources in many cases, because it does not require individual replication results. It is generally accepted that a significant positive correlation of yields between two planting systems suggests a minimal genotype by cropping system interaction and thus a minimal need for separate breeding strategies for the two systems.

Extreme caution must be exercised, however, in the use of results from correlation analysis in this fashion. Table 10.7 summarizes data from selected studies with maize using this type of correlation analysis. In some cases, correlation coefficients as low as 0.35 are highly significant. This means that only $(0.35)^2$ or about 12 percent of the variability in maize yields in the second cropping system can be attributed to their linear regression on yields in the first cropping system, and 88 percent of the variation is independent of that regression. Even with a correlation coefficient of 0.70, half of the variation in yields in one cropping system is independent of the yields obtained in the other. Clearly the correlation should be not only significant and positive, but also quite high to justify the inference that genotype by cropping system interactions are minimal. A glance at some scatter diagrams with high, significant correlation coefficients reveals that even under these conditions, a fair number of outlying points may occur which could represent useful genotypes that would not get picked up in monoculture selection. Thus, care should be taken in the interpretation of correlation coefficients between cropping systems.

It has been suggested that correlations of the rank orders of genotypes between two cropping systems may be more relevant than correlations of the actual yields. Since a selection scheme usually is designed to keep a certain proportion of the genotypes tested, the rank order correlation would reflect more accurately the similarity of the selections that would be made in two systems. The correlation of rank orders, however, has the same limitations as those discussed above for the correlation of absolute yields, and should be used with caution.

Analysis of Variance

Perhaps the most widely used approach to evaluating genotype by cropping system interactions is the analysis of variance procedure. Its use was reviewed in detail recently by Gomez and Gomez (1983), and hence it will not be discussed here in depth. Studies employing this technique to detect genotype by cropping

Table 10.7 Correlation Coefficients for Maize Yields between Pairs of Cropping Systems

Number of genotypes tested	Cropping system 1	Cropping system 2	Correlation coefficient	Reference
20	Monoculture	Maize + bush beans	0.90**	Francis (1981)
20	Monoculture	Maize + bush beans	0.40	Francis (1981)
20	Monoculture	Maize + climbing beans	0.89**	Francis (1981)
20	Monoculture	Maize + climbing beans	0.73**	Francis (1981)
38	Monoculture	40% shade	0.37*	Gomez (1976)
10	Monoculture	40% shade	0.12	Gomez (1977)
58	Monoculture	40% shade	0.35**	Gomez (1977)

Note: *, ** = statistically significant at the 0.05 and 0.01 probability levels, respectively.

system interactions are becoming more and more common. Significant interactions have been detected for soybeans in monoculture and intercrops with three different cereals (Makena and Doto, 1980), for pigeon peas in monoculture and associated with sorghum (Nerkar, 1980), for maize in monoculture and intercropped with bush and climbing beans (Francis et al., 1983), and for maize in conventional and minimum tillage systems (Brakke et al., 1983).

Francis (1985) has summarized the results of analyses of variance for genotype by cropping system interactions for a number of species. He points out that the majority of the variation observed in all cases is due to the effects of the different cropping systems. Of the 14 experiments included in the summary, the genotype by cropping system interaction was significant in 11. This interaction was consistently significant in experiments with climbing beans, with maize, and with soybeans, whereas bush bean and mung bean experiments showed a significant interaction in some cases but not in others. This may be attributable to the short growing cycles of these legumes, which minimize the period during which interaction with associated crops might occur.

Results from these sorts of experiments suggest that genotype by cropping system interactions are not necessarily universal, and whether they are significant or not may vary from one crop species to another. It appears that selection of maize, sorghum, climbing beans, soybeans, and pigeon peas for specific cropping systems could capitalize on genotype by cropping system interactions, and thus lead to the release of varieties of these crops with specific adaptation to the chosen system. Analysis of variance studies require complete data by replication, and also require a fair amount of statistical calculating, but their accuracy in detecting genotype interactions with cropping systems is greater than that for the other methods discussed.

Regression Analysis

The calculation of a stability index using regression analysis as proposed by Finlay and Wilkinson (1963) and Eberhart and Russell (1966) is another route to comparison of genotypes and their reactions to the environment or to cropping patterns. Both Gomez and Gomez (1983) and Francis (1985) discuss the use of the regression method in evaluating genotype stability. The former authors presented data for 24 soybean varieties grown in five different cropping systems. The regression indices for the varieties showed significant differences, indicating that the varieties responded in different ways to the range of cropping systems used. This, again, is evidence for genotype by cropping system interactions and, as Gomez and Gomez (1983) indicate, these data suggest that some varieties that performed well in one cropping system might have been eliminated from a breeding program if testing were done in a single different cropping system. This points out again the importance of multisystem testing in breeding programs for crops grown in a range of different cropping systems.

Davis and Garcia (1983) used regression analysis in a different way, to facilitate selection of genotypes based on the performance of both crops in an intercrop. For a study of bean genotypes in association with three maize genotypes, they calculated and plotted a regression of maize yield against bean yield. In this case the yields of the two crops were inversely proportional: for each additional 1 kg/ha of bean yield, 2 kg/ha of maize yield were lost. With plots such as this, points lying above and to the right of the regression line represent crop combinations that are more favorable than the average. Depending on the relative importance or market value of the two crops, combinations can be identified that will produce more than the average in combined yield, and favor the preferred species.

Methods of Selection

Methodology for breeding for multiple cropping systems is an area where much remains to be investigated (Francis, 1981). The complexities of competition, both intraspecific and interspecific, create a very difficult situation for the plant breeder. Desirable plant types in an intercrop may look very different when grown in a monoculture for selection (Wien and Smithson, 1981). The need for testing in several systems or in several environments at some stage of the breeding process is obvious, however the cost and effort involved in testing materials in two or more systems are such that testing clearly cannot be done at every generation of the breeding program.

Not only is the identification of breeding objectives difficult, but so is the choice of the agronomic conditions under which screening is carried out. Multiple cropping systems are often concentrated on marginal lands and grown with relatively limited inputs. Selection should be done under conditions that are representative of the target farm environment, yet the marginal conditions found on many farms give rise to a level of environmental variability that obscures genetic differences. Some balance must be sought between the degree of similarity between farmer conditions and the screening environment and the need for reasonably uniform screening conditions (Woolley, 1982). Wien and Smithson (1981) suggest using median stress level, achieved by manipulations of cultural variables such as planting time and fertility.

Standard breeding procedures are probably adequate for handling genetic material, but considerations such as those discussed in the preceding paragraphs must be taken into account in setting up a breeding program. Finlay (1976) recommends cyclical population improvement methods for improving the many traits that are important in multiple cropping systems. Screening early generations in monoculture has been suggested, with emphasis on negative selection against undesirable, simply inherited traits (e.g., those controlled by a small number of genes such as seed color and type, plant growth habit, adaptation to day length, temperatures, and rainfall). In later generations, positive screening for adaptation within the multiple cropping system (or systems) of interest should be done. At this stage, multilocation testing and screening for quantitative traits should be incorporated (Francis, 1985; Wien and Smithson, 1981). Hamblin et al. (1976) recommend simultaneous selection in two crops, a comprehensive yet expensive approach that would not allow large numbers of progeny to be evaluated. Finlay (1976) points out that in the latter stage of breeding programs, yield should not necessarily be the only, or even the primary, selection criterion. Stability, economic value, nutritional value, total system productivity, or some combination of these may be more appropriate.

One scheme proposed for simultaneous selection uses the principles of reciprocal recurrent selection (RRS) for cross-pollinated crops such as maize. The adaptation of this well-known method considers the ability of two species to combine well with each other in an intercrop, just as the RRS method seeks two populations that combine with each other genetically. This scheme was called reciprocal recurrent selection for compatibility (Francis et al., 1985). Two varieties or populations of species that are to be intercropped are chosen (called cycle zero of species $A = C_{A0}$, and cycle zero of species $B = C_{B0}$) which have genetic variability and favorable traits for production in the mixture. Individual plants of species A are self-pollinated and selected while grown with species $B(C_{B0})$. Concurrently, individuals from species B are self-pollinated while intercropped with species $A(C_{A0})$. In the next step, selected plants of species A are intercrossed to form the first improved cycle (C_{A1}); selected plants of species B are intercrossed to form C_{B1}. The next cycle is identical to the first, with C_{A1} and C_{B1} used as testers and the selected plants recombined to form C_{A2} and C_{B2}. After several selection cycles, the two resulting varieties or populations (C_{An} and C_{Bn}) should be finely tuned to grow together in the conditions under which the selection was done. Similar schemes were presented for self-pollinated and vegetatively propagated crops. This scheme has not been tested in the field.

As the preceding discussion points out, breeding for multiple cropping systems is a complicated process. The researcher must strive to define clear objectives, and limit the number of variables included in any one experiment. Testing is important, and at this stage the appropriate choice of testing environments is crucial. Only through a steady buildup of specific information on these complex systems will the level of understanding necessary to design and develop better systems, and better crop varieties for them, be achieved.

ACKNOWLEDGMENTS

The authors would like to express their sincere appreciation to many colleagues for helpful dialogue and suggestions, particularly to Charles Hall for his comments on agroecological theory. The conscientious efforts of Steffie David in typing the manuscript are gratefully acknowledged, as are those of Christopher McVoy in improving and drafting the figures.

REFERENCES

Adams, M. W., Ellingboe, A. H., and Rossman, E. C. 1971. Biological uniformity and disease epidemics, *Bioscience* 21:1067–1070.

Altieri, M. A. 1977. Ecological regulation of pests in tropical ecosystems: Bean and maize mono- and polycultures diversified with weeds, M.S. thesis, Univ. Nacional de Colombia, Instituto Colombiano Agropecuario, Bogota.

Altieri, M. A., Francis, C. A., van Schoonhoven, A., and Doll, J. D. 1978. A review of insect prevalence in maize (*Zea mays* L.) and bean (*Phaseolus vulgaris* L.) polycultural systems, *Field Crops Res.* 1:33–49.

Anand Reddy, K., Raj Reddy, K., and Devender Reddy, M. 1980. Effects of intercropping on yield and returns in corn and sorghum, *Exp. Agric.* 16:179–184.

Andrews, D. J. 1972a. Intercropping sorghum, in: *Sorghum in the Seventies,* (N. G. P. Rao and L. R. House, eds.), Oxford and IBH Publishing Co., New Delhi, India, pp. 545–556.

———. 1972b. Intercropping with sorghum in Nigeria, *Exp. Agric.* 8:139–150.

AVRDC (Asian Vegetable Research and Development Center). 1983. Annual Report. Tainan, Taiwan.

Baker, E. F. I. 1978. Mixed cropping in northern Nigeria. I. Cereals and groundnuts, *Exp. Agric.* 14:293–298.

———. 1979. Mixed cropping in northern Nigeria, III. Mixtures of cereals, *Exp. Agric.* 15:41–48.

Borlaug, N. E. 1959. The use of multilineal or composite varieties to control airborne epidemic diseases of self-pollinated crop plants, in: *Proc. First International Wheat Genetics Symp.* University of Manitoba, Winnipeg, Canada, pp. 12–27.

Brakke, J. P., Francis, C. A., Nelson, L. A., and Gardner, C. O. 1983. Genotype by cropping system interactions in maize grown in a short season environment, *Crop Sci.* 23:868–870.

Browning, J. A., and Frey, K. J. 1969. Multiline cultivars as a means of disease control, *Annu. Rev. Phytopathol.* 7:355–382.

Carballo, M., and Peairs, F. B. 1984. Effects of a maize cassava association on insect populations, in: *Progr. Rep. Int. Fund for Agricultural Development,* Centro Agronómico Tropical de Investigación y Enseñanza (CATIE), Turrialba, Costa Rica, pp. 110–113.

Carter, D. C., Francis, C. A., Pavlish, L. A., Heinrich, G. M., and Matthews, R. V. 1983. Sorghum and soybean density interactions in one intercrop pattern. *Agron. Abstr.* 1983:43.

Carter, D. C., Francis, C. A., Youngquist, W. C., and Pavlish, L. A. 1985. Evaluation of a constant soybean:sorghum plant density ratio for intercropping experiments conducted in a 'replacement series' design, (unpublished manuscript).

Catedral, I. G., and Lantican, R. M. 1978. Mung bean breeding program of the University of the Philippines, in: *First International Mungbean Symp.,* AVRDC, Tainan, Taiwan, pp. 225–227.

Chan, L. M., Johnson, R. R., and Brown, C. M. 1980. Relay intercropping soybeans into winter wheat and spring oats, *Agron. J.* 72:35–39.

Chui, J. A. N., and Shibles, R. 1984. Influence of spatial arrangements of maize on performance of an associated soybean intercrop, *Field Crops Res.* 8:187–198.

CIAT (Centro Internacional de Agricultura Tropical). 1980. Catálogo Descriptivo del Germoplasma de Frijol Común *Phaseolus vulgaris* L., Cali, Colombia.

———. 1983. *Annual Report,* Cali, Colombia.

CIMMYT (Centro Internactional de Majoramiento de Maiz y Trigo). 1983. *Annual Report,* El Batan, Mexico.

CIP (International Potato Center). (Centro Internacional de la Papa). 1983. *Annual Report,* Lima, Peru.

Clements, R. H. G., and Morris, R. A. 1982. Adaptation of soybeans to pre- and post-rice environments, Saturday seminar, 27 February 1982, International Rice Research Institute (IRRI), Los Baños, Philippines.

Coffman, W. R. 1977. Rice varietal development for cropping systems at IRRI, in: *Proc. Symp. on Cropping Systems Research and Development for the Asian Rice Farmer,* Los Baños, Philippines, pp. 359–371.

Davis, J. H. C., and Garcia, S. 1983. Competitive ability and growth habit of indeterminate beans and maize for intercropping, *Field Crops Res.* 6:59–75.

Davis, J. H. C., van Beuningen, L., Ortiz, M. V., and Pino, C. 1984. Effect of growth habit of beans on tolerance to competition from maize when intercropped, *Crop Sci.* 24:751–755.

del Rosario, D. A., and Santos, P. J. A. 1982. Screening for flooding tolerance, Paper presented at the Workshop on Varietal Improvement of Upland Crops for Intensive Cropping, 15–17 April 1982, IRRI, Los Baños Philippines.

Eberhart, S. A. and Russell, W. A. 1966. Stability parameters for comparing varieties, *Crop Sci.* 6:36–40.

Finlay, R. C. 1976. Selection criteria in intercrop breeding, in: *Intercropping in Semi-Arid Areas,* (J. H. Monyo, A. D. R. Ker, and M. Campbell, eds.), Univ. Dar es Salaam, Morogoro, Tanzania, IDRC-076e, pp. 33–36.

Finlay, K. W., and Wilkinson, G. N. 1963. The analysis of adaptation in a plant breeding programme, *Austr. J. Agric. Res.* 14:742–754.

Francis, C. A. 1978. Multiple cropping potentials of beans and maize, *Hortscience* 13:12–17.

———. 1980. Developing hybrids of corn and sorghum for future cropping systems, in: *Proc. 35th Ann. Corn Sorghum Research Conf.,* Amer. Seed Trade Assoc., Washington, D.C., pp. 32–47.

———. 1981a. Development of plant genotypes for multiple cropping systems, in: *Plant Breeding Symposium II,* (K. J. Frey, ed.), Iowa State University Press, Ames, Iowa, pp. 179–231.

———. 1981b. Rationality of small farm cropping systems, Paper presented to Farming Systems Workshop, Kansas State University, Manhattan, Kansas, November 1981.

———. 1985. Variety development for multiple cropping systems, *Crit. Rev. Plant Sci.* CRC Press, Boca Raton, Florida, 3:133–168.

Francis, C. A., Barker, T. C., and Smith, M. E. 1985. Multiple cropping for resource efficient production: A plant breeding perspective, Plant Breeding Dep. Seminar, Cornell University, Ithaca, New York, March 15.

Francis, C. A., Flor, C. A., and Temple, S. R. 1976. Adapting varieties for intercropping systems in the tropics, in: *Multiple Cropping,* (R. I. Papendick, P. A. Sanchez, and G. B. Triplett, eds.), Amer. Soc. Agron. Spec. Publ. 27, pp. 235–253.

Francis, C. A., Prager, M., and Tejada, G. 1982. Effects of relative planting dates in bean (*Phaseolus vulgaris*) and maize (*Zea mays* L.) intercropping patterns, *Field Crops Res.* 5:45–54.

Francis, C. A., Prager, M., Tejada, G., and Laing, D. R. 1983. Maize genotype by cropping pattern interactions: Monoculture vs. intercropping, *Crop Sci.* 23:302–306.

Francis, C. A., and Sanders, J. H. 1978. Economic analysis of bean and maize systems: Monoculture versus associated cropping, *Field Crops Res.* 1:319–335.

Gomez, A. A. 1976. Varietal screening for intensive cropping, Progress report II, Univ. Philippines, Los Baños, Philippines.

———. 1977. Varietal screening for intensive cropping, Progress report III, Univ. Philippines, Los Baños, Philippines.

Gomez, A. A., and Gomez, K. A. 1983. *Multiple Cropping in the Humid Tropics of Asia,* International Development Research Center, Ottawa, Canada, IDRC-176e.

Goodman, D. 1975. The theory of diversity-stability relationships in ecology, *Quart. Rev. Biol.* 50:237–266.

Gregory, P. J., and Reddy, M. S. 1982. Root growth in an intercrop of pearl millet/groundnut, *Field Crops Res.* 5:241–252.

Hamblin, J., Rowell, J. G., and Redden, R. 1976. Selection for mixed cropping, *Euphytica* 25:97–105.

Hart, R. D. 1977. Caracteristicas de variedades que pueden tener potencial como componentes de los sistemas de cultivos en Yojoa, Honduras, in: *Reunión Internacional de Colaboración Técnica,* CATIE, CIMMYT, Inst. Interamer. de Ciencias Agric., Turrialba, Costa Rica (mimeograph).

Harwood, R. R., and Price, E. C. 1976. Multiple cropping in tropical Asia, in: *Multiple Cropping,* (R. I. Papendick, P. A. Sanchez, and G. B. Triplett, eds.), Amer. Soc. Agron. Spec. Publ. 27, pp. 11–40.

Hawkins, R. 1983. Maiz asociado con sorgo en Centroamerica: Importancia, localización, y caracterización. CATIE, Turrialba, Costa Rica, mimeograph.

Heinrich, G. M., Francis, C. A., and Eastin, J. D. 1983. Stability of grain sorghum yield components across diverse environments, *Crop Sci.* 23:209–212.

Hookstra, G. H., and Ross, W. M. 1982. Comparison of F_1's and lines as female parents for hybrid sorghum seed production, *Crop. Sci.* 22:147–150.

ICRISAT (International Crops Research Institute for the Semi-Arid Tropics). 1977. Report of the cropping systems research carried out during the Kharif (monsoon) and Rabi (post-monsoon) season of 1976, Farming Systems Research Program (mimeograph).

———.1983. *Annual Report,* Hyderabad, India.

IITA (International Institute for Tropical Agriculture). 1983. *Annual Report,* Ibadan, Nigeria.

INTSOY (International Soybean Program). 1983. *Annual Report,* University of Illinois, Urbana, Illinois.

IRRI (International Rice Research Institute). 1973. *Annual Report,* Los Baños, Philippines.

———.1974. Cropping systems, in: *IRRI Annual Report for 1974,* Los Baños, Philippines, pp. 323–347.

———.1975. *Cropping Systems Program Report for 1975,* Los Baños, Philippines.

———.1983. *Annual Report,* Los Baños, Philippines.

Jennings, P. R., and Herrera, R. M. 1968. Studies on competition. II. Competition in segregating populations, *Evolution* 22:332–336.

Jensen, N. F. 1952. Intra-varietal diversification in oat breeding, *Agron. J.* 44:30–34.

Kass, D. C. L. 1984. Tolerance to flooding, in: *Progress Report to the International Fund for Agricultural Development,* CATIE, Turrialba, Costa Rica, pp. 35–41.

Krantz, B. A., Virmani, S. M., Singh, S., and Rao, M. R. 1976. Intercropping for increased and more stable agricultural production in the semi-arid tropics, in: *Intercropping in Semi-Arid Areas,* (J. H. Monyo, A. D. R. Ker, and M. Campbell, eds.), Univ. Dar es Salaam, Morogoro, Tanzania, 10–12 May 1976, IDRC-076e.

Lantican, R. M. 1977. Field crops breeding for multiple cropping patterns, in: *Proc. Symp. on Cropping Systems Research and Development for the Asian Rice Farmer,* IRRI, Los Baños, Philippines, pp. 349–357.

———.1980. Desirable characteristics of dryland crops for pre- and post-rice planting, Paper presented to Cropping Systems Conference, IRRI, Los Baños, Philippines, 3–7 March 1980.

Litsinger, J. A. 1982. Plant breeding priorities for developing insect resistant legume, maize, and sorghum varieties for rice-based cropping systems in Asia. Paper presented to Workshop on Varietal Improvement of Upland Crops for Intensive Cropping, Institute of Plant Breeding, Univ. Philippines, IRRI, Los Baños, Philippines, 15–17 April 1982.

Long, S. P., and Woolhouse, H. W. 1977. Primary production in *Spartina* marshes, in: *Ecological Processes in Coastal Environments,* (R. L. Jefferies and A. J. Davy, eds.), Blackwell Scientific Publications, Oxford, U.K., pp. 333–352.

Makena, M. M., and Doto, A. L. 1980. Soybean-cereal intercropping and its implications in soybean breeding, in: *Second Symp. on Intercropping in Semi-Arid Areas,* (Keswari and Ndungura, eds.), Univ. Dar es Salaam, Morogoro, Tanzania, 4–7 August 1980.

Meckenstock, D. 1984. Breeding aspects in the Honduran case of farming systems work, Paper presented to Workshop on Sorghum and Millets in Latin American Farming Systems, CIMMYT, El Batan, Mexico, 16–22 September 1984.

Mora, L. E. 1978. Effect of soil tillage on incidence and severity of maize and bean leaf diseases in several cultivation systems, M.S. thesis, Univ. Costa Rica and CATIE, Turrialba, Costa Rica.

Moreno, R. A. 1977. Effect of different cropping systems on severity of angular leaf spot of beans caused by *Isariopsis quiseola, Agron. Costaricense* 1:39–42.

Moseman, A. H. 1966. International needs in plant breeding research, in: *Plant Breeding,* (K. J. Frey, ed.), Iowa State University Press, Ames, Iowa, pp. 409–420.

NAS (National Academy of Sciences). 1975. Underexploited tropical plants with promising economic value, Washington, D.C.

Nerkar, Y. S. 1980. Performance of early generation lines under different cropping systems and its bearing on the selection procedure in pigeonpea breeding. In: *Proc. Int. Workshop on Pigeonpeas,* vol. 2, ICRISAT, Patancheru, India, 15–19 December 1980, pp. 159–163.

Ojomo, O. A. 1976. Development of cowpea ideotypes for farming systems in Western Nigeria, in: *Intercropping in Semi-Arid Areas,* (J. H. Monyo, A. D. R. Ker, and M. Campbell, eds.), Univ. Dar es Salaam, Morogoro, Tanzania, 10–12 May 1976. IDRC-076e, p. 30.

Okigbo, B. N., and Greenland, D. J. 1976. Intercropping systems in tropical Africa, in: *Multiple Cropping,* (R. I. Papendick, P. A. Sanchez, and G. B. Triplett, eds.), Amer. Soc. Agron. Spec. Publ. 27, pp. 63–101.

Osiru, D. S. O., and Willey, R. W. 1972. Studies on mixtures of dwarf sorghum and beans (*Phaseolus vulgaris*) with special reference to plant population, *J. Agric. Sci. Camb.* 79:531–540.

Perrin, R. M. 1977. Pest management in multiple cropping systems, *Agro-ecosystems* 3:93–118.

Poehlman, J. M. 1978. What we have learned from the international mungbean nurseries, in: *First Int. Mungbean Symp.,* AVRDC, Tainan, Taiwan, pp. 97–100.

Rao, M. R., and Willey, R. W. 1980. Preliminary studies of intercropping combinations based on pigeonpea or sorghum, *Exp. Agric.* 16:29–39.

———. 1983. Effects of genotype in cereal/pigeonpea intercropping on the alfisols of the semi-arid tropics of India, *Exp. Agric.* 19:67–78.

Rodriguez, M. W., Kass, D. C. L., and Oñoro, P. 1982. Performance in association of cultivars of cassava (*Manihot esculenta* Crantz) and cowpea (*Vigna unguiculata* Walp.) of different growth habits. CATIE, Turrialba, Costa Rica (mimeograph).

Saeed, M., and Francis, C. A. 1983. Yield stability in relation to maturity in grain sorghum, *Crop Sci.* 23:683–687.

———. 1984. Association of weather variables with genotype x environment interactions in grain sorghum, *Crop Sci.* 24:13–16.

Saeed, M., Francis, C. A., and Rajewski, J. F. 1984. Maturity effects on genotype x environment interactions in grain sorghum, *Agron. J.* 76:55–58.

Shackel, K. A., and Hall, A. E. 1984. Effect of intercropping on the water relations of sorghum and cowpea, *Field Crops Res.* 8:381–387.

Swaminathan, M. S. 1970. New varieties for multiple cropping, *Indian Farm.* 20:9–13.

Tarhalkar, P. P. and Rao, N. G. P. 1975. Changing concepts and practices of cropping systems, *Indian Farm.* 25:3–7,15.

Teare, I. D., Kanemasu, E. T., Powers, W. L., and Jacobs, H. S. 1973. Water-use efficiency and its relation to crop canopy area, stomatal regulation, and root distribution, *Agron. J.* 65:207–211.

Thompson, D. R., Monyo, J. H., and Finlay, R. C. 1976. Effects of maize height difference on the growth and yield of intercropped soybeans, in: *Intercropping in Semi-Arid Areas,* (J. H. Monyo, A. D. R. Ker, and M. Campbell, eds.), Univ. Dar es Salaam, Morogoro, Tanzania, 10–12 May 1976, IDRC-076e, p. 29.

Tiwari, A. S. 1978. Mungbean varietal requirements in relation to cropping seasons in India, in: *First Int. Mungbean Symp.,* AVRDC, Tainan, Taiwan, pp. 129–131.

Trenbath, B. R. 1975. Diversify or be damned?, *Ecologist* 5:76–83.

van Rheenen, H. A., Hasselbach, O. E., and Muigai, S. G. S. 1981. The effect of growing beans together with maize on the incidence of bean diseases and pests, *Neth. J. Plant Path.* 87:193–199.

Villareal, R., and Lai, S. H. 1977. Developing vegetable crop varieties for intensive cropping systems, in: *Proc. Symp. on Cropping Systems Research and Development for the Asian Rice Farmer,* IRRI, Los Baños, Philippines, pp. 373–390.

Wanki, S. B. C., Fawusi, M. O. A., and Nangju, D. 1982. Pod and grain yields from intercropping maize and *Vigna unguiculata* (L.) Walp. in Nigeria. *J. Agric. Sci. Camb.* 99:13–17.

Wien, H. C., and Nangju, D. 1976. The cowpea as an intercrop under cereals, in: *Intercropping in Semi-Arid Areas,* (J. H. Monyo, A. D. R. Ker, and M. Campbell, eds.), Univ. Dar es Salaam, Morogoro, Tanzania, 10–12 May 1976. IDRC-076e, p. 32.

Wien, H. C., and Smithson, J. B. 1981. The evaluation of genotypes for intercropping, in: *Proc. Int. Workshop on Intercropping,* Hyderabad, India, 10–13 Jan. 1979, pp. 105–116.

Willey, R. W. 1979a. Intercropping—Its importance and research needs. Part 1. Competition and yield advantages, *Field Crop Abstr.* 32:1–10.

———. 1979b. Intercropping—Its importance and research needs. Part 2. Agronomy and research approaches, *Field Crop Abstr.* 32:73–85.

Willey, R. W., and Osiru, D. S. O. 1972. Studies of mixtures of maize and beans (*Phaseolus vulgaris*) with particular relation to plant population, *J. Agr. Sci. Camb.* 79:517–529.

Willey, R. W., and Reddy, M. S. 1981. A field technique for separating above- and below-ground interactions in intercropping: An experiment with pearl millet/groundnut, *Exp. Agric.* 17:257–264.

Woolley, J. 1982. Selección e identificación de variedades apropiadas para los pequeños agricultores, in: *Semilla Mejorada para el Pequeño Agricultor,* CIAT, Cali, Colombia, 9–13 August 1982, pp. 75–80.

Zandstra, H. G., and Carangal, V. R. 1977. Crop intensification for the Asian rice farmer, *Agric. Mech. in Asia,* vol. 8(3):21–30.

Chapter 11

Economics and Risk in Multiple Cropping

John K. Lynam
John H. Sanders
Stephen C. Mason

Historically, agricultural researchers and administrators have neglected the small farm production systems in developing countries because they were viewed as being commercially unimportant. Recently, social concerns about the welfare of small farmers and the extremely rapid urbanization process in most developing countries have led to development projects that focus on small-farm production systems. A primary feature of these production systems is multiple cropping. However, a comprehensive understanding of the biological and economic factors involved in multiple cropping systems remains incomplete. An economic evaluation of two principal types of multiple cropping systems is presented in this chapter.

Economics provides a 2-fold input into multiple cropping research. First, a large body of literature characterizing farmer decision making in developing countries has been published since T. W. Schultz's *Transforming Traditional Agriculture* (1964). Understanding the rationale and decision criteria determining the farmers' choices of cropping systems is an essential element in the design of alternative systems. Collection and analysis of farm-level data provide diagnostic input into the experimental design of improved cropping systems. Second, the acid test of an improved cropping system is whether it is adopted, yet prior to release and promotion of new activities, discrimination among alternatives requires the use of performance criteria that approximate the farmers' methods of decision making. Economics is thus used in the evaluation of alternative cropping systems during the testing phase of the research process.

Another classic economic study was Ruthenberg's *Farming Systems in the Tropics* (1980). The first edition came out in 1956, and Ruthenberg continued to revise and expand it, turning out a third edition in 1980 shortly before his death. This chapter, following the Ruthenberg tradition, attempts not only to offer some explanations for multiple cropping, but also to stress those factors causing the systems to be dynamic, and to project the directions in which these farming systems appear to be evolving. After considering some regional differences and the dynamics of intercropping and sequential cropping, some concepts and methods for improving these systems are suggested. Finally, the conclusions synthesize the principal results and suggest several research areas.

WHY STUDY MULTIPLE CROPPING?

One reason for the interest in multiple cropping by policy makers and researchers is the assumption that multiple cropping can be a means to increased productivity, principally land productivity. Multiple cropping can lead to increases in total productivity by making fuller use of seasonally underemployed labor or capital resources. Multiple cropping may make more productive use of underemployed resources. Alternatively, it may be able to exploit some biological complementarity that comes from associating different crop species, either spatially or temporally. Nevertheless, this does not imply that there is a "free lunch" to be plucked in tropical agricultural systems; the cost side of multiple cropping is often not as rigorously considered as the benefit side.

The principal motivation for research and policy interest in multiple cropping results from the empirical observation that a significant proportion of the farmer population in developing countries utilizes multiple cropping. To produce new technology for these farmers requires an understanding of the functioning and the rationale of their present systems.

People design farming systems to satisfy basic objectives. Which objectives do farmers in developing countries pursue in the management of their farms? The issue is generally set in two principal dimensions: (1) the degree of market integration of the farmer, i.e., whether production objectives can be analyzed independently of consumption goals; and (2) the degree of certainty of outcomes in the farmer's pursuit of objectives.

In small farm agriculture in developing countries there is a continuum from subsistence farms to farms that market most of their output. There are few pure subsistence farms, even in Africa, or many purely commercial farms, at least for small-scale farmers. The average farmer in the developing world has some monetary income objectives and some subsistence objectives, and therefore consumption and production decisions are interrelated.

The other dimension is that agriculture is an uncertain enterprise, where crop output is subject to the vagaries of rainfall, temperature, pests, and diseases. As the farmer's resource base becomes more limited and the environment in which crops grow becomes more variable, security objectives and the maintenance of subsistence needs are put increasingly at risk. How semisubsistence farmers evaluate risky outcomes and their preferences for risk-taking then become elements in how farmers decide to meet their goals.

In summary, farmers, especially small farmers, often are concerned with home consumption objectives and with diversification for protection against natural and economic risks. Given these multiple objectives of farmers, one consequence is the utilization of multiple cropping. Multiple cropping is one method of crop diversification. As with a stock portfolio, there can be gains to diversification. Moreover, with multiple cropping there are often physical complementarities in the use of light, water, or nutrients (Fussell and Serafini, 1985).

The two types of multiple cropping to be considered in the next two sections are (1) intercropping (two or more crops planted at approximately the same time), and (2) sequential cropping (crops planted in sequence after the previous crop is harvested) (Sanchez, 1976). Cereal/legume intercropping is pervasive in Latin America and in sub-Saharan Africa. Sequential cropping is more characteristic of Asian rice production and has been concentrated in the irrigated river valleys with high population pressures. Double cropping in rice systems has been reported in China from the eleventh century (Beets, 1982). Although there are many other types of intercropping systems in addition to the cereal/legume combinations and the rice-based sequential cropping systems, this chapter focuses on these systems. Other issues for future study are the multiple crop systems of the humid tropics, the cereal mixtures, and the combinations of bushes or trees with cereals or grasses.

ECONOMIC EVALUATION OF INTERCROPPING SYSTEMS

Maize (*Zea mays*) and field beans (*Phaseolus vulgaris*) are found intercropped throughout the mountain highlands of Latin America and in many small farms in Brazil. In the semiarid regions of the Brazilian northeast, field beans are replaced by cowpeas (*Vigna unguiculata*) (Sanders and de Hollanda, 1979). In semiarid West Africa the cowpeas and peanuts are combined with millet, sorghum, or maize. An estimated 80 percent of the cultivated area in tropical West Africa is intercropped (Stern, 1984). Where rainfall is more evenly distributed through the year, as in Antioquia, Colombia, a relay system spreads the maize and field beans out more during the year, reducing competition and increasing yields (Sanders and Johnson, 1982). The relay is an intermediate system between intercropping and sequential cropping found where water is less limiting but full water control is not available. Surveys conducted by CIAT and published reports lead to an estimate that 60 percent of the maize and 80 percent of the field beans are produced in associated cropping systems in Latin America (Francis, 1978).

Why do small farmers in most of the developing world intercrop? Intercropping systems are not consistently more profitable than monoculture. The profitability depends on relative prices, costs, and the degree of complementarity or competition of the activities; hence, other factors affecting profitability are specific to site, time, and input level. Figure 11.1 compares the profitability of associated and monoculture bean/maize systems for the prices received by farmers during 2 crop years. In 1979 monoculture bean systems were more profitable than the associated systems, whereas these results were reversed with 1980 field bean and maize prices.

In intercropping systems net income advantages appear to be secondary to risk reduction, particularly in farm systems where subsistence is an important objective. The sources of reduced risk in intercropping as compared to sole cropping are usually attributed to both a reduced variance in output and/or net income and a higher probability of avoiding complete crop failure. In regions where water is limiting, intercropping is extremely widespread among small farmers (Sanders and de Hollanda, 1979; Norman et al., 1981) and may be the most efficient strategy for reducing the probability of crop failure. Norman (1974) and Abalu (1976) present indirect evidence for the use of intercropping in northern Nigeria as a means of reducing variability in net income. A mix of four crops was found to be the most profitable pattern, yet two-crop patterns were more

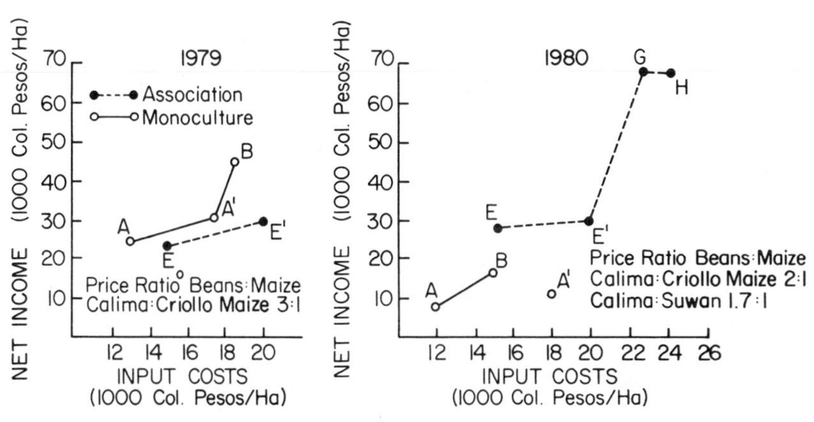

a: Mean ratio of the prices received by farmers.
A: Practices of average farmer in monoculture.
A': Best farmer practices in monoculture. Better farmers utilize fertilizer and spraying for disease and insect control.
B: Farmers' field bean variety (Calima) with improved agronomy including higher density, weed control and curative spraying.
E: Average farmer practices in association. Calima x local ("Criollo") maize variety.
E': Better farmer practices in association. Calima x hybrid maize variety with fertilizer and pesticide on the maize.
G: Calima x Suwan I without chemicals at high bean density.
H: Calima x Suwan I with chemicals at high density.

Figure 11.1 Income and input costs for farmers' fields and new technology treatments in monoculture and association, 1979 and 1980. (Reprinted by permission, J. H. Sanders and D. V. Johnson, "Selecting and Evaluating New Technology for Small Farmers in the Colombian Andes," *Mountain Research and Development* 2(3), 1982, p. 313.)

frequently observed on farmers' fields. Abalu (1976) showed that the two-crop patterns had a lower variance in net income. In semiarid areas of India, plot yield surveys demonstrated that intercropping could reduce the probability of absolute crop failure as compared with sole cropping (Singh, 1981). Farmers apparently used intercropping to reduce the risk of failure to meet subsistence needs.

Some distinctions need to be made between crop diversification and multiple cropping as potential strategies in meeting farmer objectives. Crop diversification is defined here as planting various crops on the same farm without them being interplanted. Crop diversification also would reduce yield and price uncertainty, as the weather and economic conditions would have different effects on the different crops. Walker et al. (1983) demonstrated that crop diversification significantly increases crop income stability in semiarid India. Intercropping may give more yield stability than diversification due to the interaction effects of more effectively utilizing available light, water, or nutrients or by the insulation of multiple crops to the spread of crop specific pathogens or insects. To the extent that these types of complementary effects of intercropping reduce the variance in output, crop diversification would not be a substitute for this feature of intercropping.

Crop diversification, however, can also arise as a result of within-farm variations in soil quality and/or slope (Stoop et al., 1982) and variation between planting seasons in rainfall and/or temperature. The latter type of diversification rests on different adaptation characteristics of alternative crops and is a means of maximizing net income rather than reducing risks.

One way of analyzing the explanations for intercropping is by comparing the different types of intercropping and monoculture systems. Three generalizations are proposed from observing differences in farmers' practices in developing countries:

1 Quality of the resource base, especially its homogeneity and the certainty of the correspondence between output and resource input, seems to be an important determinant of intercropping. The more variable the production conditions, due either to climate, soil variation within the farm, or pests and diseases, the more likely that intercropping will be practiced.
2 Intercropping can provide a more varied production of food from a limited land area, and most evidence suggests that intercropping produces a lower variance in per area unit net income. However, the greater the marketing of farm output, the less likely that crops will be intercropped.
3 The greater the complementarity in yields between crops in association, the more likely that intercropping will occur (Flinn, 1979).

Small farmers are expected to be more concerned with consumption or subsistence requirements and to be more risk-averse than large farmers; hence intercropping would be expected to be more common on small farms. Jodha (1977) presents data for semiarid regions of India to show that intercropping is practiced at a proportionally higher rate on small farms as compared to large farms. McIntire (1983) found no relation between farm size and intercropping in the semiarid areas of Upper Volta. In Latin America the differences in farm size are much more extreme than in Africa or Asia. In El Salvador farms with 0.5 to 2 ha had over one-half of their cultivated area in intercropping, whereas only 5 percent of the cultivated area in farms more than 50 ha in size was in intercropping (Table 11.1).

There is proportionately less intercropping of purely commercial crops, such as cotton (*Gossypium sativum*) or wheat (*Triticum* sp.). Legumes are intercropped more than other crops. Maize is widely intercropped either with a legume or a crop with a similar growth habit, such as sorghum (*Sorghum bicolor*) in El Salvador. In this case the sorghum gives the system more drought resistance in drier years. Where there is a complementary yield relationship in the mixture, large farmers are also observed to intercrop.

Irrigation leads to a significant increase in sole cropping (Jodha, 1977). Although there are farms where intercropping occurs on irrigated land in Asia, for example in China (FAO, 1980), these instances are relatively rare. The homogeneity that comes from irrigation and the control over environmental variables allow the farmer to maximize profit by growing the most profitable crop.

A distinguishing feature of multiple cropping systems is their diversity. Even within a relatively homogeneous region, cropping systems are still marked by significant diversity. Matlon (1977) found that 35 farmers identified 305 different intercropping systems in a region of northern Nigeria. Farmers exploit and adapt the inherent flexibility in multiple cropping systems to their particular agronomic, resource, and price conditions. Farmers are aware of the microvariation, whether due to soil type, soil fertility, cropping history, fertilization

Table 11.1 Cropping Pattern by Farm Size, El Salvador, 1970 to 1971

						Intercropping			
	Total cultivated	Monoculture					Maize and	Maize and	Percentage of crop area in
Farm size	area	Maize	Beans	Rice	Millet	Total	beans	millet	intercropping
0.0–0.5	13.8	6.5	0.6	—	0.2	6.1	1.2	4.9	44%
0.5–1.0	42.8	15.3	1.9	0.5	0.7	22.6	3.1	19.5	53
1–2	64.4	19.7	4.0	1.4	1.5	32.2	4.1	28.1	50
2–3	42.8	12.0	2.6	1.4	1.0	17.7	2.2	15.5	41
3–4	21.3	4.9	1.3	0.8	0.5	7.7	1.0	6.7	36
4–5	19.8	4.1	1.1	0.7	0.4	6.3	0.8	5.5	32
5–10	49.8	11.4	2.4	1.6	0.8	13.5	2.1	11.4	27
10–20	42.9	7.8	1.6	1.2	0.6	9.0	1.3	7.7	21
20–50	48.8	7.4	1.3	1.4	0.5	7.5	1.1	6.4	15
Over 50	142.0	15.0	1.1	4.8	1.2	7.2	1.4	5.8	5
Total	488.4	107.3	17.9	13.8	7.4	129.8	18.3	111.5	24

Source: Data from Dirección General de Estatísticas y Censos, 1974.

history, or soilborne pests and pathogens. Also, farmers can favor the yield of one crop over another in response to changes in relative prices by varying population, arrangement, and combination. In contrast to U.S. agriculture where specialization is the rule, in tropical agricultural systems diversification and heterogeneity are the predominant characteristics. Part of this difference results from the increased variability (both agronomic and economic) that farmers in the tropics face and the decreased control that these same farmers are able to exert over this variation. The competitive nature of purely commercialized agriculture and the well-integrated markets, typical of U.S. agriculture, induce the search for income maximization of one or a few activities for which there is a comparative advantage.

The spatial diversity of cropping systems suggests significant scope for modification of multiple cropping systems. Factors that may induce such changes may be agronomic in nature, especially changes in soil fertility. This factor is dominant in the changing crop rotations observed in shifting agricultural systems. It is found particularly where increasing population pressure on the land causes a decrease in the fallow period, a decline in fertility, and even shifts into crops with lower soil fertility requirements, such as cassava. This process is frequently observed in African farming systems (see Lagemann, 1977; Lang et al., 1984).

The dynamics of multiple cropping systems, however, are also influenced by changing economic factors. Structural changes in the agricultural sector and the economy in general will result in increasing commercialization of farm production and increasing responsiveness to changing input and output prices. The net income from intercropping field beans with maize or producing either in monoculture varies substantially with the variation in the bean/maize price ratios and the yields that can be obtained (Fig. 11.2). Since the agronomic focus is often on optimizing a land equivalent ratio, this variation in net income with the relative prices indicates the importance of the incorporation of economic analysis into evaluation of alternative farming systems. Note that the comparisons in Fig. 11.2 are with net income rather than gross income since it is unlikely that the costs of alternative activities would be the same.

Intercropping can reduce labor input as compared to sole cropping. Crops with different times to full canopy can be combined to reduce the weed problem, as for example in cassava/legume systems (Leihner, 1983). Moreover, different intercropping systems can have vastly different labor requirements. Norman (1974) found in northern Nigeria that intercropping systems had a higher total labor requirement compared to sole cropping but that intercropping gave a significantly higher return to the labor input in the peak season. There appears to be significant flexibility in adapting intercropping systems to labor availability, more so than in sole cropping systems.

A central factor determining the future of intercropping in developing countries is the impact of mechanization on these systems. The impact will be related to the evolutionary pattern of mechanization of those activities with rather general applicability (e.g., land preparation) to those with very specific applicability (e.g., cotton harvesters). Mechanization of cultural practices in developing coun-

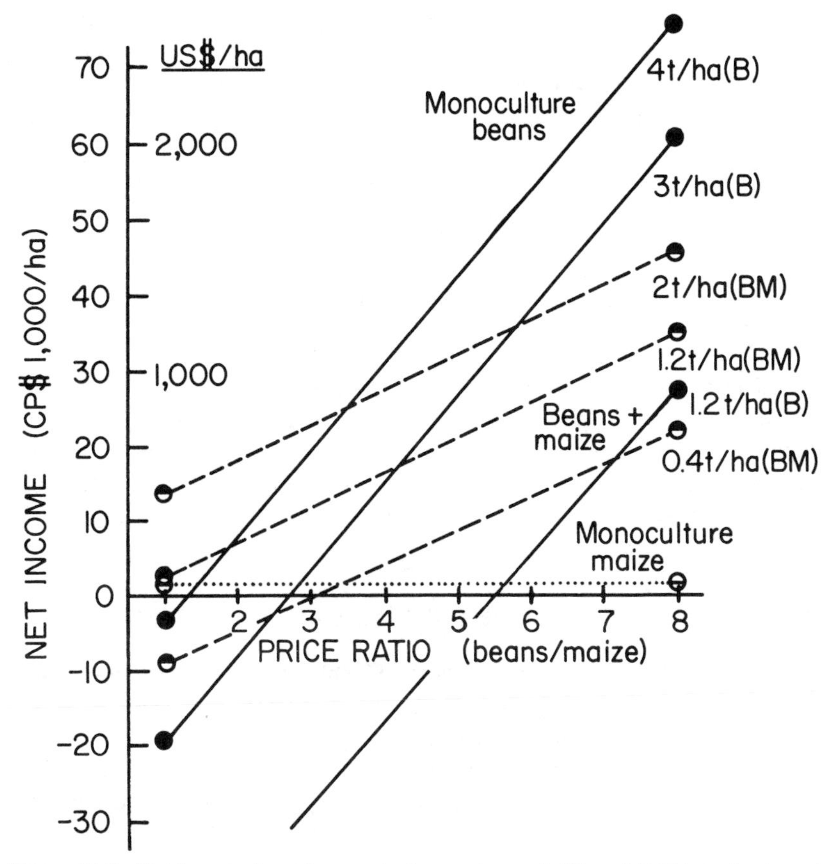

Figure 11.2 Net income from three cropping systems at different field bean/maize price ratios with different field bean yields. (Reprinted by permission from C. A. Francis and J. H. Sanders, "Economic analysis of bean and maize systems: monoculture versus associated cropping, *Field Crops Res.* 1, 1978, p. 328.)

tries frequently proceeds in the following order: land preparation or interrow cultivation, seeding, and harvesting (Binswanger, 1984).

The principal impact of mechanization of the first two operations above is to expand the area cultivated per laborer. Where it is possible to expand the cultivated area, the impact of mechanization will in most cases be positive in its effects on employment. McIntire (1983) finds in Upper Volta a significant increase in area planted but little proportional change in cropping patterns with the shift to animal traction. When seasonal labor availability becomes constraining, intercropping systems can be organized for mechanizing the between-row cultivation practice. As farmers move toward complete mechanization, however, task specialization and the uniformity of timing in the harvest will move farmers out of most intercropping systems, apart from strip cropping. In the later stages of mechanization of the seeding and harvesting, farmers generally shift to sole cropping. The employment reduction of mechanization can be substantial.

Intercropping systems appear to be a response to environmental diversity and to the multiple objectives of small farmers including subsistence and risk reduction. Shifts occur out of intercropping to sole—but often diversified—cropping systems, as the following factors occur:

1. More control over environmental risk is obtained, such as the availability of irrigation
2. Farmers become more market oriented, i.e., sell a higher proportion of their output
3. The cost of labor increases sufficiently relative to the machinery price so that the farmer mechanizes a large number of his operations

The principal advantage of intercropping appears to be risk reduction. This risk reduction is made possible by diversification and, in some cases, by the complementarity or interaction effects from growing the crops together. As input levels are increased and more environmental control is obtained, gradual shifts to more specialized production activities are anticipated and have been observed in developing and developed countries.

ECONOMIC EVALUATION OF SEQUENTIAL CROPPING SYSTEMS

Crop sequencing requires adequate moisture, sunlight, and temperature regimes during the growing season to support back-to-back activities. This system is most common in Asian irrigated systems. However, ratooning of crops, such as sorghum or sugarcane, and finding a shorter-season crop to fit into a rotation are both sequential activities found throughout the developing world including areas with seasonal water deficits. Sequential cropping of winter wheat and soybeans is an increasingly common practice in the United States (Lewis and Phillips, 1976).

An assured moisture supply provides a powerful inducement for sequential cropping. In India where agroclimatic conditions are very heterogeneous, irrigation explained over half the interstate variation in sequential cropping (Narian and Roy, 1980). The impact of irrigation on sequential cropping was moderated by the amount and distribution of rainfall. In the low rainfall zones of India the correlation coefficient between irrigation and cropping intensity was 0.71; in the median rainfall area, 0.44; and in the high rainfall areas, 0.22 (Narian and Roy, 1980).

Small farms are more likely than large farms to adopt sequential cropping because of the greater relative availability of family labor on small farms and the higher labor requirements of sequential cropping systems. In the Indian study, labor availability explained 30 percent of the variation in cropping intensity between states (Narian and Roy, 1980). Not only can these systems be constrained by labor availability, but sequential cropping significantly increases the demand for management skills (Harwood and Price, 1976). Timing of activities in the more intensive systems is crucial to the productivity of the whole system, which in turn requires accurate scheduling of labor activities and assurance of labor availability.

These features of sequential cropping provide some explanation for the greater adoption of these systems by small-scale farmers. Larger farms are constrained by the availability of management skills. For larger farms sequential cropping would require a very intensive use of the limited stock of management skills for labor recruitment and supervision. Small farms are thus able to marshall a more secure supply of labor in relation to the land area and use it more effectively. Thus, both higher labor inputs per unit area and a greater employment of sequential cropping were found on small farms in Bangladesh (Table 11.2), Taiwan, and Indonesia (Table 11.3).

The scope for increasing sequential cropping in Asia is linked very closely to expansion of irrigation. China has promoted triple cropping in many of its principal rice growing regions to replace existing double cropping systems. Critics have argued that in many regions, while gross yields have increased modestly, there has been no increase in net production and certainly no increase in labor productivity (Wiens, 1982). However, if increasing area in sequential cropping is costly, the Taiwanese experience (Mao, 1975) suggests that individual crop yields can be increased within very intensive agricultural cropping systems. The most dramatic case, however, comes from the "green revolution" in the Indian Punjab (Bartsch, 1977; Rao, 1975). A synergism between the high yielding wheat varieties and investment in tube wells for irrigation led to significant increases in cropping intensity. The profitability of the wheat varieties justified the investment in irrigation, which in turn motivated an expansion in sequential cropping.

Sequential cropping is a clear means of increasing net income over sole cropping, as long as the value of output of each successive crop covers costs and there are no critical resource constraints. Sequential cropping systems are driven essentially by environmental variation during the year and by labor availability. As stresses due to temperature and/or lack of rainfall increase in the

Table 11.2 Land Use and Cropping Intensity by Farm Size in Bangladesh, 1975 to 1976

Farm size (ha)	Cropping intensity index	Labor use per hectare (workday/ ha)	Labor use per cropped hectare (workday/ha)
0.01–0.40	133.5	222.3	168.1
0.40–0.81	123.4	199.9	162.0
0.81–1.42	131.5	216.0	164.2
1.42–2.02	130.4	213.5	163.5
2.02–3.04	121.0	196.6	159.9
More than 3.04	116.2	177.9	153.1

Source: Data from Ahmed, 1981.

Table 11.3 Cropping Intensity by Farm Size in Taiwan and Indonesia

	Taiwan (1970)		Indonesia (1963)	
Farm size (ha)	Farm size distribution	Cropping intensity index	Farm size distribution	Cropping intensity index
0.1–0.3	21.8%	206.6	18.8%	149.8
0.3–0.5	18.8	203.7	24.8	140.4
0.5–1.0	27.6	196.9	26.5	132.5
1.0–1.5	13.1	187.5	12.7	122.9
1.5–2.0	6.9	180.0	5.5	115.4
2.0–3.0	5.3	170.0	5.7	104.9
3.0–5.0	2.5	154.9	3.5	93.5
5.0 and over	0.7	131.6	2.5	59.3

Source: Data from Wang and Yu, 1975; Birowo, 1975.

second or succeeding season, there is a tendency first to plant more stress-tolerant crops and, in more extreme cases, to fallow. Where timing of a second crop is critical for yield, labor constraints for land preparation, cultivation, or harvesting may limit the extent of sequential cropping. Hence, there would be economic presssure to mechanize these critical operations if the potential profits were sufficiently high and the machinery were available. In the initial stages, mechanization could facilitate sequential cropping and increased labor utilization by spreading the labor requirements more evenly throughout the year. As the mechanization process continues, release of labor and farm consolidation would occur (Binswanger, 1984).

In summary, sequential cropping has been principally encountered in those Asian small farms that are labor intensive systems without serious limitations of water availability. Continuing investments in irrigation lead to further sequential cropping. Ultimately, nonagricultural employment opportunities will attract small farmers and farm laborers. As shortages of farm labor occur, the importance of sequential cropping will decline, as has been occurring in Taiwan in the seventies (Beets, 1982).

LOCATION AND METHODS FOR IMPROVING MULTIPLE CROPPING SYSTEMS

The significant increase in multiple cropping research in developing countries over the last two decades represents both a broadening of the scope of traditional agronomic research and, more importantly, a significant change in the research process itself. Multiple cropping research has been the precursor to the more holistic approach now termed "farming systems research" (Technical Advisory Committee, 1978; Gilbert et al., 1980), under which multiple cropping research is now a specialized component. The dominant element in this process change was the forging of an interactive link between experimental design and understanding of farmer conditions, where in many instances the experimental research was actually carried out in the farmers' fields. Multiple cropping research stresses the whole system and its interactions, thereby moving away from the component emphasis predominantly found on the experiment station.

Why multiple cropping research should have been the cutting edge in this change probably lies in two characteristic features. First, the content of the research questions forced an integration of various crop components and disciplines, since an understanding of interactions was essential. This integrative element is key to farming systems research. Second, the combinational arithmetic of factorial experiments in multiple cropping research saw the number of treatments, increasing as they do at an exponential rate, reach unmanageable size without exhausting combinational possibilities. One IRRI cropping system experiment had 750 possible treatment combinations (Beets, 1982). Exogenous decision criteria had to be imposed on experimental design to constrain the treatment number. What more logical criteria than those constraints faced by the farmer? This link between experimental design and farmers' circumstances, while initially forged by expediency, has come to be seen as an integral part of the farming systems research process.

Multiple cropping, thus, represents a fundamental change in the organization of agricultural research. At the agricultural experiment station, the research task is broken down into parts, usually crop-specific programs at the first level and component or disciplinary research at a lower level. Moreover research using *ceteris paribus* conditions implies that nonexperimental sources of variance need to be controlled. Such an organizational structure, however, raises fundamental problems in multiple cropping research.

Multiple cropping research must be concerned with system diversity and temporal stability. Diversity in multiple cropping systems arises from variation in edaphic, climatic, biological, market, and farming systems factors across the research target area. Multiple cropping research must therefore cope with a significant location-specificity problem. Not only are there more components in the cropping system, but there are also significant interactions between components, many of which are influenced by environmental factors. Therefore, breaking the cropping system down into components and then researching those components misses the point of multiple cropping systems, which is their malleability as a system. New approaches to breeding and agronomic research adopting a systems perspective are necessary if improved multiple cropping technologies are to be developed (e.g., see Rao and Willey, 1980).

Multiple cropping research must deal not only with significant spatial variation but also with temporal stability which is another key objective, especially in intercropping research. As previously emphasized, intercropping systems are especially dominant in savanna and semiarid agricultural environments of Latin America and sub-Saharan Africa. With greater risks of crop failure, temporal yield stability becomes a critical factor in selecting improved technologies. The interactions between the components of the cropping systems will be important for evaluating both productivity and stability over time (Fussell and Serafini,

1985; Walker and Rao, 1982a, 1982b). Maintaining or improving stability parameters will require innovative approaches in methodology and experimental design.

The principal lines along which multiple cropping needs to be organized are then the following: (1) a systems perspective in experimental design and testing, (2) decentralization of most of the research to farmers' fields with only the purely basic research carried out at the experiment station, and (3) priorities being set principally by field research.

On-farm research will be necessary to combine the component pieces from the experiment station and to evaluate them under the diverse conditions of farmers' fields. One key principle is that a certain degree of plasticity needs to be built into intercropping systems. Farmers will make their own adjustments in cultural practices, input levels, relative population densities, and spatial arrangements on the basis of relative prices, resource availability, and agroclimatic conditions. A clear advantage of intercropping systems over monoculture is that they can be adapted to a range of economic and environmental conditions by the farmer. This plasticity should not be sacrificed in an attempt to optimize "yield" potential of the system. Understanding how the cropping system responds over ranges of key variables such as fertilization, water availability, and plant populations will be critical to maintaining this plasticity.

Due to their greater regional diversity and the importance of risk to farmers the initial concentration in intercropping research may need to be agronomic improvements. In the improved agronomic environment, it should be easier to define the breeding requirements for new varieties (Roth and Sanders, 1985) as it already is in the sequential cropping systems. Once an improved agronomic environment is attained for intercropping systems, there will be much higher returns to breeding activities.

Sequential cropping research is easier to undertake than intercropping research because of the greater environmental control usually attained in this higher rainfall (or irrigated) and higher cash input system. However, the process of combining component parts from the experiment station for on-farm systems experiments has been the same as recommended for intercropping (Zandstra et al., 1981).

Breeding for sequential cropping systems relies on developing links between the cropping systems research programs with their on-farm, agronomic focus and the relevant crop breeding program. This linkage may be based on no more than information feedback to the breeding program, or there may be more structured linkages where advanced selection is actually done within the cropping system research sites. The principal point, however, is that the identification of a breeding role in developing a sequential cropping system must be done by the cropping system research team. The essential parameters—the specific crop, maturity length, stress tolerance, environmental conditions—are defined by the cropping system program, and where these parameters are recognized to have relatively widespread applicability, they are incorporated into a breeding project.

In summary, multiple cropping represents a change in how agricultural research is organized (i.e., a movement away from the breakdown of the whole into components to a systems focus). The emphasis is on on-farm testing of the new multiple cropping systems to resolve the location specific requirements of these complex systems and to obtain more farm level input into the research design at the experiment station.

CONCLUSIONS

Intercropping systems tend to be low input, risk-reducing approaches that enable crop diversification and the fulfillment of subsistence objectives. At higher input levels it will be necessary to reevaluate and recombine various activities. Simple economic evaluation with partial budgeting and profitability comparisons is a minimum extension to the various measures of total crop production frequently utilized to compare multiple cropping systems. Better understanding of how farmers' objectives change over time with higher incomes and increased marketed surplus appears to be a necessary input into more complex economic evaluation of new technology systems (Norman, 1982; Balcet and Candler, 1981). Those developing new technology systems have to anticipate changes in the agronomic and economic environments and evolving farmers' objectives.

Sequential cropping systems are customarily encountered where resource endowments, especially water availability, are more adequate than in intercropping systems. These sequential cropping systems utilize higher inputs, and income maximization appears to be a much more important objective than in the case of intercropping. Economic evaluation of the potential profitability of alternative activities is thus considerably easier. The very high labor input requirements during the crop season with sequential cropping indicate the potential for employment creation in these systems. However, in the long run many of the smallest farms will find more remunerative activities outside agriculture as has been occurring in the Japanese, South Korean, and Taiwanese systems.

Any technology evaluation technique should be first concerned with adequately defining farmer objectives. For example, yield stability analysis across sites is not the concept of risk appropriate to farmers. Yield variance over time on the same site would be useful to a farmer especially if the mean yield of the new activity were higher than that of the farmer's present activities. Yield variance between sites has been shown to be a poor proxy for yield variance over time (Evenson et al., 1978; Watson and Anderson, 1977). Low cost methods for obtaining improved estimates of yield variance over time are badly needed.

Multiple cropping research has helped focus the agricultural research process more on understanding the farmers' production systems and defining the farmers' constraints to productivity improvement. The importance of the combined effects of various changes in the component parts of new technology and the location specific nature of most agricultural technology are strong forces pushing toward doing more agricultural research on farmers' fields. The systems approach to technology development attempts to integrate the component parts, to evaluate their combined effects, and to study the interactions. The systems approach and the on-farm trials will supplement and help orient experiment station research. On-farm trials are a complement not a substitute for the more basic research on the experiment station.

To analyze the performance of multiple cropping research attention will need to be focused on profitability. Agronomists object that profitability can be a poor performance indicator due to seasonal and annual crop price variability and the difficulty of adequately estimating certain costs such as the returns to family labor at different times of the crop season (Beets, 1982). These are valid objections and an adequate response is necessary to improve interdisciplinary collaboration. Sensitivity analysis of the variation in net income over the crop price ranges expected by farmers and programming techniques to obtain shadow prices of the on-farm resources can handle these objections. The fundamental point is that farmers are more interested in raising profits than in maximizing yields so it is important to move rapidly to economic analysis of the on-farm trials. Moreover, low-income farmers are expected to be concerned with minimum consumption requirements and with the temporal stability of profits, especially in intercropping systems. Another part of the evaluation will be in simulations to consider alternative states of nature for which there are data gaps from the on-farm and experiment station data. This modeling will help supplement and organize the data from on-farm trials and give insights into future research areas on farms and on the experiment station.

REFERENCES

Abalu, G. 1976. A note on crop mixtures under indigenous conditions in northern Nigeria. *J. Develop. Stud.* 12:212–220.

Ahmed, I. 1981. Farm size and labor use: Some alternative explanations, *Oxford Bull. Econ. Statist.*, 43:73–88.

Balcet, J. C., and Candler, W. 1981. *Farm technology adoption in northern Nigeria*, World Bank, Washington, D.C. (mimeograph).

Bartsch, W. H. 1977. *Employment and Technology Choice in Asian Agriculture*, Praeger Publishers, New York.

Beets, W. C. 1982. *Multiple Cropping and Tropical Farming Systems*, Westview Press, Boulder, Colorado.

Binswanger, H. P. 1984. *Agricultural mechanization: A comparative historical perspective*, World Bank Staff Working Paper 673, World Bank, Washington, D.C.

Birowo, A. 1975. Employment and income aspects of the cropping system in Indonesia, *Philippine Econ. J.*, 14:272–278.

Dirección General de Estatística y Censos. 1974. *Tercer Censo Nacional Agropecuario*, 1971, San Salvador, El Salvador.

Evenson, R. E., O'Toole, J. C., Herdt, R. W., Coffman, W. R., and Kauffman, H. E. 1978. Risk and uncertainty as factors in crop improvement research, IRRI Research Paper Series 15, IRRI Manila, Philippines, 19 pp.

FAO (Food and Agriculture Organization of the United Nations). 1980. *China: Multiple Cropping and Related Crop Production Technology*, FAO Plant Production and Protection Paper, Rome.

Flinn, J. 1979. Agronomic considerations in cassava intercropping research, in: *Intercropping with Cassava: Proc. International Workshop*, Trivandrum, India,

(E. Weber, B. Nestle, M. Campbell, eds.) International Development Research Center, Ottawa, Canada.

Francis, C. A. 1978. Multiple cropping potentials of beans and maize, *Hortscience,* 13(1):12–17.

Francis, C. A., and Sanders, J. H. 1978. Economic analysis of bean and maize systems: Monoculture versus associated cropping, *Field Crops Res.* 1:319–335.

Fussell, L. K., and Serafini, P. G. 1985. Crop associations in the semi-arid tropics of West Africa: Research strategies past and future, seminar paper presented at the Workshop on Technologies Appropriate for Farmers in Semi-Arid West Africa, Ougadougou, Burkina Faso, April 1985.

Gilbert, E. H., Norman, D. W., and Winch, F. E. 1980. *Farming Systems Research: A Critical Appraisal,* Rural Development Paper 6, Michigan State University, East Lansing, Michigan.

Harwood, R. R., and Price, E. C. 1976. Multiple cropping in tropical Asia, in *Multiple Cropping,* (R. I. Papendick, P. A. Sanchez, and G. B. Triplett, eds.), Amer. Soc. Agron. Spec. Publ. 27, Madison, Wisconsin.

Jodha, N. S. 1977. Resource base as a determinant of cropping patterns, in: *Symposium on Cropping Systems Research and Development for the Asian Rice Farmer,* IRRI, Los Baños, Philippines.

Lagemann, J. 1977. *Traditional African Farming Systems under Conditions of Increasing Population Pressure,* African-Studen des Ifo-Institutes, Munich.

Lang, M., Cantrell, R., and Sanders, J. 1984. Identifying farm level constraints and evaluating new technology in the Purdue Farming Systems Project in Upper Volta, in: *Animals in the Farming System,* (C. B. Flora, ed.), Kansas State University, Manhattan, Kansas.

Leihner, D. 1983. *Management and Evaluation of Intercropping Systems with Cassava,* Centro Internacional de Agriculture Tropical, Cali, Colombia.

Lewis, W. M., and Phillips, J. A. 1976. Double cropping in eastern United States, in *Multiple Cropping,* (R. I. Papendick, P. A. Sanchez, and G. B. Triplett, eds.), Amer. Soc. Agron. Spec. Publ. 27, Madison, Wisconsin.

Mao, Y. 1975. Implications of Taiwan's experience in multiple-crop diversification for other Asian countries, *Philippine Econ. J.* 14:216–234.

Matlon, P. J. 1977. The size distribution, structure, and determinants of personal income among farmers in the north of Nigeria, Ph.D. thesis, Cornell University, Ithaca, New York.

McIntire, J. 1983. Two aspects of farming in SAT Upper Volta: Animal traction and mixed cropping, West Africa Economics Program Progress Report 7, International Crops Research Institute for the Semi-Arid Tropics (ICRISAT), Ougadougou, Upper Volta.

Narain, D., and Roy, S. 1980. *Impact of Irrigation and Labor Availability on Multiple Cropping: A Case Study of India,* Research Report 20, International Food Policy Research Institute, Washington, D.C.

Norman, D. W. 1974. Rationalizing mixed cropping under indigenous conditions: The example of north Nigeria, *J. Develop. Stud.* 11:3–21.

———. 1982. Socio-economic considerations in sorghum farming systems, in *Sorghum in the Eighties,* vol. 2, ICRISAT, Andhra Pradesh, India, pp. 633–646.

Norman, D. W., Newman, M. D., and Oedraogo, I. 1981. *Farm and Village Production Systems in the Semi-Arid Tropics of West Africa: An Interpretive Review of Research,* Research Bull. 4, ICRISAT, Patancheru, India.

Rao, C. H. H. 1975. *Technological Change and Distribution of Gains in Indian Agriculture,* The Macmillan Publishing Company of India Limited, Delhi.

Rao, M. R., and Willey, R. W. 1980. Evaluation of yield stability in intercropping studies on sorghum and pigeon pea, *Exp. Agric.* 16:105–116.

Roth, M., and Sanders, J. H. 1985. An economic evaluation of selected agricultural technologies with implications for development strategies in Burkina Faso, International Programs in Agriculture (IPIA), Staff Paper, Purdue University, West Lafayette, Indiana.

Ruthenberg, H. 1980. *Farming Systems in the Tropics,* 3d ed., Oxford University Press, Cambridge, U.K.

Sanchez, P. A. 1976. *Properties and Management of Soils in the Tropics,* John Wiley & Sons, New York.

Sanders, J. H., and de Hollanda, A. D. 1979. Technology design for semi-arid Northeast Brazil, in: *Economics and the Design of Small Farmer Technology,* (A. Valdes, G. M. Scobie, and J. L. Dillon, eds.), Iowa State University Press, Ames, Iowa, pp. 102–111.

Sanders, J. H., and Johnson, D. 1982. Selecting and evaluating new technology for small farmers in the Colombian Andes, *Mountain Res. Develop.* 2:307–316.

Schultz, T. W. 1964. *Transforming Traditional Agriculture,* Yale University Press, New Haven, Connecticut.

Singh, R. P. 1981. Crop failure and intercropping in the semi-arid tropics of India, Economics Program Progress Report 21, ICRISAT, Patancheru, India.

Stern, K. G. 1984. *Intercropping in Tropical Smallholder Agriculture with Special Reference to West Africa,* German Agency for Technical Cooperation (GTZ), D-6236, Eschborn, Germany. 310 pp.

Stoop, W. A., Pattanayak, C. M., Matlon, P. J., and Root, W. R. 1982. A strategy to raise the productivity of subsistence farming systems in the West African semi-arid tropics, *In: Sorghum in the Eighties,* (L. R. House, L. K. Mughogo, and J. M Peacock, eds., ICRISAT, Andhra Pradesh, India, pp. 519–526.

Technical Advisory Committee, Consultative Group on International Agricultural Research. 1978. Farming systems research at the international agricultural research centres, FAO of the United Nations, Rome, Italy.

Walker, T. S., and Rao, K. V. S. 1982a. Yield and net return distributions in common village cropping systems in the semi-arid tropics of India, Economics Program Progress Report 41, ICRISAT, Patancheru, India.

———. 1982b. Risk and the choice of cropping systems: Hybrid sorghum and cotton in the Akola region of central peninsular India, Economics Program Progress Report 43, ICRISAT, Patancheru, India.

Walker, T. S., Singh, R. P., and Jodha, N. S. 1983. Dimensions of farm-level diversification in the semi-arid tropics of rural south India, Economics Program Progress Report 51, ICRISAT, Patancheru, India.

Wang, Y. T., and Yu, T. Y. H. 1975. Historical evolution and future prospects of multiple-crop diversification in Taiwan, *Philippine Econ. J.* 14:26–46.

Watson, W. D., and Anderson, J. R. 1977. Spatial versus time series data for assessing response risk, *Rev. Market. Agric. Econ.* 45:80–84.

Wiens, T. B. 1982. The limits to agricultural intensification: The Suzhou experience, in China under the Four Modernizations. Part 1. U.S. Congress, Joint Economic Committee, Washington, D.C.

Zandstra, H. G., Price, F. C., Litsinger, J. A., and Norris, R. A. 1981. *A Methodology for On-Farm Cropping Systems Research,* IRRI, Los Baños, Philippines.

Chapter 12

Sociocultural Factors in Multiple Cropping

Stillman Bradfield

A computer-assisted library search on "sociocultural factors in multiple cropping" produces no results. Even changing "multiple cropping" to "cropping systems" doesn't improve the yield. Yet, in fact, there is a large amount of literature in the social sciences having to do with sociocultural factors in multiple cropping. In anthropology, one can find ethnographies of societies all over the world that are engaged in multiple cropping. Some of the most interesting include studies by Carneiro (1961), Conklin (1957), and Rappaport (1968) on slash-and-burn horticulture in the tropics. One of the advantages of reading ethnographies of peoples in an area of interest to agricultural researchers is they not only provide descriptions of the farming systems, but attempt to provide a functional analysis of different aspects of the system. Usually, the ethnographer tries to get the farmer's point of view as to why he or she makes certain decisions.

There is another large body of literature in the behavioral sciences that deals with peasant cultures in many parts of the world. Peasants are usually considered to be a different population from slash-and-burn horticulturalists, in that peasants by definition are part of a larger state and produce a surplus for market and are subject to taxation by outsiders. While sociologists and anthropologists have predominated in their descriptions of peasant societies, others such as the political scientist Banfield (1958) have also provided valuable descriptions of peasant society. The relatively small number of behavioral scientists working in this area in this century has led them to spread out to many different areas of the world in order to record indigenous cultures before they have been radically altered by contact with the rest of the world. One unfortunate side effect of this tendency is that we do not have many cases where we have either continuous observation and recording or repeated revisits to the same village. Exceptions to this rule are Foster's (1967) work in Tzintzuntzan in Mexico, and Robert Redfield (1930) and his student Oscar Lewis (1960) who studied Tepoztlan in Mexico some years apart and provided very different interpretations of the village.

One of the key differences between behavioral sciences and physical sciences lies in the inability of behavioral scientists to keep any type of controlled experiments with people going over a considerable period of time. Therefore, we will never be able to produce a social science equivalent of the Rothamsted Experimental Station (1981) experiments in England where, every year since 1843, wheat (*Triticum* sp.) has been sown and harvested on all or part of the same field. Their continuous evaluation of organic vs. inorganic fertilization and other treatments simply has no parallel in social science. Perhaps the most important pioneer work in multiple cropping where we have good descriptive accounts as well as photographs is that contained in F. H. King, *Farmers of Forty Centuries* (1911). King toured China, Manchuria, Korea, and Japan in 1908 and recorded his observations on multiple cropping at that time. Apparently, this book is ancestral to all modern multiple cropping research. It would be fascinating if a team consisting of an agronomist and a social scientist were to retrace King's journey and bring his descriptions up to date.

Most of the early collaboration between behavioral scientists and agriculturalists was achieved through the extension services of various countries. Since the extension services were focused mainly on monocrop agriculture, there was little if any impact of this collaboration on multiple cropping work. After World War II, there was a considerable movement to integrate both cropping research and the work of social scientists interested in cultural change through what was then called the community development programs, particularly in the 1950s and 1960s. Systematic research in multiple cropping was started at the International Rice Research Institute (IRRI) in the 1960s and has subsequently spread to some of the other international centers such as Centro Agronómico Tropical de Investigación y Enseñanza (CATIE) and some of those in the Consultative Group in International Agricultural Research (CGIAR) network. By the 1970s, some national research organizations such as ICTA in Guatemala had also moved in this direction.

Once multiple cropping research was moved from the experiment stations to farmers' fields, social scientists began to be incorporated into the research process. In some places, this emphasis has been dropped, and in others has been expanded as a result of the successful collaboration between social scientists and agricultural scientists in these teams. For example, Hildebrand's (1978) work at ICTA in Guatemala yielded a number of novel approaches to research with farmers, including the *sondeo,* a joint reconnaissance survey carried out by teams consisting of one agricultural scientist and one social scientist. Perhaps the most successful and consistent collaboration between social scientists and agricultural scientists has been at the International Potato Center (CIP) in Peru, where Horton (1984) has recently summarized the impact of the social sciences on agricultural research.

In addition to reports published at the various centers where multiple cropping research is going on, reports by social scientists can also be found in some of the newer journals and newletters, such as *Culture and Agriculture,* published by the Department of Anthropology at the University of Arizona, and a new journal, *Mountain Research and Development.* Topics of special interest to social scientists are now being collected into edited volumes such as that by Barlett (1980) on agricultural decision making. Other volumes such as those by Whyte and Boynton (1983), Wagley (1974), and Moran (1981) pull together the work of both social scientists and agricultural scientists working on the same project or in the same area. Similarly, there are a number of detailed accounts of some of the better-known development programs such as Comilla, the Puebla Project, and the Caqueza Project.

BACKGROUND OF COLLABORATION BETWEEN SOCIAL AND AGRICULTURAL SCIENTISTS

There appear to be two main forces responsible for the incorporation of social scientists into agricultural development since the 1970s. The first of these is the Foreign Assistance Act of 1974, variously known as "New Directions" or "Congressional Mandate," which requires the Agency for International Development (AID) to "conduct its programs as if poor people mattered," to paraphrase E. F. Schumaker (Steinburg, 1980–1981). AID is now required to consider the effects of its programs on the lives of the people in the affected area. Social analyses are required of every project, and social science participation in the Country Development Strategy Statement and Project Evaluation is also built in. Apparently, by law, social scientists will have to be involved from the earliest planning stages of a project right through to evaluation. In the past, social scientists were frequently brought in to do a postmortem after everything had gone wrong with a project, or at best to help write the final reports and recommendations that had already been decided by specialists in biological sciences.

The second main thrust has been a Rockefeller Foundation postdoctoral research program in agricultural and rural development for the social sciences, also started in 1974. After completing their program, some of these fellows have accepted employment at the CGIAR centers but few, if any, are in core staff positions which are permanently funded.

DIFFICULTIES OF COLLABORATION BETWEEN SOCIAL AND AGRICULTURAL SCIENCES

At first blush, it would appear that all sciences should be able to collaborate easily with one another since they are all concerned with the same two basic sorts of questions. The first of these could be called static analysis, having to do with issues of structure and function. That is, all sciences are concerned with explaining the relationship between different parts of a system, and all are concerned with explaining how any given part contributes to the maintenance of the system. The second question has to do with dynamic analysis, or how systems or phenomena change over time. However, in the real world multidisciplinary research teams have a great deal of trouble starting with the most basic problem of defining the role and purpose of the team.

Even such apparently simple goals as increasing productivity can have many meanings. Does the team wish to increase productivity per unit of land? This is usually readily attainable by lavishing more labor on small plots of land. Although economists may argue that it pays to increase the amount of labor until the marginal product falls to zero, most people don't want to earn zero wages for their efforts. On the other hand, W. Arthur Lewis (1977) argued that none of the industrialized countries achieved industrialization without a prior increase in productivity in agriculture. He pointed out that human productivity in agriculture must increase in order to raise the real wages of farmers and thereby provide a market for the products of industry. Should, therefore, the research team be concerned with increasing the productivity of labor? If so, should we be trying to raise the productivity of labor only on a particular crop for a particular season or should we be looking at the productivity of labor of the entire farm family throughout the year? If the latter is the case, then we would have to consider not only off-farm labor, but modifications in the production system that would smooth out the peaks and valleys of labor demand on the farm to enable family members to optimize their productivity over the full agricultural cycle.

In the past, most commodity programs have emphasized increasing the productivity per acre of that commodity. Clearly, in multiple cropping research, one is constantly thinking of the productivity of the total system and not of a single commodity. But even when the systematic approach is taken, are we still to be mainly concerned with productivity either measured in terms of land or labor, or should we also be thinking about the profitability of the system? At a time of sharply rising input prices and relatively constant product prices, there should be a search for systems of production that use less in the way of purchased inputs in order to improve profitability, even if production declines. Harwood and Banta (1974) called this "substitutive technology" and argued that it would be appropriate for small farmers who could not afford the cash outlays required to adopt high technology.

Beyond the problem of defining the general goals of the research program, we find that team ideas as to how to best realize these goals will be at least as varied as the number of disciplines represented on the team. Disciplinary training sharpens the sensitivities of its members to the existence of certain problems and possibly supplies blinders that makes it difficult to recognize other problems. A weed specialist visiting a farm sees weeds and thinks of ways to control them, whereas a plant breeder thinks in terms of genetic improvements that can be made in the crop of his or her specialty. Social scientists, on the other hand, have a much wider and less specific area of focus, and are inclined to see more problems than other members of the team really want to hear about. Moreover, the most intractible of these problems frequently are in the areas of social institutions, where a research team has little or no influence. Quantitatively oriented scientists are particularly frustrated when many of the most serious problems prove difficult, if not impossible, to quantify in terms of their existing models or computer programs. Scientists used to having tight control over experimental procedures find it difficult to deal with all of the factors that affect a farmer's decisions with respect to technology. Similarly, the considerations a farmer has to keep in mind with respect to all of his or her various crops and animals are far more numerous than the considerations occupying the mind of a maize (*Zea mays* L.) specialist who is accustomed to thinking only of the variables that affect the physical production of one crop of interest.

Economists were the first of the social scientists to be fitted into multidisciplinary agricultural teams, probably because of their quantitative orientation and ability to provide useful techniques of analysis which incorporated both the physical and the economic factors. Hildebrand (1977) summarized the dilemma as follows:

> The problem stems from having most top level technology 'generators' who are agriculturally trained and 'product'-oriented, working on experiment stations or other highly controlled conditions where they consider only a limited number of variables; most of the 'transfer-mechanism generators', who are trained in the social sciences are not 'cause' but product oriented, struggling with the vast quantity of variables which condition acceptance or rejection of the technology at the farm level; and 'goal'-oriented agricultural economists in the middle complaining that the agricultural scientists do not consider enough of the variables in their work, but ignoring the pleas of the social scientists that including just the quantifiable variables is not sufficient either.

Once the goals of the multiple cropping team have been adequately defined, new problems of research methodology immediately appear. CGIAR, which now has 13 widely scattered research stations, has organized most of its work on a commodity (crop or animal) program basis. Focusing the work of many disciplines on the problems of a particular crop or animal has found increasing acceptance in agricultural research both at the international level and in national agricultural research programs. The model is a direct outgrowth of its successful application in the Rockefeller Foundation program on maize and wheat in Mexico where both the International Center for the Improvement of Maize and Wheat and the National Institute for Agricultural Research continue this emphasis today. Much of the success of the so-called green revolution is traceable to this concentration of effort on a particular commodity. No doubt the methodology will continue to be of great value for the study of certain kinds of problems. However, the strategy is based on fairly traditional reductionist thinking, which seeks to limit the variables under consideration to as few as possible, and to those most amenable to control by the various disciplines involved. The procedure requires identification of one or more limiting factors which, at a given point in time, are preventing productivity increases. As one limiting factor is overcome, another is identified, and the research proceeds in an orderly, linear fashion. The remarkable advances in wheat and rice breeding in recent years indicate that the tool is indeed powerful.

However, those teams that are now beginning to focus on multiple cropping research are finding the model not to be transferable to their concerns. Work on whole-farm systems requires holistic thinking of the ecology or systems theory type which, rather than proceeding linearly takes account of positive and negative feedback and multiple causation, and represents a very different way of thinking about agricultural production. One of the most serious problems of any multidisciplinary team is trying to get everybody on the team thinking in terms of the entire system, and not just the components they are used to dealing with. Economists have proven themselves useful in both kinds of research effort, but I suspect that interest by sociologists and anthropologists in national and international centers will be focused more on whole-farm systems programs. Nevertheless, the International Potato Center has recently reported on cases where consultations between potato storage specialists and anthropologists led to radical changes in the definition of technical problems. Similar stories have been coming out of IRRI since it acquired its first anthropologist.

Agronomists have always recognized the need to adapt general crop recommendations to such local conditions as variations in soil conditions, slope, and distribution of rainfall. They have not always recognized the simplifying and homogenizing effects of tractor-based technology on farming systems. Large-scale production for the market leads to specialization in one or at best a few commodities.

By way of contrast, small farms in the Third World are generally not mechanized, are not exclusively oriented to the market, and therefore grow many commodities in multiple cropping combinations throughout the year. Their systems are complicated from an agronomic point of view, and the constraints affecting farmers' decisions are not simply removed. For example, when most of the work is done by the farm family without mechanical help, the number, age, and sex composition of the family is a crucial limiting factor which changes slowly through time, necessitating periodic adjustments in the farming system. A farmer with a mechanized, commercial farm in the American Midwest can decide to double or halve the acreage in maize without changes in the labor force or in equipment. The small, nonmechanized farmer in the Third World has to deal with a more complex set of trade-offs when considering changes in the farming system. Social scientists have made substantial contributions toward understanding these systems (e.g., see Barlett, 1980; Roumasset et al., 1979.)

Another consequence of the commodity-based research organization is that the scientist's attention is focused on the plant itself and the problem in overcoming obstacles to increasing the productivity of that plant. Thus it is possible to have a very successful commodity program in terms of increasing genetic capacity, breeding for resistance to pests or disease, and still have little or no effect on production in a given area because the plants developed simply do not fit into the multiple cropping systems of small farmers. A very sophisticated maize breeding program in Mexico for the past 40 years has had very little impact on the small-farm sector. On the other hand, the wheat program was very similar in its basic strategies but was spectacularly successful, in large part because Mexican wheat farmers were large-scale commercial farmers with all the necessary infrastructure support to take advantage of the new wheat. Similar cases of a lack of fit, or of a spectacular fit, have been noted in India and other parts of the Far East.

The multiple cropping approach, on the other hand, requires that problems be defined in terms of a whole production system and not simply one crop. Moreover it requires realistic attention to constraints on labor, financial resources, input availability, markets, and problems of this sort which are not considered in traditional experiment station work. Indeed one of the most vexing aspects of multiple cropping research on the farm is the lack of control by either the farmer or the researcher over crucial variables that limit production.

One of the first requests that the social scientist is likely to get from an agricultural research team is to get out to the villages and find out what the farmers are doing and why they are doing it. Scientists want to know why the superior technological packages developed over the past few decades have not been acceptable to small farmers. If the economic benefits of the technological package have also been proven to the scientist's satisfaction, why are the farmers not adopting the recommended technology? This is certainly a legitimate request, and it would be easy for the social scientist to become isolated from the rest of the team and occupy all of the time doing field studies of this sort, as well as studies of the impact of outside institutions, such as credit agencies, input suppliers, and marketing organizations, on farmers' decisions. In many cases, this could well be the major contribution of the behavioral scientist. However, the danger of succumbing to this temptation is that the social scientist would probably not have very much impact on the decisions made by the rest of the team. It is essential that multidisciplinary teams consist not only of specialists in various disciplines, but that each member of the team perform as a multidisciplinary person, with primary responsibility in one particular area, but with rights and obligations to participate in the disciplines of others. If this does not happen, the team will not develop the esprit de corps and loyalty to the team objectives that is required. Hildebrand (1977) has found it most useful to send an agronomist and social scientist out together to interview farmers. This ensures not only that members of different disciplines are getting the same information from farmers, but also that questions relevant to each discipline get asked. Moreover, no team is going to include all possible specialities within agricultural research, so various members are going to have to risk opinions in areas where they have no formal training. For example, none of the international centers has poultry specialists on its staff, yet small farmers all over the world raise poultry as an integral part of their farm operations.

Since World War II, behavioral scientists have been involved in programs of culture change and development, with one of their main tasks being the identification of cultural constraints that hinder change. At times, they have been justly accused of showing how all sorts of exotic behaviors were functional in a particular society. This kind of emphasis led to a long list of possible constraints with which change agents must deal. Some anthropologists have argued that in many cases a culture should not be tampered with at all for fear that changing cultural traits in one area might lead to the disruption of functional linkages that would bring on a total social collapse.

Rather than present a list of individual cultural factors that should be taken into account in attempting to bring about agricultural development (Foster, 1973), we may simply group them into three major categories in order to examine the kinds of problems these present to the change agent. Since our final goal involves changing human behavior, our immediate task must be to ask the question, "How do we explain human behavior?"

PSYCHOLOGICAL FACTORS

The study of diffusion of culture traits has always been a favorite topic of the social sciences. Similarly, this discipline has given a great deal of attention to the processes of invention and innovation in general. Barnett (1953) presents a thorough synthesis of this research. What are the conditions that favor adoption of new technology from other areas, or the development of new technology within a given culture (Rogers and Shoemaker, 1971)? Many researchers have sought the explanation of human behavior by looking to the internal states of the individual, viewing behavior as a function of attitudes, values, beliefs, and knowledge. This being the case, if behavioral change were to take place, changes would have to be sought in these mental states.

This was the approach of many social scientists during the community development phase of international development in the 1950s. It has always been the strategy of missionaries and other agents of change who find themselves in a rather powerless position. The general assumption is that, since people are not doing those things that lead to optimal rates of development, and since they act according to their state of knowledge, attitudes, beliefs, values, etc., these psychological or mental factors must be changed first in order to change behavior.

The appeal of this strategy to relatively powerless change agents lies in its rational approach to individuals and groups and the fact that it does not require institutional change. It appeared reasonable at the time, yet we found in the massive community development program that it was not very effective. Later, psychologist David McClelland (1961) sought to explain differences in rates of development in terms of the amount of need-achievement which had been inculcated into individuals as small children. This approach also was found to have rather hopeless policy conclusions and has been largely abandoned. We now suspect that the reason it did not work is that people's behavior is in fact normally quite rational, given the circumstances of their lives. New behaviors that lack the institutional and environmental supports to make them pay off will not be adopted, or if adopted will not persist for lack of support. This realization has led some behavioral scientists to concentrate their attention in other areas.

INSTITUTIONAL CONSTRAINTS

The assumption here is that behavior is rewarded or punished by a system of incentives built into the basic institutions of the society. If behavior does not lead to optimal development, it must be because the institutional structure encourages less than optimal behaviors. This realization led to the institution-building phase in development work. Sometimes, it may be that a simple restructuring of the price system is all that is needed to reward newly desired practices and/or to punish old practices. Other approaches require a massive overhauling of virtually all of the social institutions impinging on farmers' behavior (Mosher, 1969). This approach is more attuned to behavioral psychology, whereas the emphasis on internal states elaborated above is associated with the psychodynamic theories of psychology. The principal problems of trying to pursue the behavioral approach derive from the outsider's lack of power to change the institutions that shape the behaviors of farmers. Moreover, since the structure of advantage has built up winning groups who have been able to rise to the top and stay there, vested interests are normally arrayed against changes that would jeopardize their favored position. Although this is sometimes interpreted as the inherent conservatism of Third World societies, except under extreme circumstances, life under the present institutional structure is predictable and the rewards are reasonably certain. Most people have some hope of being able to achieve satisfactorily within the system and therefore are not ordinarily inclined toward revolutionary change. If most behavior makes sense most of the time, we must look at some other factors that may explain why people do what they do.

ECOLOGICAL AND ENVIRONMENTAL FACTORS

Marvin Harris (1979) calls this strategy "cultural materialism," and argues for focusing attention first on basic infrastructural variables such as climate, resource base, and population characteristics. Theorists using this approach also include technology as part of the basic infrastructure as technology shapes the institutional structure of the society. Harris argues that in order to understand the constraints on development, one must look first at the infrastructural level since, "It's a good bet that these constraints are passed on to the structural and superstructural components."

All three of these strategies have some validity and are therefore useful. Many of our past failures are due to our tendency to quickly identify one or a few components in a problem and to seek a solution based on dealing with those particular components, while simply assuming away the relevance of other variables, or assuming that factors required in the other two strategies are in place, when in fact they are not. For example, we might recommend a new practice, including a particular seed variety plus cultivation practices while overlooking the absence of the necessary institutional structure to deliver the inputs and credit needed to apply that particular technology. Or, we might ignore tastes and other market considerations that might affect the adoption of new varieties.

We have found that little can be assumed to be in place, and that we have failed regularly in the Third World whenever we allow ourselves the luxury of the unconscious assumption that the requisite institutions function there as they do in the United States. For example, an extension agent may feel that the physical conditons warrant specific soil conservation practices, or a particular cropping pattern or technology, and make recommendations accordingly. But such things as land ownership patterns, tenancy conditions, lack of credit, markets, and other necessary institutional supports may be lacking to the point where such advice is useless. Therefore, we may assume, with Schultz (1964), that farmers usually make efficient use of the combination of resources that are available to them in the institutional climate in which they operate. To make sense of farmer behavior we need to be much more specific as to how these environmental, institutional, and psychological variables shape their decisions on how to manage their farms.

It is not surprising that people from different disciplines would have difficulty in communicating with one another, nor is it surprising that their priorities may differ considerably. It may be surprising to some, however, to learn that people within the same discipline frequently differ markedly in their approach to any given problem. Pablo Gonzalez Casanova (cited by Kahl, 1976) has made the claim that, generally speaking, quantitatively oriented sociologists tend to be conservative in their outlook, whereas qualitatively oriented sociologists tend to be more radical. This seems to fit with the experience of many American sociologists in Latin America where the North Americans find themselves out of tune with Latin American sociologists who are more qualitatively and historically oriented.

The new technological advances represented by the term "green revolution" were presented to the public as "scale-neutral." That is, since the technology consisted of new genetic material plus chemical inputs, but did not require mechanization on any particular scale, it was argued that it was scale-neutral and would be just as useful for small farmers as for large farmers. In practice, of course, it didn't work out that way. As is normally the case, those farmers with the most land, machinery, capital, and knowledge were those best able to take advantage of the new technology. Moreover, in many areas, farmers who used to rent their land out to small farmers found that it was economically worth while to take over all of the production themselves. So scale-neutrality did not prove to be the boon for small farmers that it was hoped to be. Now, with multiple cropping research to update many of the primitive small-scale systems of production that have been around for thousands of years, we do have a possibility of a "scale-specific" technology for small farmers. To the extent that the multiple cropping involves growing two or more plant species in the field at the same time, mechanization will be difficult at best, and small farmers using hand tools and/or small cultivation machinery should not be at a disadvantage when compared to large farmers.

Even if researchers recognized that the large farmers would be the first to benefit from the new green revolution technology, it was hoped that the benefits of this technology would "trickle down" to the smaller farmers. In retrospect, we apparently had the wrong hydraulic metaphor. Rather than trickle down, wealth and power seemed to move by "capillary action" in an upward direction. As with most technological change, those who are in the best position to take advantage of innovation can increase the gap between themselves and their poorer colleagues.

Although in general multiple cropping research seems to be most adapted to the small farm, there are some intriguing possibilities of combining multiple cropping technology on large commercial estates. Plucknett (1979) has pulled together the evidence for cattle raising under coconut (*Coco* sp.) trees in the tropics. Presumably one could have a large coconut plantation with individual workers having pastorage rights under the trees. Similarly, it has been known for some time that in many areas where sugarcane (*Saccharum spontaneum*) is grown, it is possible to interplant maize right after the cane is cut in order to take off a crop of maize before the cane ratoons back. During the early 1970s, technicians in the Peruvian ministry of agriculture carried out some experiments on 300 ha on large sugarcane plantations on the Peruvian coast. They reported getting a $1\frac{1}{2}$ t crop of "free" maize per hectare with no influence on sugar yields. The fertilizer and crop protection chemicals applied to the sugarcane also benefitted the maize. They were not interested, however, in allowing plantation workers to enjoy private interplanting rights on the cooperative cane fields, as the military government of the time did not favor individual initiative. Similarly, in countries where sugarcane plantations are in the hands of large private companies or individuals, these landowners see intercropping rights by labor as a dangerous first step toward land reform. Some multiple cropping techniques, such as those mentioned above, may be technically possible and are certainly desirable from the point of view of farm labor, but they may be politically inconvenient from both the left and right ends of the political spectrum.

ROLE OF SOCIAL SCIENCE IN FUTURE MULTIPLE CROPPING RESEARCH

There are a number of works published in recent years that detail specific contributions to be made by social scientists in working with small farmers in the Third World that apply directly to any future research on multiple cropping. Shaner et al. (1982) provide a condensed list of information factors affecting small farmers that need to be investigated. Some of the topics fall clearly within the traditional interests of economists, whereas others are of economic importance but are not necessarily investigated only by economists. In any multiple cropping system we would have to know what commodities are traded in the market and which ones are primarily for home consumption. Can farmers sell their products directly, or must they go through intermediaries? If market prices are satisfactory and they normally sell to truckers at the farm gate, when prices drop can they do some of the processing, such as washing potatoes, or carrots, and sell directly to the consumer? What is the nature of the contract between the farmer and the intermediary: does the intermediary pay cash on delivery, or after he or she in turn has sold the produce? Are there government purchasing agents for some commodities that tend to put a floor under the market price, and are these used by farmers in the region? These are just some of the marketing questions that affect what kinds of crops are raised for sale.

Transportation facilities and costs are another major factor affecting which crops are feasible in a given area. Quite typically, if roads are either unavailable or passable only at certain seasons of the year, farmers may adapt their cropping system to crops that mature during the dry period of the year when the roads are passable. If there are no roads at all, the farmer usually opts for a combination of subsistence crops plus animals that can be walked out to the nearest road. If transportation is relatively good and the farms are near urban markets, then more perishable products can be incorporated into the cropping system. Availability of storage and processing facilities both on and off the farm both have a direct effect as to what is feasible to include in a cropping system.

The national government frequently provides some information about market conditions and frequently establishes regulations having to do with weights and measures and acceptable grades for certain commodities. Many governments in Latin America have their own purchasing agencies for basic grains and other storable commodities such as beans. This enables them to establish a floor under the prices of these commodities for the entire market, even if their own purchases are relatively small.

Information that is always urgently needed by any research team in agriculture has to do with those things that are directly needed by farmers, such as all purchased inputs, tools, equipment, and credit. Investigators frequently find that while, in principle, all of these things are available in a country, there may be large areas of the country that are not well served in any sense. Starting with credit, the official conditions may appear quite reasonable both in terms of interest rate and the size of the loans. Yet these may fail in practice because of the red tape involved that requires that a farmer make a number of trips to the bank, thereby incurring transportation costs as well as lost labor time, only to find that the loan becomes available too late in the cropping cycle to be of any use.

More general cultural traits that are frequently cited as worthy of investigation would include any kinds of norms and customs, whether religiously based or not, that define what is proper behavior. Religion in most societies affects farming systems directly by both taboos against certain kinds of activities, as well as taking time and resources away from the farm operations in order to carry out required religious obligations, such as service on various committees or accepting posts that require considerable expenditure of personal resources. Where heavy ceremonial expenditures are required, religion serves as a leveling device to convert dangerous economic power into harmless social prestige. Whether or not redistribution is affected by religious institutions, there is usually a certain movement of both commodities and labor outside market channels in any village as people help each other out when the need arises, and pay back obligations for help received in the past.

Most authors would agree that one of the main tasks of social scientists working in agriculture would be to study various aspects of the social structure. Starting at the level of the individual household, social scientists could produce a good estimate of the range in household size and the customary division of labor by sex and age. A brief census of a village done with village leaders can provide a lot of information not only on the families in the village, but on the size of their holdings, conditions of land tenure, and crops grown.

We always need to understand the land tenure system with its different categories, such as full title to land, working family-owned land or lineage-owned land, communally owned land that may be assigned to a particular family for an indefinite period, renting, or sharecropping. The land tenure system is particularly important for any multiple cropping research team because the varying conditions have markedly different effects on the incentives of farmers to invest inputs as well as labor. Where a farmer has secure title, tenure is not a limiting factor for what is possible to do. At the other extreme, there are systems of tenure that provide powerful disincentives on the part of the renter. For example, I have found systems in eastern El Salvador where the owner of a large amount of scrub pasture would rent out small parcels to landless farmers with the rent paid in cash in advance. The motives of the landowner are several, in addition to the cash payment. First, he wants to get the land cleared of scrub and the soil loosened so that it can be reestablished in pasture after one cropping cycle. Second, he will rent only a small patch in order to establish an obligation on the part of the small farmer to work for him when needed. Third, renting the land to a small farmer does not deprive the owner's cattle of forage, since the farmer is allowed to take only the grain, leaving all stalks for the owner's animals. In this particular village, there were 3 years of almost total crop failure in maize and beans. As a result, the only product the renter could get out of the land to compensate for the rent payment was to take whatever soil he could and make adobe bricks out of it and haul it away. Under such extreme conditions of tenancy, it is difficult to see how cropping research can make much of an improvement since the disincentives for investment by small farmers are so strong. CENTA, the National Agricultural Research Institute, had done the only feasible thing by showing that a sorghum (*Sorghum bicolor*)/pigeon pea (*Cajanus cajan*) association produced better than the preferred maize/bean association under drought conditions.

Inheritance patterns and family organization also have considerable effect on farming systems. For example, do parents hold title to all of their land up to their death, or do they distribute land to their children when the children begin to establish their own families? Is farm land divided equally among all descendants, whether or not they are farmers, male or female, or is it kept in viable farm units for one or a few of the descendants? In cases where a father or mother holds title to all lands up to the time of death, does he or she also try to retain all authority with respect to the farming operations?

Social classes are frequently delineated partly on the basis of land ownership, but education and nonfarm occupations also play a large part in the layering of rural society. Wealth differences, as well as power differences, are important determinants of access to credit and inputs, and therefore must be thoroughly understood.

Whyte and Boynton (1983) stress the need to study formal organizations. At the local level, local government and particularly any organization, such as a cooperative, which deals directly with agriculture should be thoroughly investigated not only to understand the capabilities it may have in promoting a research project, but also to see if it constitutes an obstacle to research and development activities in an area. The role of the ministry of agriculture in general and its extension services in particular, requires some study. In some areas extension agents have proven to be effective in the provision of credit and inputs, whereas in other areas they may be regarded as completely useless by local farmers.

In order to understand the present land tenure system and the farming system, it is necessary to see present practices and conditions in historical context. We cannot understand "what is" without understanding the process of how it came into being. Even if the data are only oral history from some of the older farmers in the area, it is extremely important to understand what changes have taken place over time as access to roads and markets have led to changes in the cropping system. The present-day situation needs to be seen as part of a continuous process of change that is going on in a given area. In areas where there has been a recent land reform or any substantial change in land ownership patterns, it is common to see a general evolutionary process in which farmers go first into crops to help sustain themselves and pay debts, and later move more into livestock and tree crops as their accumulation of capital permits.

Multiple cropping is often practiced on slopes that are too steep for large-scale cultivation with machinery. Where minor differences in altitude provide significant differences in microclimates, it is common for farmers to try to have plots at varying altitudes, which help to safeguard against losses due to climate in one area and provide a more varied calendar of planting and harvesting dates, thereby spreading family labor and permitting a wider association of crops (Mayer, 1979).

A survey research team with behavioral scientists could be used in a variety of ways. For example, periodic small-scale surveys could be taken to monitor the reaction of farmers to the research and extension efforts of the multiple cropping research program. Similarly, sometime after the conclusion of any experimental intervention, a survey would help to evaluate the impact of the research program.

Before beginning active experimentation in farmers' fields, it might be well to conduct a series of small group meetings with farmers and their spouses to get their reactions to the research plans. Researchers would be expected to spell out in considerable detail what they plan to do and what results they hope for. An economic analysis of those results in relationship to costs should be presented even though at this stage it is understood by all that it is clearly hypothetical. The reaction of farmers and their families to these various alternatives could be assessed with a view to choosing to begin with those alternatives that appear most feasible and most important to the farmers. As a result of these meetings, the team will know which of the proposed innovations are of interest and the reasons for that interest. Farmers expressing most interest in particular experiments will in all probability be the cooperators who will want to see those experiments done on their fields. From the team point of view, time and resources will not be wasted on trials that are of no interest to the local farmers. In this kind of experimental design, rather than carrying out a matrix type experiment on experiment fields, the selection of experiments is done at the mental level through joint farmer-researcher discussion.

FUTURE DILEMMAS IN MULTIPLE CROPPING RESEARCH

Multiple cropping research has been complicated from the outset, and appears to grow more complicated with each new step toward realism. The tremendous potential of multiple cropping was demonstrated conclusively on the fields of IRRI (Streeter, 1972). The emphasis there was in finding the right combinations of crops that would yield well and produce a reasonably well-balanced diet. No attempt was made to measure the costs of production under this system either in labor or monetary terms. When the research was subsequently moved off the experimental fields and onto farmers' fields, new methodologies had to be invented to cope with farmer-managed experiments (Harwood, 1979). Under these circumstances, not only did researchers have to consider the interests of the farmer, but also the real limitations in terms of labor and capital that could be invested in the experimental program.

The next logical step in achieving the maximum realism in multiple cropping research would be to move toward some form of complete experimental farm. I have heard agronomists talk about this possibility for many years, but at this point I don't know of any cases where it has been carried out. One design involved an agronomist who intended to retire to the tropics, buy a small farm, and hire a farm family to manage it with a view to optimizing production. A second version by another agronomist involved establishing one or more experimental farms each with its own family within the confines of one of the CGIAR stations in the tropics. This idea presented the opportunity of optimal access by the research staff but a rather unsatisfactory life for the farm families who would be living in a "zoo" rather than in a village. Other possibilities would include purchasing a small farm near a research station and renting it at zero or nominal rent to a young farmer on the condition that he or she keep accurate records of all operations. Other variations on this theme include having a cooperative or the extension service own the land. As agricultural research moves further from the tightly controlled manipulation of a relatively few experimental variables in the laboratory or on the experimental field toward the more realistic conditions of small farmers in the Third World, the question comes up, "Where do we stop?" Is there a point beyond which agricultural researchers can and should say that it is now up to the extension service or farmer organizations or some other group to take over? To what extent is this experimental farm also a demonstration farm?

Moving beyond the individual farm level to that of the local farmer organization or cooperative, should we be content with an assessment of the problems faced by these organizations, or should we, as Whyte and Boynton (1983) argue, get actively involved in finding ways to make these organizations more effective in promoting agricultural development in their spheres of influence? As we move up the line toward the national structure, we face the same kinds of dilemmas. Should we simply recognize an inadequate extension service when we see it and avoid it, or should we be trying to find ways to improve its performance?

McDermott (1982) has argued that we need to look at various governmental and market structures in order to find blockages to development. Should we merely identify these blockages in our reports or should we be trying to find ways to improve the functioning of these various institutions? If we choose to do the latter we will necessarily wind up paying some attention to national policy and planning efforts. This is not to say that each multiple cropping research team should have a representative in the capital city trying to infiltrate the highest policy and planning groups. However, major research institutions could benefit by having detailed knowledge of, and contact with, these agencies in the hopes of influencing them as new policy is put into effect.

There is a considerable body of research indicating that the principal constraints limiting agricultural production in the Third World are those of a social, political, and economic nature. Biological research will not remove those constraints. As long as these constraints are operating, the biological potential of multiple cropping research will be severely limited. Arnon (1981) points out that planners have always had the dilemma of choosing policies somewhere between two extremes. At one extreme the goal is economic efficiency, and limited resources are invested in those sectors of agriculture that are already commercialized, mechanized, and best able to take advantage of the resources. At the other extreme is a goal of increasing equity, which would argue for focusing on small farmers, assuming that the large commerical farmers can look after themselves. The evidence from multiple cropping research to date suggests that the most efficient farmers, as measured by productivity per hectare, are those practicing multiple cropping on small-scale farms. This being the case, it could be argued that efficiency in the use of land resources as well as greater social equity could be best served by focusing more national attention on the small-farm sector in agriculture. Proving this point through intellectual debate does little to win power struggles (or change priorities) at the national level.

REFERENCES

Arnon, I. 1981. *Modernization of Agriculture in Developing Countries: Resources, Potentials, and Problems,* John Wiley & Sons, New York.

Banfield, E. C. 1958. *The Moral Basis of a Backward Society,* The Free Press, New York.

Barlett, P. R. 1980. *Agricultural Decision-Making: Anthropological Contributions to Rural Development,* Academic Press, New York.

Barnett, H. G. 1953. *Innovation: The Basis of Culture Change,* McGraw-Hill, New York.

Carneiro, R. L. 1961. Slash-and-Burn Cultivation Among the Kuikuru and Its Implications for Cultural Development in the Amazon Basin, *Antropologica* (Suppl. 2) (reprinted in: Cohen, Y. A. 1968, *Man in Adaptation: The Cultural Present,* Aldine, Chicago, p. 131.

Conklin, H. C. 1957. *Hanunóo Agriculture in the Philippines,* FAO Forestry Development Paper 12, Rome, Italy.

Foster, G. M. 1967. *Tzintzuntzan: Mexican Peasants in a Chaning World,* Little, Brown, & Co., Boston, Massachusetts.

———. 1973. *Traditional Societies and the Impact of Technological Change,* 2nd ed., Harper & Row, New York.

Harris, M. 1979. *Cultural Materialism: A Struggle for a Science of Culture,* Random House, New York.

Harwood, R. R. 1979. *Small Farm Development: Understanding and Improving Farming Systems in the Humid Tropics,* Westview, Boulder, Colorado.

Harwood, R. R. and Banta, G. 1974. Intercropping and Its Place in Southeast Asia, *Agron. Abstr.* 1974, p. 45.

Hildebrand, P. E. 1977. Generating Small Farm Technology: An Integrated Multidisciplinary System, presented at the 12th West Indian Agricultural Economics Conference, Caribbean Agro-Economics Society, 24–30 April 1977, Antigua.

———. 1978. Motivating small farmers to accept change, paper presented at the Conference on Integrated Crop and Animal Production to Optimize Resource Information on Small Farms in Developing Countries, Bellagio, Italy, October 1978.

Horton, D. E. 1984. *Social Scientists in Agricultural Research: Lessons from the Mantaro Valley Project,* International Development Research Center, Ottawa, Canada.

Kahl, J. A. 1976. *Modernization, Exploitation, and Dependency in Latin America,* Transaction Books, New Brunswick, New Jersey.

King, F. H. 1911. *Farmers of Forty Centuries,* Madison, Wisconsin, (reprinted in 1973 by Rodale Press, Emmaus, Pennsylvania).

Lewis, O. 1960. *Tepoztlan: A Village in Mexico,* Holt, Rinehart, & Winston, New York.

Lewis, W. A. 1977. *The Evolution of the International Economic Order,* Princeton University Press, Princeton, New Jersey.

Mayer, E. 1979. *Land Use in the Andes: Ecology and Agriculture in the Mantaro Valley of Peru, with Special Reference to Potatoes,* International Potato Center, Lima, Peru.

McClelland, D. 1961. *The Achieving Society,* D. Van Nostrand Co., Princeton, New Jersey.

McDermott, J. K. 1982. Social Science Perspectives on Agricultural Development, in: *Social, Cultural, Economic and Political Dimensions of International Agricultural Development,* (D. G. Anderson and N. E. Tooker, eds.). Proc. Faculty Development Conference, Mar. 10–11, 1982, Univ., Nebraska, Lincoln, Nebraska, pp. 3–10.

Moran, E. F. 1981. *Developing the Amazon,* Indiana University Press, Bloomington, Indiana.

Mosher, A. T. 1969. *Creating a Progressive Rural Structure,* Agricultural Development Council, New York.

Plucknett, D. L. 1979. *Managing Pastures and Cattle Under Coconuts,* Westview Tropical Agriculture Series 2, Westview, Boulder, Colorado.

Rappaport, R. A. 1968. *Pigs for the Ancestors: Ritual in the Ecology of a New Guinea People,* Yale University Press, New Haven, Connecticut.

Redfield, R. 1930. *Tepoztlan: A Mexican Village,* University of Chicago Press, Chicago, Illinois.

Rogers, E. M., and Shoemaker, F. F. 1971. *Communication of Innovations,* The Free Press, New York.

Rothamsted Experimental Station. 1981. Lawes Agricultural Trust, Harpenden, Herts, U.K.

Roumasset, J., Boussard, J. M., and Singh, I. 1979. *Risk, Uncertainty and Agricultural Development,* Agricultural Development Council, New York.

Schultz, T. W. 1964. *Transforming Traditional Agriculture,* Yale University Press, New Haven, Connecticut.

Shaner, W. W., Phillip, P. F., and Schmehl, W. R. 1982. *Farming Systems Research and Development: Guidelines for Developing Countries,* Westview, Boulder, Colorado.

Steinburg, D. 1980–1981. Problems of antropology in AID as viewed from the periphery of anthropology, *Practicing Anthropol.* 3(2).

Streeter, C. P. 1972. *Help for the World's Small Farmers,* Rockefeller Foundation (condensed), *Readers Dig.,* Oct. 1972, p. 217.

Wagley, C., ed. 1974. *Man in the Amazon,* University Presses of Florida, Gainesville, Florida.

Whyte, W. F., and Boynton, D. 1983. *Higher-Yielding Systems for Agriculture,* Cornell University Press, Ithaca, New York.

Chapter 13

Research Methods for Multiple Cropping

Anne M. Parkhurst
Charles A. Francis

Multiple cropping is both a means of subsistence and a way of life for many limited-resource farmers. Research has been limited to date on the many and complex systems or patterns we call multiple cropping. Experiments have employed classical statistical methodology and time-honored, well-tested experimental designs (Parkhurst and Francis, 1984a). Researchers have tended to use simple designs as in monoculture (Mead and Stern, 1980). Although many underlying biological and mathematical assumptions are likely to apply to more complex systems, the sheer number of factors and interactions make analysis, interpretations, and setting priorities for research much more difficult. This chapter explores the magnitude of the complexity of research questions, and how it is possible to quantify production constraints. Once this is accomplished, priorities must be established in a research program. Some methods are proposed for doing this in a systematic way.

Although it is difficult to manipulate single components of a cropping pattern without affecting many others, an efficient methodology is needed to control most factors in order to study each component of a system. This technology must be evaluated on the farm. Only if the end user is given an opportunity to view and try new alternatives can this research and development cycle be complete. This methodology complements Chap. 14 by Mead, providing a broad framework into which the analyses may be expected to fit.

Central to the theme of the chapter is the assumption that the research job is not finished if the results are not published or put together in some form that will reach the farmer and affect farming success. Whether to increase production, to reduce costs or inputs, or to improve the stability of the cropping pattern, the ultimate objective of research is to improve production practices and produce results that will effect some change on the farm. More specific objectives of research may be appropriate for basic studies or work in other areas of science, but research for complex systems needs to be carefully directed toward solving immediate food production problems and the sustainability of production systems while attempting to improve or regenerate the production resources available to the farmer.

COMPLEXITY OF RESEARCH QUESTIONS

Complex cropping systems that intensify the use of land are important in the agriculture of developing countries. Multiple cropping may take the form of sequential double or triple cropping, or may involve some degree of intercropping, where two or more species are found in the field at the same time. The former, an intensification in time only, is similar to traditional monoculture and presents no unusual design or statistical challenges that cannot be handled with traditional research methodology. Intercropping, on the other hand, represents a degree of complexity that is difficult to handle with the usual application of our current techniques.

This cropping pattern complexity is illustrated in Figs. 13.1 to 13.3, where some of the many possible interactions involved in a two-crop pattern and a three-crop pattern are compared to the better understood monoculture system (adapted from Francis, 1981a). Climatic and soil factors are relatively unaffected in the short run by intercropping, compared to monoculture. There are no doubt some microclimatic changes in humidity, wind speed, and amount of light reaching the soil surface, but these have not been quantified in convincing detail. There may also be longer-term effects on soil organic matter and structure due to the species variety and cropping intensity of intercrop patterns. More apparent differences are found in cultural practices required for these intensive systems and used by farmers, and in the genetic factors which involve not only crop genotypes but also the insects and pathogens that attack each crop. Considering Fig. 13.1, there are 15 major factors listed for the monocrop situation, and thus 105 combinations of two factors that may interact. In Fig. 13.2, a two-crop pattern and simplest of all possible intercrops, the 26 factors (15 from Fig. 13.1 plus an additional 11 factors) listed could combine in 325 possible pairs of factors. Figure 13.3 shows a three-crop pattern with 39 single factors and the resulting 741 possible two-way interactions. The increasing number of possible interactions as the number of single factors increases is illustrated in Fig. 13.4.

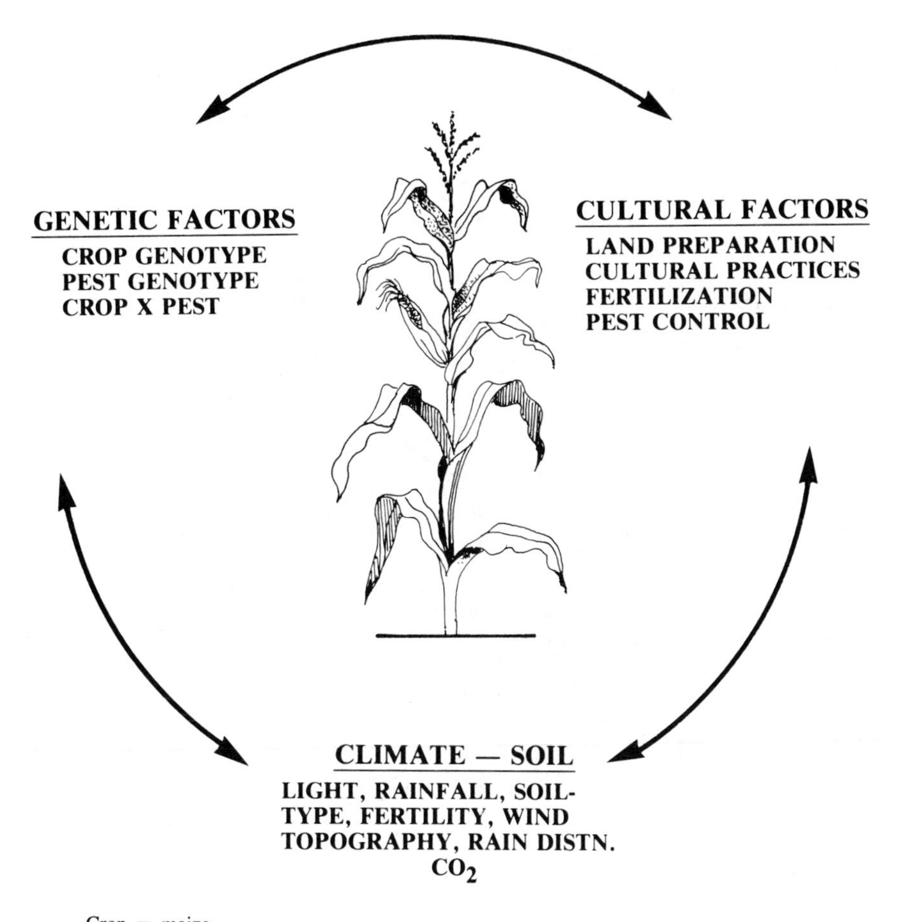

Crop = maize
Total factors = 15
Possible two-way interactions = 105

Figure 13.1 Interactions in a model monoculture cropping system. (Reprinted by permission from *Plant Breeding II*, Kenneth J. Frey, ed., © 1981 by the Iowa State University Press, 2121 South State Ave., Ames, IA 50010.

Although this oversimplifies the comparison between monoculture and intercropping, the figures attempt to quantify to some degree the complexity that must be dealt with in research of these systems.

There is a further degree of complexity found with intercropping patterns when we consider the economic and social situation in which farmers use these systems. Often found on small farms with limited resources, intercrop patterns are used to fill a series of needs for the farmer: food, income, and security. The criteria that the farmer uses to evaluate success or failure also may be different from the purely commercial/economic yardstick used for large farms. Thus, family nutrition, a range of food crops and animal species, stability and distri-

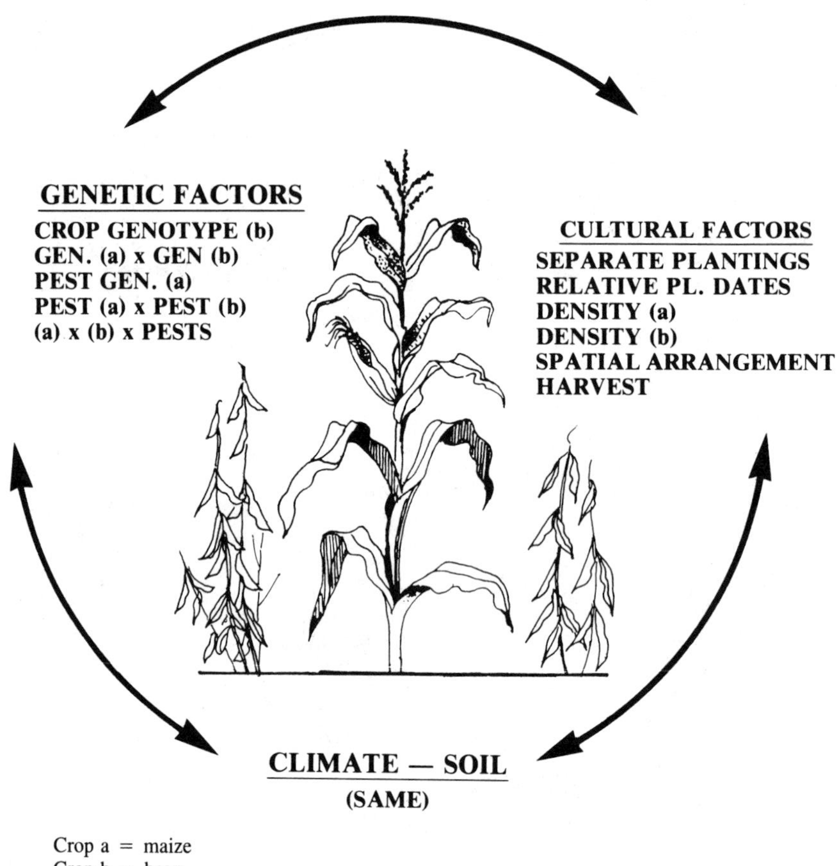

Crop a = maize
Crop b = bean
Total factors = 26
Possible two-way interactions = 325

Figure 13.2 Interactions in a model two-crop pattern, giving those factors in addition to those presented in monocrop in Fig. 13.1. (Reprinted by permission from *Plant Breeding II,* Kenneth J. Frey, ed., © 1981 by the Iowa State University Press, 2121 South State Ave., Ames, IA 50010.)

bution of production through the year, and minimizing risk may all be more important to the farmer than economic return, which is used to evaluate most commercial monocultures (Francis, 1981b; Willey, 1979a, 1979b).

EVALUATION OF CONSTRAINTS TO PRODUCTION

In production research on complex cropping systems, it is particularly difficult to evaluate what factors most constrain production in a given region (Parkhurst and Francis, 1984b). A number of factors complicate this evaluation. It may be

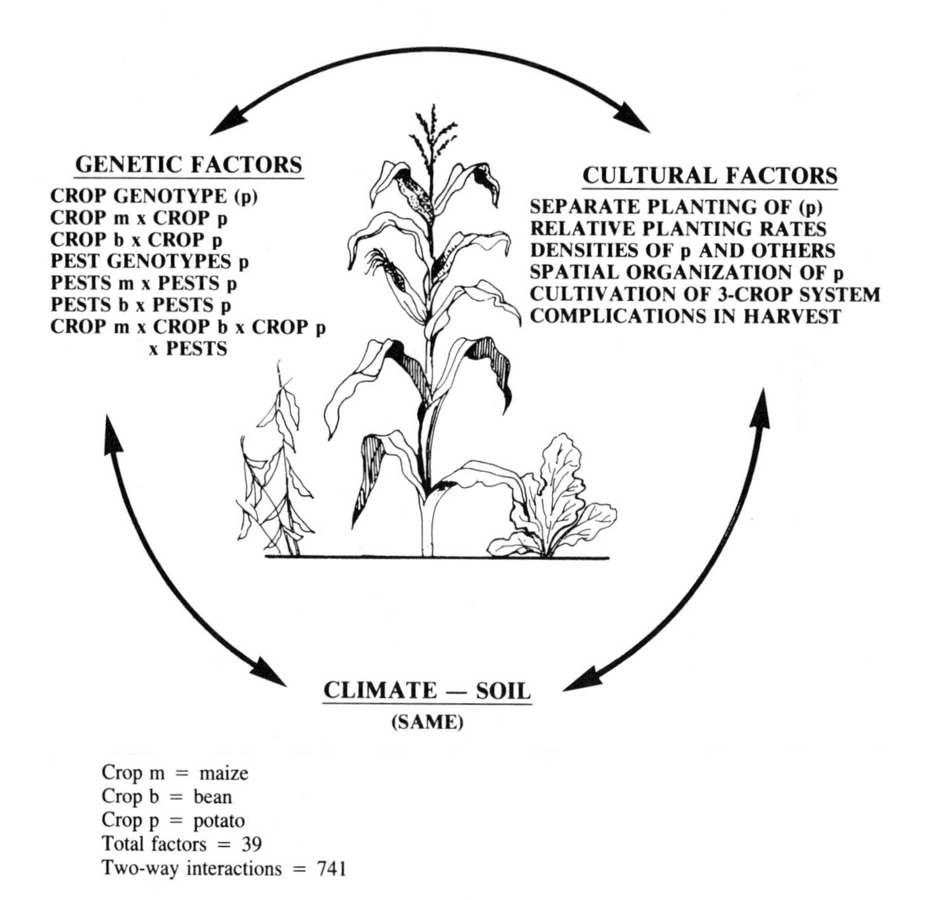

Figure 13.3 Interactions in a model three-crop pattern, giving those factors in addition to those presented in monocrop (Fig. 13.1) and intercrop (Fig. 13.2).

difficult to examine a crop growing in a mixture with other species and determine immediately if deficiency symptoms or lack of vigor are due to a soil-related problem, shortage of water, or some other type of competition inherent in the system. Since relatively little is known about crop species interactions in these systems, and since most technical agronomists are trained to observe, describe, and research monoculture systems, even the initial evaluation of limitations to production is difficult.

Further complicating the activity is a lack of understanding of the farmer's objectives in growing crops in complex mixtures. The agronomist may be thinking in terms of maximum production, optimum economic return, or how to mechanize a farm to save labor. On the other hand, the farmer may be more concerned about food production for the family, stability of production through the year, how to minimize risk of failure, or how to most profitably employ the family in this enterprise. A specialist who is trained to examine biological,

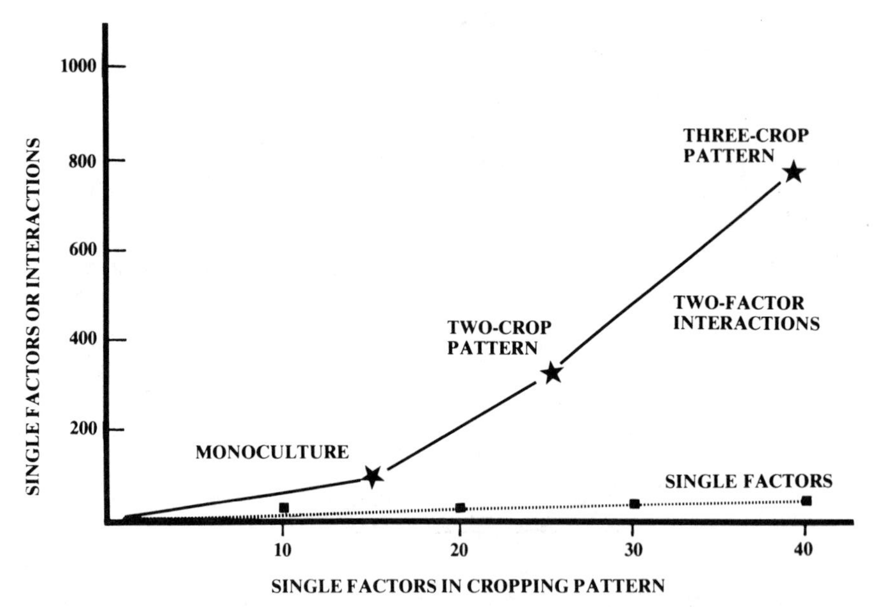

Figure 13.4 Increase in two-factor interactions as a function of increasing number of factors which may vary in cropping patterns.

climatic, or economic constraints in commercial monoculture may be unprepared to use the appropriate yardstick to evaluate an intercrop system. A frequent recommendation by specialists has been to eliminate intercropping completely, and to plant soybeans or maize as a commercial monocrop. This illustrates a lack of understanding by researchers of the farmer's goals and the critical role the complex cropping system plays in the family's survival (Parkhurst and Francis, 1984b).

Problem identification can be as simple as talking to a few established experts in the area and using this conventional wisdom to embark on a research program, or as complex as spending several years talking to farmers and using detailed questionnaires to quantify constraints. Between these two extremes are some alternatives that should be useful to the researcher.

Conventional Wisdom Approach

A method frequently employed to evaluate constraints is the collection of conventional wisdom on what is going on in prevalent cultural systems on the farm. In a semiarid zone, "Everyone knows that water is the most limiting problem!" Therefore, a program is designed to develop drought-tolerant cultivars. Minimum tillage is included in the package to reduce moisture loss and lower crop densities are recommended to make best use of available water. This is illustrated by monoculture sorghum (*Sorghum bicolor*) grown in the western part of Nebraska, an area where rainfall is marginal for sorghum.

Another example is the maize (*Zea mays*)/bean (*Phaseolus vulgaris*) intercrop found in the Andean zone. Conventional wisdom among researchers who are trained to research and extend monoculture technology may be that pure stands of soybeans (*Glycine max*) are more economically rational than the intercrop system of the farmer. Replicated experiments show that net income is greater from soybeans on the research station than the farmers appear to gain from their intercrop with low inputs. Although counter-intuitive to those who have worked in multiple crop systems, this approach to research may in fact be successful if the problems are the correct ones, and the choice of solutions is appropriate to the problems and to the farmers they are designed to help.

The approach may be to talk to researchers with extensive experience in the zone, to read the literature (if any exists), or to meet with a few key informants who farm or who are familiar with cropping in the area. The limitations of this approach are obvious: the information base is limited to those consulted, and to their biases about what is actually happening in the zone.

The range of options that is suggested to improve current systems also may be limited, since these same sources will have some conventional wisdom ideas about how to solve production constraints. In the sorghum program in Mali, local and international researchers have assumed that water was limiting in a zone where rainfall was between 400 and 700 mm per year. A recent study by a Dutch team showed that nitrogen is often the most limiting factor in any zone with a good distribution of at least 300 mm of rainfall on those soils (Art Onken, personal communication). Thus, research programs designed around moisture as the most limiting factor may have ignored or at least deemphasized the importance of nitrogen in this area. In summary, we cannot afford to ignore ideas of those with experience in a region. It is equally risky to depend on this as the only source of information.

Research and Literature Evaluation

A more comprehensive approach to evaluation of limiting factors includes the above collection of conventional wisdom plus the careful study of research results in the literature of the region and of areas with similar climate and crops. This information may be available in local libraries, in annual reports, in summaries of research that do not have wide circulation, or in the files of other researchers who are active in the area. Although we need to take into account these results and build any new research program on past results rather than start from zero, we must keep in mind that this prior research may have suffered from the same problems outlined above. It is both difficult and necessary to evaluate not only the results but the research questions they were designed to answer and the conclusions that were drawn. What has been the impact of this past research? What were the farmer's reactions to the results, or have results reached the farmer? To the extent possible, the conditions under which research was conducted must be evaluated, and the creditability of the results needs to be considered.

In many situations where intercropping is important, there is a dearth of information from the research stations. Since most emphasis has been placed on plantation crops for export and on commercial monocultures, there may be little that is helpful on multiple cropping. Nevertheless, there will be data and recommendations on monoculture on some of the same crops that are or might be included in the intercrop, and these data need to be evaluated. For example, genetic resistance to a specific pathogen that causes stalk rot in maize will work equally well in monoculture or in an intensive cropping pattern, even though the incidence and relative importance of stalk rot may be different between the two systems. On the other hand, a specific fertility recommendation for a cereal crop in monoculture may have to be modified drastically to make that recommendation appropriate to a system that mixes the cereal with one or more legume species. These pieces of information have to be sorted out to improve the knowledge base for intercropping.

Observational Experiments on Farms

Another approach to learning about constraints is to plant observational trials. These preliminary experiments both on the experiment station and on the farm can combine careful study of farmer's systems with the introduction and observation of some types of improved technology. This may include some changes in cultural practices identified by either the researcher or the farmer, new crop varieties, some new inputs, or a combination of these (Parkhurst and Francis, 1984b). These types of observational plots bring the researcher into close communication with the farmer, and help to build an appreciation of the farming systems in the area. This approach is being developed under the general umbrella of "farming systems research," although the total activity of FSR is more complex than what is suggested here. Recent publications that detail some of the work done in this area include those by Garrity et al. (1979) and Harwood (1979). The recent book by Gomez and Gomez (1983) and the methodology published by Centro Internacional de Mejoramiento de Maiz y Trigo (CIMMYT) (Byerlee and Collinson, 1980; Perrin et al., 1979) on planning technologies both developed this theme of using farm-generated results for making recommendations.

Detailed Limiting-Factor Surveys

Detailed surveys of crop production on specific crops or on cropping systems can lead to precise definition of constraints, although the cost both in time and resources may be prohibitive. A survey may be as simple as a mail questionnaire, such as that used to evaluate the incidence of the downy mildew disease in the Philippines on maize (Francis, 1967). The results from this survey showed quantitatively that the incidence of the disease was greater in the rainy than in the dry season, and that there was a heavy occurrence in four centers in the country. The data were used to locate screening trials to select for resistance.

A more complex study was done on the factors limiting bean production in three departments in Colombia (Gutierrez et al., 1975). This consisted of visits by trained agronomists to more than 150 farmers during the growing season, with at least three visits to each farm per season. A detailed questionnaire, completed during each visit, gave information on farm size, level of technology, and sociological aspects of the farm environment. It detailed land preparation, planting, protection practices, irrigation (if any), fertilization, other cultural details during the season, harvest and yield data, and disposition of the crop. The analysis of these questionnaires gave a broad look at the problems in bean production in these departments, and provided the basis for evaluation of different research strategies directed toward solving these constraints. Given the costs for doing research, the value of crops harvested, and the financial advantage of controlling one specific problem (e.g., resistance to rust disease), the basis was given for a cost-benefit analysis of different research strategies.

This type of approach is rarely feasible before embarking on a research project. In addition, the evaluation of a multiple cropping system is far more complex than that for a single crop species. Nevertheless, there are some questions and a part of this methodology that could be very useful in an attempt to quantify problems in a multiple cropping system.

These approaches to identification of constraints are not mutually exclusive. There is a need to spend some resources and some time on this evaluation, and a need to bring from each of these approaches any information that can be useful. The time spent on this part of a project usually is minimal, and it would be advisable to err on the side of more time spent evaluating constraints than to follow the time-honored path of conventional wisdom.

ESTABLISHING RESEARCH PRIORITIES

Priorities in research may be established after the constraints outlined above have been carefully and comprehensively evaluated. More often, some abbreviated process is used to quickly arrive at a set of objectives, and the researcher is off at full speed with the program. This is illustrated by the development projects that put a plant breeder and a soil fertility expert in the field to work in their discipline-specific areas, since "everyone knows that new varieties and good fertility recommendations are needed before progress can be made." Several years later, an agronomist and agricultural economist may be called upon to evaluate the results of the work, and may in fact find that neither genetic potential nor soil fertility were and are the most limiting factors. Thus, some careful consideration given to choice of research priorities and design of a research program is essential to ensure that this program will in fact have a payoff in the future.

Some of the factors that need to be integrated into this decision include a comprehensive knowledge of the crops grown and the constraints to their production, the probabilities of solving these problems through research, the chance of adoption of new solutions once they are available, the costs in time and resources of reaching these recommendations, and the total cultural and economic milieu into which the recommendations are to be made. What is the chance for adoption? Do these solutions provide an alternative to solve some problem which is actually perceived by the farmer as a limitation to production? What is the cost of adoption, and what risk is involved? And if the new technology is adopted on a wide scale what will be the effect on total production in the region and on product prices? These questions are rarely contemplated during the design of a research program.

Priorities often are set according to one or several criteria. Usually they are set by crops, since most research programs in the world are organized by crops and not by cropping systems. This immediately hampers the solution of problems that are complex and involve more than one species. This is both an institutional constraint, since departments, research projects, and budgets are organized along crop and discipline lines, and a barrier confronted by the individual researcher who has been trained to look at one crop at a time. In addition to the focus on single crops, the priorities may be set by geographic area, or areas that are defined by some political boundary such as a state, department, province, or country. These boundaries rarely correspond to any climatic or cropping system region. Priorities may be set by farm size, where large farmers have the economic influence and interest in research to stimulate scale-specific research that is more useful to the large than to the small farm. Research plans frequently are drawn up by one academic discipline, since the departments and budgets are organized this way and since we have learned to communicate best with those who speak the same specific dialect in the technical community. Most of these approaches to setting priorities ignore the need to study several crop species at once, the need to look at integration of disciplines, and the importance of a holistic approach to solving production problems in complex systems. Infrequently, there will be a team or a rare individual who looks at the entire farm and evaluates the limiting factors to production on that farm and how alternative strategies might help the farmer to improve production, food balance for the family, or total income and distribution of that income through the year.

One approach that can be used to quantify priorities is a process that puts the available information into a matrix of (1) limiting factors, (2) probabilities of solving those limitations, (3) importance of each factor vis-à-vis the others, (4) probability of adoption of new technology, and (5) some composite of these data to give a ranking of research priorities. This approach is illustrated with two examples, one for monoculture sorghum (*Sorghum bicolor*) in Nebraska, and another for a maize/bean intercrop in the Andean zone.

Monocrop Example

The simplest case is illustrated with a monoculture crop in the temperate zone. This is done to explain the method by using a crop that has a relatively well-known set of constraints, on which considerable research has been done, and where the probabilities of success of alternative approaches to increase production have been established. An example for sorghum in Nebraska is given in Table 13.1 (adapted from lecture notes of second author). Eight limiting constraints to sorghum production are listed, along with their relative importance in limiting yields of the crop. Next, alternative solutions are presented that appear to be feasible either through plant breeding or agronomy or a combination of the two. In the case of drought, either tolerant hybrids or irrigation could help solve this constraint. The probability of developing more tolerant hybrids is relatively low

Table 13.1 Stepwise Method of Calculating Research Priorities for Sorghum in Nebraska
Numbers Based on Survey of Active Research Workers in State

Limiting factor	Importance of factor,[a] X	Potential solution	Probability of solution, Y[b]	Probability of adoption, Z[c]	Index of priority, XYZ[d]	Order of priority
Drought	9	Tolerant hybrids	0.3	0.9	2.43	2
		Irrigation	1.0	0.1	0.90	10
High temperature	3	Tolerant hybrids	0.4	0.9	1.08	9
		Planting date	0.2	0.3	0.18	14
Low temperature	2	Tolerant hybrids	0.8	0.7	1.12	8
		Planting date	0.5	0.6	0.60	12
Greenbug	6	Chemical treatments	1.0	0.2	1.20	6
		Resistant hybrids	0.8	1.0	4.80	1
Stalk rot	4	Rotation/management	0.7	0.6	1.68	5
		Chemical treatment	0.1	0.1	0.04	16
Fertilizer cost	7	More efficient hybrids	0.3	0.9	1.89	4
		Rotate legumes	0.8	0.4	2.24	3
Chinch bug	3	Resistant hybrids	0.2	0.8	0.48	13
		Chemical treatment	0.8	0.5	1.20	6
Stalk borer	2	Resistant hybrids	0.5	0.6	0.60	11
		Chemical treatment	0.4	0.2	0.16	15

[a] 10 = most important, 1 = least important.
[b] Probability based on prior research on this crop and experience with other crops.
[c] Chance of wide adoption of new technology in state.
[d] Product of these three items, to give a weighted priority for each factor.

based on research to date, while the probability of correcting this deficiency with irrigation is high.

The next column lists these probabilities of finding solutions through research. Some types of technology have been successful on this crop or another cereal, and are given a high chance of success. Greenbug (*Schizaphis graminum*) resistance has been found before, and no doubt will be found for new races that emerge. Chemical treatment, if applied in a timely way, can control greenbugs with complete success, just as irrigation can solve the drought problem. At the other extreme, tolerance to high temperature, efficiency of fertilizer use, and resistance to chinch bugs have been difficult research problems and these solutions are less likely.

Once these solutions are available, if the research is successful, there is a question of adoption. Any factor that can be incorporated into the seed as a genetic trait, such as greenbug resistance or more efficient use of nitrogen by the crop, has a high probability of success if the farmer in fact perceives this constraint and is willing to buy a hybrid with that characteristic. On the other hand, use of irrigation on a traditionally dryland crop such as sorghum or use of chemicals to control greenbug at a prohibitive cost would have a low probability of adoption.

If these three factors are multiplied,

$$\text{Importance } X \times \text{ probability of solution } Y \times \text{ probability of adoption } Z = \text{index of priority } XYZ$$

the resulting index of priority gives a quantitative measure of where research emphasis should be placed. Needless to say, the results of this exercise are only as good as the information that is used and the assumptions that go into the determination of probabilities. This is a way to quantify conventional wisdom, or to further process data that comes out of a limiting-factor survey of the type described above for beans.

Intercrop Example

A parallel analysis of priorities for an intercrop situation is presented for a maize/bean cropping pattern in the intermediate elevation in Colombia in the Andean zone. This exercise is more complicated for a number of reasons. First, one needs to deal with two crop species and their associated disease and insect problems. These pest problems not only affect each crop, but the intercrop pattern may influence the severity of each problem and their relative importance may be different between this pattern and the better-studied monoculture. It is more difficult to assess the relative importance of each factor, since these will depend on specific rainfall patterns in the area, number of potential crops per year, and the objectives of the farmer. Even within the same cropping pattern, these objectives may vary depending on the need for maize and for beans, and the orientation toward sale of some excess production of each. Probabilities of solution of the problems can be extrapolated from monoculture experience on each crop and from past research on this important intercrop pattern, but the probabilities of adoption are less confident.

Table 13.2 lists several constraints that may limit maize/bean production in the area, along with some potential solutions to these problems. Some of the solutions that are obvious to those who have worked in developed countries in the temperate zone (irrigation to avoid drought stress, chemical control of anthracnose disease, purchased chemical fertilizers) are either very expensive or not available to the farmer with limited resources. A number of problems might be solved by use of new varieties or hybrids with disease resistance or different morphology. This is an excellent solution for the farmer who cannot adopt more expensive alternatives, but as shown later this approach does not have the high probability of adoption that the genetic package enjoys in a developed country. The management solutions to several of these constraints are designed to be minimal in cost, but their adoption will depend on how the farmer perceives both the problem and the potential solution, and whether this solution appears to meet one of the objectives of the farmer.

This information is used to calculate a priority index. The importance assigned to each factor is a general value for this intermediate elevation cropping region (1500 to 2200 m above sea level), and the number will vary with specific location, elevation, soil type and fertility, current cropping systems practiced by farmers, varieties, and the level of technology. Probabilities of solution are based on experience with monoculture on the same crops, or on research that has been conducted in the zone (see Parkhurst and Francis, 1984a, 1984b, for references to this research). The adoption probabilities are scale-specific to small and medium sized farms in this area, which are most likely to use the intercrop pattern of maize and beans.

Based on this analysis, the highest priority should be given to breeding for anthracnose resistance in beans. Anthracnose disease is a large problem with the currently used varieties of beans but is likely to be solved by research. The adoption level is high because it is easy to demonstrate the difference between resistant and traditional susceptible varieties in the field, and after a one-time purchase or trade for seed, the farmer can save seed for the next planting.

The next priority is use of manure or compost on maize. The practice would employ materials that are available locally to the farmer and involve family labor, while the other solution, using chemical fertilizer, would be difficult to implement in spite of its high probability of success.

The third priority is to breed maize with a stronger stalk to resist lodging when planted with the beans. This type of goal has been reached before by maize breeders, and the adoption of a new variety or hybrid based on visual comparisons by the farmer should have a high success rate. Lower priority is placed on such changes in technology as irrigation of the intercrop for obvious reasons, and complicated changes in management that require fine tuning a system when the results are not obvious.

This exercise is only as useful as the assumptions made in generating the priority index. Some of the estimates of importance are based on observations

Table 13.2 Stepwise Calculation of Research Priorities for Maize/Bean Cropping Pattern in the Andean Zone at Medium Elevation

Limiting factor	Importance of factor, *X*	Potential solution	Probability of solution, *Y*	Probability of adoption, *Z*	Index of priority, *XYZ*
Anthracnose (in beans)	9	Resistant variety	0.7	0.8	5.04
		Chemical treatment	1.0	0.1	0.90
Lodging in maize	7	New variety	0.9	0.6	3.78
		Less aggressive bean	0.5	0.2	0.70
Competition in maize	5	Taller variety	0.7	0.5	1.75
		Changed management	0.4	0.3	0.60
Rust (in maize)	3	Resistant variety	0.9	0.3	0.81
		Chemical treatment	0.8	0.1	0.24
Water stress	6	Irrigation	1.0	0	0
		Reduced densities	0.7	0.4	1.68
Fertility (for maize)	8	Chemical fertilizer	1.0	0.2	1.60
		Compost/manure	0.7	0.8	4.48

Note: See Table 13.1 for explanation of numbers.
Source: Data from Parkhurst and Francis, 1984b.

and discussions with farmers, plus experience in research in the zone. Others are based on conventional wisdom. The better the farmers' systems are known, the better these estimates will be. Probabilities of solution of constraints by research are well established, based on previous research on these crops and others in the zone. The probabilities of adoption are less secure, since there is little experience in extension work with intercrop patterns, and it is difficult to extrapolate from experience with large farmers to those with limited land and resources. More important than the absolute numbers given in the example is the approach. With these priorities set for the pattern or patterns of interest, it is possible to proceed with design of research to reach solutions to limiting constraints.

METHODOLOGY FOR COMPONENT RESEARCH

Given the complexity of the questions to be resolved in the experimental search for new components of technology for multiple cropping, a careful evaluation must be made of the methodology available for research. Treatment designs available for this work range from the simplest of single-factor experiments with several levels or varieties, to the most complex complete or incomplete factorial designs (Parkhurst and Francis, 1984b). There is a need for balance between unreplicated observational plantings (to get general ideas about combinations of practices and to bracket treatments of interest) and more detailed and replicated experiments (from which careful analyses can be made of main effects and interactions). The complexity of research and design questions was addressed in a recent paper (Francis, 1983) in which some of the alternative designs were evaluated in a qualitative way. Huxley and Maingu (1978) and Mead and Stern (1980) have proposed use of systematic designs. These options are further described by Mead in Chap. 14.

First, it is necessary to consider the numerous interactions outlined in the figures. The two-crop pattern illustrated in Fig. 13.2 is used as an example. Although there are 325 possible two-way interactions between pairs of factors, some of these are much more likely to occur than others. For example, genotype of bean is more likely to interact with the morphological plant types of associated maize than with topography. Relative planting dates of maize and beans are more apt to interact with weed growth and success of different herbicide mixtures than with the type of land preparation practiced. It is possible to predict from prior experience with monoculture and from limited research on intercrops which of these interactions are most likely to be important, and which are unlikely to affect results. This exercise can greatly simplify the process of experimentation, since the researcher can focus on those main effects and interactions that are most likely to be important in the design of components of new technology. Another useful focus is to concentrate on those factors over which the farmer can exercise some control.

If rainfall is limited in a region, there is little the farmer can do to increase the supply of water to crops if irrigation is impossible. It may be possible for the farmer to adjust the planting date to take advantage of available rainfall, to choose different crop species or varieties that are more tolerant to drought conditions or that mature more rapidly and thus avoid drought. Thus, the potential adoption of new technology is considered from the very beginning as the researcher evaluates what factors are important in the system and which constraints should be addressed through research.

An Intercrop Example

In this context, a consideration of important interactions is a crucial step. An example of these interactions for a two-crop pattern is given in Fig. 13.5. The eight main effects are described briefly to show why these were chosen out of the 26 factors given in Figs. 13.1 and 13.2, because of their importance as main effects or because of the predicted interactions between pairs or among more of these factors in the intercrop system.

Maize Variety and Bean Variety Choice of crop variety is under direct control of the farmer, and limited-resource farmers often save their own seed from one harvest to the next planting. Conventional wisdom in the international centers and in many national programs is that introduction of new crop varieties can be one of the most cost-effective ways to increase production potential in agriculture. For this reason, there has been an emphasis on plant breeding in the centers. Tolerance to insects and pathogens, drought and other stress conditions, and better nutritional quality can all be built into the genetic package, and this component is one in which research can make a substantial contribution.

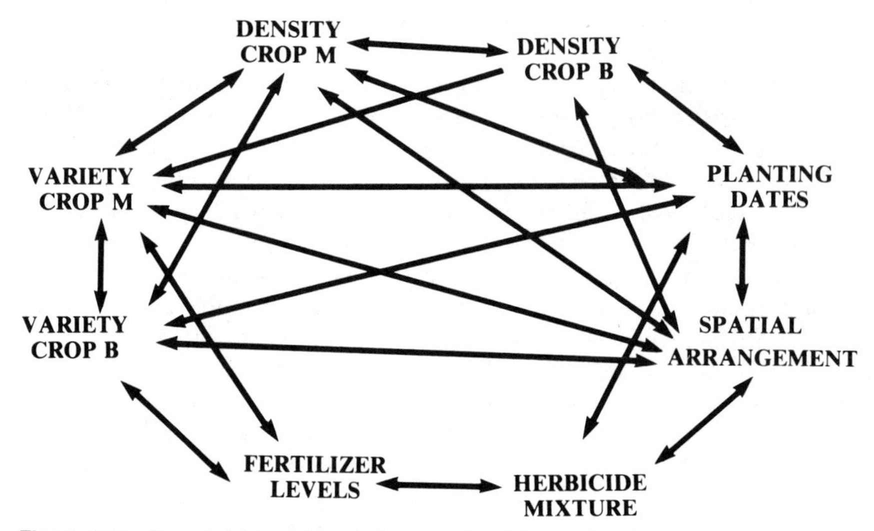

Figure 13.5 Expected interactions between pairs of factors in a two-crop pattern, maize (M) and bean (B).

Fertilizer Levels Fertility often is limiting where intensive cropping is practiced, and a strong emphasis has been placed on fertility trials over the past several decades. The use of soil tests and recommendation of chemical fertilizers has been central to the contributions of the agricultural research establishment. The principal constraint to the farmer of limited resources in adopting this technology has been its cost and the availability of fertilizers in the immediate farm environment.

Herbicide Mixtures Use of chemical weed control in multiple species combinations is both difficult, due to the potential phytotoxicity of compounds on one or more crops, and exciting, in the potentials this offers to the farmer whose crops are reduced in yield by competitive weed growth.

Spatial Arrangement The physical organization of crops in multiple species systems is under direct control of the farmer, and can drastically affect the relative performances of component crops. As new varieties are developed for testing in cropping patterns, alternative arrangements of these crops need to be studied, and these options need to be consistent with the farmer's objectives in planting two or more species in the field.

Planting Dates Relative dates of planting can be varied to favor one component or the other, and to shift both the competitive advantages and the eventual growth and productivity of the crops.

Bean and Maize Density Also important to the relative production levels of the component crops are the densities at which they are planted. This factor also is important in relation to the amount of rainfall available, the types of crops in a mixture, the relative competitiveness of the crops, and the objectives of the farmer.

Choice of these eight factors in the example does not imply that other factors cannot be manipulated by the farmer in the design and choice of a cropping pattern and production strategy. Factors in Fig. 13.5 represent those that have shown importance in experiments and on the farm, and those which have been shown to interact with each other in past experiments. The relative magnitude of these interactions is important to evaluate before considering choice of an experimental design. Again, based on monoculture experience and some research results, these can be compared.

Since it is not feasible to study all interactions simultaneously, it is useful to build a priority index in the manner described above for prioritizing constraints. To get a priority index for interactions, PI_i, each interaction is assigned a probability of existence, $PROB_e$, and rank, RK, according to influence from least (1) to most important (9). The priority index is

$$PI_i = (PROB_e)(RK)$$

Table 13.3 illustrates which of the factors are likely to interact with which other factors ($PROB_e$), and a ranking (RK) is given to the expected importance of the interaction. As shown in Table 13.3, interactions are predicted to be highest between varieties of the two cultivars and bean variety with maize density (rating = 9). Bean variety is expected to interact with relative planting dates and bean density, and maize variety with maize density (rating = 8). Other important though less critical interactions are predicted between varieties and several cultural practices, and between pairs of these cultural practices (ratings = 4 to 7). Other interactions (ratings = 3 or less) are expected to be nonexistent or small enough to be ignored in the research process. This identification of most-probable significant interactions allows the design of an experiment or series of experiments that can explore effectively the main effects of these factors and the interactions that need to be known to design new technology for the farmer.

Research on these factors and their manipulation can confirm the initial predictions and identify which factors are sensitive to changes in cropping pattern and which ones are not. For example, experience and results may show that the

Table 13.3 Probability of Two-Way Interactions, Importance of Interactions, and Priority Indices for Two-Way Interactions for Maize/Bean Intercrop Pattern in the Andean Zone

Factors	Bean var.	Maize var.	Fert. level	Herb. mix	Spatial arrange.	Planting dates	Bean density	Bean density
Bean variety		0.8[a] 9[b]	0.2 3	0.2 5	0.6 4	0.4 8	0.7 8	0.6 9
Maize variety	7.2[c]		0.3 6	0.2 4	0.4 3	0.3 6	0.5 5	0.6 8
Fertility level	0.6	1.8		0.1 5	0.2 3	0.1 2	0.4 4	0.6 4
Herbicide mix	1.0	0.8	0.5		0.5 4	0.6 3	0.3 2	0.4 3
Spatial arrangement	2.4	1.2	0.6	2.0		0.7 7	0.5 6	0.6 4
Planting dates	3.2	1.8	0.2	1.8	4.9		0.6 5	0.7 3
Bean density	5.6	2.5	1.6	0.6	3.0	3.0		0.9 6
Maize density	5.4	4.8	2.4	1.2	2.4	2.1	5.4	

[a]Probability that two-way interaction exists.
[b]Influence ranking: Importance of interaction in intercrop success.
[c]Priority indices for two-way interactions.
Source: Data from Parkhurst and Francis, 1984b.

bean/maize cropping pattern performance is strongly dependent on the plant type of the taller maize component, and thus any new maize suggested for the system must be tested carefully in conjunction with other components of technology to avoid those types that would provide extreme competition for the lower story bean crop. On the other hand, any bean variety will be shaded to some degree by the maize, and if all bean varieties are reduced equally in yield (e.g., no bean variety by cropping pattern interaction) there is limited need to test new bean varieties in combinations with other components. The bean component of the mixture can be changed with a minimum amount of testing. The upper story crop (maize) may be relatively unaffected by weed competition, and thus not interact significantly with herbicide treatment, while the lower story bean crop may be drastically influenced by weed growth and the ability of different herbicide mixtures to control unwanted weeds in the cropping pattern. These types of information will result from the testing of component technology, and will lead to the most efficient possible testing scheme.

In order to design efficient experiments to evaluate these main effects and interactions, the importance of higher-order interactions must be evaluated. If there are three-way, four-way, or other interactions that have been shown in the past to be important, their activity in the field will confound or obscure the main effects or two-way interactions that might be expected to have importance in this bean/maize intercrop pattern (Fig. 13.5). The probability of each three-way interaction and its predicted importance shown in Table 13.4 is multiplied to determine the priority index for research on each factor. The same was done for a few important four-way interactions (Table 13.5). Higher-order interactions were ignored because of the difficulties in visualizing their effects in a practical field setting. Of the 56 possible three-way interactions, there were 11 that had a research priority greater than 1. Of these, 7 included the effect of maize variety, an overriding factor due to differences in plant height and the dominance of this factor in the system. Of the 70 possible four-way interactions, only 6 had a research priority equal to or greater than 1. All these included maize or bean variety, maize or bean density, or both. These interactions are most likely to influence results of studies of main effects and lower-order interactions, and are the ones most logical to include in the first cycle of research.

What are the design options? There is an on-going debate between those who advocate large complex multifactor designs (Mead and Riley, 1981; Mead, Chap. 14) and those who insist on simplicity and designs that study one to three factors at a time (Francis, 1983). In general, the approach using larger complete or incomplete factorial designs may be the appropriate one when there are many interactions of interest and the researcher is anxious to get as much information as efficiently as possible on both main effects and interactions.

To compare three of these options, Table 13.6 shows a situation where eight factors are to be tested: (1) four bean varieties, (2) three maize varieties, (3) six bean density levels, (4) eight maize density levels, (5) three herbicide mixtures (including checks), (6) three fertilizer treatments, (7) four alternative spatial arrangements of the two crops, and (8) three relative planting dates of

Table 13.4 Probability of Three-Way Interactions, Importance of Interactions, and Priority Indices for Three-Way Interaction for Maize/Bean Intercrop Pattern in the Andean Zone

	Var M	Fert.	Herb	Spat.	Pl.Dt.	Dns B	Dns M
Var B × Var M		0.2[a]/2[b]	0.1/3	0.6/5	0.4/5	0.3/4	0.4/5
Var B × Fert	0.4[c]		0.1/1	0.2/2	0.1/3	0.2/2	0.2/2
Var B × Herb	0.3	0.1		0.2/3	0.2/3	0.1/2	0.2/2
Var B × Spat	3.0	0.4	0.6		0.3/2	0.2/3	0.4/3
Var B × Pl.Dt.	2.0	0.3	0.6	0.6		0.2/4	0.2/4
Var B × Dns B	1.2	0.4	0.2	0.6	0.8		0.3/3
Var B × Dns M	2.0	0.4	0.4	1.2	0.8	0.9	
Var M × Fert			0.1/3	0.1/3	0.1/2	0.2/3	0.3/4
Var M × Herb		0.3		0.3/3	0.2/3	0.2/2	0.3/3
Var M × Spat		0.3	0.9		0.4/4	0.2/2	0.3/3
Var M × Pl.Dt.		0.2	0.6	1.6		0.2/3	0.3/4
Var M × Dns B		0.6	0.4	0.8	0.6		0.2/3
Var M × Dns M		1.2	0.9	0.9	1.2	0.6	
Fert × Herb				0.1/3	0.2/3	0.2/2	0.2/2
Fert × Spat			0.3		0.3/2	0.2/2	0.2/2
Fert × Pl.Dt.			0.6	0.6		0.3/3	0.3/3
Fert × Dns B			0.4	0.4	0.9		0.4/4
Fert × Dns M			0.4	0.4	0.9	1.6	
Herb × Spat					0.3/3	0.2/2	0.2/2
Herb × Pl.Dt.				0.9		0.2/2	0.3/2
Herb × Dns B				0.4	0.4		0.4/3
Herb × Dns M				0.4	0.6	1.2	
Spat × Pl.Dt.						0.2/3	0.3/3
Spat × Dns B					0.6		0.4/3
Spat × Dns M					0.9	1.2	
Pl.Dt. × Dns B							0.3/3
Pl.Dt. × Dns M						0.9	

[a]Probability that three-way interaction exists.
[b]Influence ranking: importance of intersection in intercrop success.
[c]Priority indices for three-way interactions.

Table 13.5 Probability of Four-Way Interactions, Importance of Interactions, and Priority Indices for Four-Way Interaction for Maize/Bean Intercrop Pattern in the Andean Zone

Interaction	Probability of occurrence	Influence ranking	Priority index
Var B × Var M × Dns B × Dns M	0.3	5	1.5
Var B × Var M × Pl.Dt. × Dns M	0.2	5	1.0
Var B × Var M × Spat Arr × Dns M	0.2	5	1.0
Var B × Spat Arr × Pl.Dt. × Dns B	0.3	4	1.2
Var B × Spat Arr × Dns B × Dns M	0.4	3	1.2
Spat Arr × Pl.Dt. × Dns B × Dns M	0.5	2	1.0

Table 13.6 Comparison of Three Designs for Several Levels of Eight Factors in Maize/Bean Intercrop Pattern

Factor	Levels	Factor	Levels
Bean varieties	4	Bean densities	6
Maize varieties	3	Maize densities	8
Fertility levels	3	Spatial arrangement	4
Herbicide mixtures	3	Planting dates	3

Option I: Complete factorial (4 × 3 × 3 × 3 × 6 × 8 × 4 × 3), 2 replications
Treatment combinations: 62,208
Size of experiment: 684,288 m^2 or 68 ha

Option II: Fractional replication (3^8 factorial, $\frac{1}{3}$ replicate)
Treatment combinations: 6561
Size of experiment: 72,171 m^2 or 7 ha

Option III: Eight small experiments

1. Maize (2)/bean (2) varieties and maize (2)/bean (2) densities split-plot with maize variety/density as whole plot treatment combinations and bean variety/density as subplots with whole plots in randomized complete blocks (2 replications), RCB.
   Treatment combinations: 16
2. Bean varieties (2), spatial arrangement (4) and maize (2)/bean (2) densities. Factorial treatment design in RCB.
   Treatment combinations: 32
3. Bean varieties (2), densities (3), spatial arrangements (4), and planting dates (3).
   Factorial treatment design in RCB.
   Treatment combinations: 72
4. Maize varieties (2), spatial arrangements (4), and planting dates (3).
   Factorial treatment design in RCB.
   Treatment combinations: 24
5. Maize (3)/bean (3) densities and fertility (3), split split plot in RCB with fertilizer as whole plot, maize density as subplot.
   Treatment combinations: 27
6. Variety trial for maize (8) and bean (6), split plot in RCB with maize variety as whole plot and bean variety as subplot.
   Treatment combinations: 48
7. Spatial arrangement (4) and herbicides (3). Factorial treatment combinations in RCB.
   Treatment combinations: 12
8. Maize density (3) and planting date (3). Factorial treatment combinations in RCB.
   Treatment combinations: 9

Total treatment combinations: 240

Size of experiments: 3120 m^2 or $\frac{1}{3}$ ha with 30 percent borders

Note: Minimum replications: 2. Minimum plot size is 5 m^2 plus 10 percent borders; option III has 30 percent borders.

Source: Adapted from Parkhurst and Francis, 1984b.

the two components. In option I, when eight factors with these levels are placed in a complete factorial with all combinations, there are 62,208 combinations or 124,416 plots in a two-replication experiment; using 5 m^2 per plot requires an area of 684,228 m^2 including an additional 10 percent for borders. This one experiment covers more than 68 ha, clearly an unmanageable monster of a trial. Option I allows evaluation of all main effects and all interactions.

In option II, a factorial replication with only three levels of each factor reduces treatment combinations to 6561 and total size of a two-replication experiment to about 7 ha. This is still a large experiment; it does allow evaluation of all main effects and interactions, although all specific combinations of varieties are not present. Other types of incomplete factorial designs could be utilized, and each would incur loss of some information. Finally, option III illustrates the use of eight small experiments concentrating on varieties of maize with varieties of bean (number 6), interactions of densities of both crops with varieties (number 1), herbicide mixtures with spatial arrangements (number 7), planting dates with spatial arrangements (number 4), fertilizers with both crop densities (number 5), and other factors that would give information on main effects and the most important interactions, but not all the interactions studied in options I and II. Even with two replications of these experiments, the number of plots is reduced to 240 and the size of the experimental area is only 3120 m^2 with 30 percent added for borders. These are the types of practical trade-offs that the researcher is forced to make.

There are obvious advantages both to the large and to the small experiments. Practical experience in the field has shown that a larger number of small experiments is easier to handle than one or two larger factorial treatment designs. Since operations at planting, data collection time, and harvest are complicated, it is more expeditious to handle each experiment as a unit and complete the field operations in 1 day when possible, then move on to another experiment.

Field Organization

Additional details on the organization of the trials in split-plot designs in the field may be helpful. The typical and logical hierarchy of split-plot assignments is to put the factor of greatest interest in the smallest subplot. For example, a trial may include bean densities, maize densities, bean varieties, and maize varieties, and the researcher attaches the following priorities to these factors: (1) maize densities, (2) bean varieties, (3) maize varieties, and (4) bean densities. The theoretical design approach would place bean densities in main plots, maize varieties in split plots, bean varieties in split split plots, and maize densities in the smallest subunits. The most replication, greatest number of degrees of freedom, and most powerful testing would be directed at maize densities and bean varieties.

In the practical reality of intercropping and field research, there are other concerns that strongly influence organization of treatments. The dominant nature of the maize overstory crop and competitive influence of both maize variety and maize density strongly suggest that these factors be placed in as large a plot as possible. Since the competitive effect of maize variety or density is so much greater than bean variety or density, many plots of the latter can be placed in large plots of the former with a minimal need for border rows. When plot size and border area can be minimized, a given number of treatment combinations can be fit into a smaller experimental area and this is more likely to be uniform with fewer large variations in soil fertility. A number of experiments on bean varieties with maize have been conducted with single row bean plots, since the principal competition is from the maize and there is limited influence of one bean row on the neighboring bean rows.

In the above example, a logical order for the field organization might be (1) maize variety in main plots, (2) maize densities, (3) bean densities, and (4) bean varieties in the smallest subplots. Although the precision of specific comparisons may differ from what was intended by the researcher at the outset, the overall quality of information from the experiment will be higher and the logical arrangements of treatments make this a smaller, more manageable, and more valuable experiment. Other practical considerations on borders needed to separate planting dates or spatial treatments, and minimal plot size needed for fertilizer or herbicide applications, can also influence the field use of designs (Davis et al., 1981). Of interest to the plant breeder is the testing of large numbers of new materials coming from a crossing program. It generally is much more useful to test twice as many new crosses in plots half as large, then to use larger plots (e.g., one-row versus two-row plots), if this plot size is consistant with the levels of differences one wants to detect. These are practical conclusions based on dozens of field experiments on these types of genetic and management factors.

Use of Multiple Environments

In order to move confidently from this first cycle into more refined experiments with fewer factors under study and more precise levels of factors, such as density or fertility, the researcher could make use of either multiple locations or multiple seasons in a single location. If the range of rainfall and temperature fluctuation from one year to another can be approximated by using different locations within a region, it is more time-efficient to work across locations to get information as quickly as possible and move on to subsequent cycles of research and testing on the farm. There are some important trade-offs between years, locations, and replications in experiments. These were evaluated in recent studies on sorghum (Saeed et al., 1984) and on maize (Brakke et al., 1983). Figure 13.6 shows the expected standard error of a sorghum genotype mean when a series of cultivars is tested over a range in environments and over 1, 2, or 3 years (from Saeed et al., 1984). About five environments in 1 year would give the same degree of precision in measuring a genotype mean as 2 years' testing in two environments, and about eight environments give the same precision as 3 years' testing in two environments. Economics of research may dictate multiple years of testing in one site, but the cost is a delay in getting credible results for moving on to another cycle, or a delay in getting results out to farmers. Figure 13.7 shows the improved precision as a result of increasing replications vs. testing over more

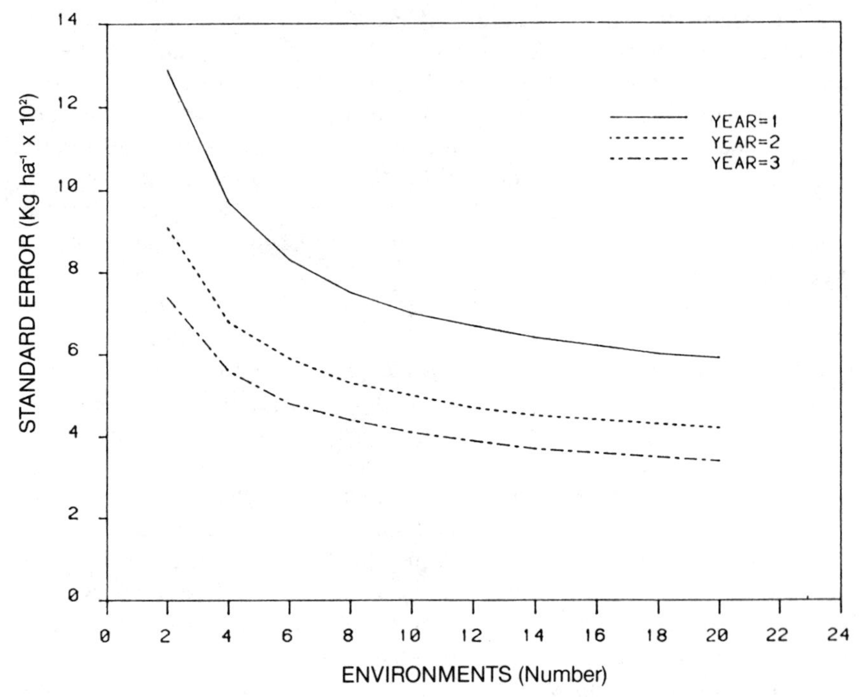

Figure 13.6 Expected standard error of a genotype mean yield for various assumed numbers of years and environments within years when replications = 2. (From Saeed et al., *Agron. J.*, vol. 76, 1984, pp. 55–58. Reprinted by permission of the American Society of Agronomy.)

locations. It is obvious that two replications are desirable to increase precision, allow valid statistical comparison of cultivars, and provide a low-cost method of producing more valid results. However, more than two replications are of limited value, and the advantages of additional locations are clear from the figure. The maize data (from Brakke et al., 1983) demonstrated similar trends. Commercial hybrid maize and sorghum breeders follow a methodology of minimal replication and maximum number of field locations.

Before proceeding to the next cycle, it is assumed that the trials in the first cycle have been run in at least two seasons or years, or that two or more locations have been used as a proxy for seasons to gain reliable results. A minimum of three locations is desirable, since this gives a better measure of treatment by location interaction, and gives an indication of most important trends if at least two of the three locations agree in results. More than two locations provides insurance, since loss of one location would not hold back the program for a season. This information leads to conclusions about interactions and main effects from the experiments outlined above, and lays the foundation for succeeding cycles of experiments.

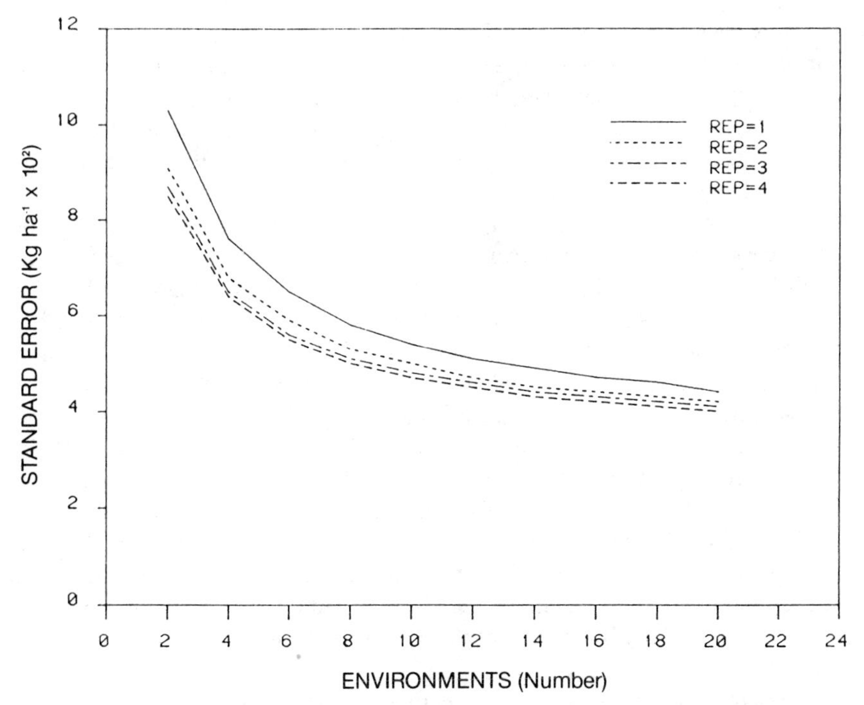

Figure 13.7 Expected standard error of a genotype mean yield for various assumed numbers of replications and environments when years = 2. (From Saeed et al., *Agron J.*, vol. 76, 1984, pp. 55–58. Reprinted by permission of the American Society of Agronomy.)

Large Factorial Designs

More complex designs are recommended by Mead and Stern (1980) to improve efficiency of research. There is no question about the theoretical validity of this approach, and many convincing arguments are presented in Chap. 14. The complications of large factorial designs can be illustrated with two examples from tropical crops research provided by Francis (unpublished data). An experiment in CATIE in Costa Rica was designed to study the main effects and interactions among five crop species and four levels of management. Rice (*Oryza sativa*), dry beans, maize, cassava (*Manihot esculenta*), and sweet potato (*Ipomoea* sp.) were planted in monoculture, in all combinations of two crops, in all combinations of three crops, in all combinations of four crops, and in the five-crop intensive intercrop pattern. Main plots were divided into four levels of management: traditional farmer treatment, use of fertilizer, use of weed control, and use of both fertilizer and weed control. Although only two replications were used, this experiment covered 12 ha (about 30 acres). Obviously, many other cultural practices were held constant to attempt to understand the effects of species, mixture, and management of weeds and fertility.

This massive undertaking was continued over several years, and a few of the component systems were evaluated, and results were published as graduate student thesis/dissertation studies. The majority of this valuable information has not been published. The magnitude of the experiment, the variability that is introduced by planting over such a large area, and many significant interactions that result in the analysis all make this type of experiment difficult to manage and the data difficult to interpret.

A second example of the massive experiment was a maize study on the north coast of Colombia that was carried out in 1971 and 1972 (Francis, unpublished data). The objective was to assess the potential impact of new technology on maize production by introducing one or more components of a new package for farmers of limited resources. Four levels of fertility, weed control (farmer practice versus herbicide), insect control (none versus insecticide), land preparation (farmer practice versus complete tillage), and maize cultivar (farmer variety versus commercial hybrid) were the five factors included in the study. These were included in all combinations in a factorial treatment design and in a split-plot field design with tillage practices as the main plots and the maize cultivars as the smallest subplots. With the five factors, four replications, and four-row plots that were 10 m long, this experiment occupied about 1 ha in each site. Six agronomists in training with CIAT carried out these experiments on six different farmers' fields, and the results were assembled for analysis in the central facility.

The most positive result of the experiments was the interaction of agronomist with farmer, as both learned about maize through the implementation of the experiment and data collection in the field. The least productive part of the exercise was the analysis and interpretation of the data. Three of the experiments produced insufficient data for useful analysis, and the other three had coefficients of variation that were higher than normally acceptable in maize research. In the analysis, most of the interactions in these three experiments were significant, making any interpretation of the main effects difficult. There was further difficulty in the interpretation of any interaction more complex than the two-way between two factors.

The end result was frustration by the young agronomists over the complexity of the experiments and disbelief by the farmers over the magnitude of the trial and why those young agronomists were so excited about putting so many pretty flags and little white stakes in their fields. It was also a learning experience for those who designed the trials about the value of complexity in farmers' field trials. Our conclusion was that large experiments are not useful for meeting this type of objective. In later trials, we invariably concentrated on two or three factors at the most in order to better understand the complexity of these components as they function in tropical cropping patterns.

Criteria for Evaluation

When analyzing the results of an experiment, decisions must be based on evaluation criteria defined by the farmers' objectives. There are several ways to evaluate cropping systems: by yields of individual component crops, comparisons to monoculture such as equivalent yields, relative yields, and LER. Examination of these criteria may lead to different conclusions (Parkhurst and Francis, 1984b).

Consider the experiments conducted on maize/bean intercropping systems in Colombia (Francis et al., 1982a, 1982b). Two varieties of beans, bush and climbing, were grown at three common bean densities and intercropped with eight maize densities. The experiments were conducted as split plots with maize density as the whole plot and bean density as the subplot. There were four replications of each whole plot.

Analysis of variance and appropriate *F* tests were performed for the individual yields, total equivalent yields, and relative yields of maize and beans, as well as for LER. A bivariate analysis (Dear and Mead, 1983) was calculated but discounted, since the correlation between maize and bean yields was inconsistent over treatments. The test of homogeneity of covariance between bush and climbing bean data indicated the two experiments should be analyzed separately.

The effect of maize density was significant in both experiments for all criteria (Table 13.7). The effect of bean density and the maize/bean density interaction was another story. Bean yield and relative bean yield gave results different from the other criteria. For bush beans, there was significant variation in bean densities; for climbers, the maize/bean interaction was significant. Bivariate plots of the yield confirmed the information provided by these criteria.

The important point is that total system yield shows a remarkable compensation in the system. The taller maize crop dominates and tends to offset the effects of the bean crop. This is one facet of the system. However, if the objective is to improve the individual components (a plant breeder's point of view), to ignore individual yields is to miss important information. For this range of bean densities, bean yield increases with increasing density. It is essential to examine the data from all angles of interest in order to reach an understanding of how a multiple cropping system works.

Table 13.7 Probability of Greater *F* Value from Analysis of Variance of Yield for Maize/Bean Cropping System in Colombia

Source of variation	df	Maize yield	Bean yield	Total equivalent yield	Relative maize yield	Relative bean yield	LER
Bush bean experiment:							
Maize density	6	0.001	0.001	0.001	0.001	0.001	0.001
Bean density	4	0.215	0.001	0.685	0.215	0.001	0.685
Maize × bean density	24	0.928	0.083	0.950	0.928	0.083	0.950
Climbing bean experiment:							
Maize density	6	0.002	0.001	0.010	0.002	0.001	0.010
Bean density	4	0.674	0.831	0.819	0.674	0.831	0.819
Maize × bean density	24	0.326	0.013	0.107	0.326	0.013	0.107

Source: Adapted from Francis et al., 1982a, 1982b.

Even more important to the success of a program is the correspondence of the evaluation criteria in these experiments to the criteria used by farmers to decide whether or not to adopt the new technology. If the researcher is fine-tuning LERs or equivalent yields, while the farmer is more interested in reducing risk, producing a range of crops for the family, or looking at stability of food and income, there is a need for better communication to bring these criteria together (Francis, 1981b). The farming systems approach that involves the farmer from the identification of constraints through the research process to the recommendation of new technologies appears to offer many advantages.

FARM TESTING AND VALIDATION OF TECHNOLOGY

Agricultural research traditionally has been conducted on the experiment station, with a second step of validation or demonstration on the farm. Where conditions on the experiment station—resource base, land quality, soil fertility, type of cropping system, and level of technology—are similar to those on farmers fields, this scheme of testing will produce credible results for farmers. This pattern of research has prevailed in temperate and developed countries, and has proven successful there. When there is a large difference between conditions on the farm and on the station, as commonly found in developing countries, this system may break down. If the experiment station research is conducted under mechanized, irrigated, pest-protected conditions, this may be very different from farms nearby where hand labor is used for tillage, farmers depend on natural rainfall, and pest problems (including weeds, insects, and plant pathogens) are severe. Under these conditions, it is important to question the conventional wisdom of doing a large amount of research and testing on the station before moving out to farmers' fields.

This concern has led researchers to greater emphasis on testing under farmers' conditions. There have been methodologies proposed by CIMMYT in Mexico and in Kenya on using research station data to make recommendations to farmers (Perrin et al., 1979; Byerlee and Collinson, 1980). Their training program has emphasized the use of on-farm trials to test components of technology under real-world conditions so that new practices will have a realistic test before any firm recommendations are organized for the farmer. These trials have assumed different forms, with some organized and carried out entirely by researchers, some done entirely by farmers, and others executed in a cooperative way. This last approach has been called the participatory model, and is being widely used in some projects and labeled "farming systems research" (Gilbert et al., 1980). Important applications have been developed and tested in Guatemala (Hildebrand, 1979).

The farming systems approach to research involves farmers in the identification of problems, the design of potential solutions, and the research carried out on farm to test alternatives and evaluate their potential to increase production. Exciting applications of this methodology are taking place in Botswana (Norman et al., 1983). This process is attractive because of the direct involvement of the farmer in all stages of the research process, and because it will help the researcher to anticipate potential problems with the adoption of new technology once it is ready. This process is a cyclical one, starting with the identification of limiting factors to production, and the design of research to solve those problems. Research may be conducted either on station or on farm or both, and the results are discussed with the farmer to determine their potential applications. Validation of results on the farm is conducted in a cooperative way, and if the package or component is successful, there is no difficulty in extension of results in that immediate area. After the new practice or package of practices is introduced into the system, an evaluation is made of the impact of this new technology, and new constraints are identified for further research. This collaborative model with farmer participation is becoming more popular in developing country research projects.

The analysis of data is complicated when research is conducted on farmers' fields under a wide range of conditions. One approach has been to use unreplicated observational plots on each farm, and to analyze these across farms confounding farm (location) with replication. Although this is valid statistically, there often are so many confounded factors with location and the error terms are so large that it is difficult to interpret data and distinguish between treatment effects. For this reason, a joint approach with some station trials including replication and researcher control and some on-farm counterpart trials for study of applications can be an efficient method. It is important to decide which of the many possible treatments of interest have potential to solve production-limiting constraints before moving to the farm. Otherwise, it would be physically impossible to conduct meaningful observational trials on the farm. The intermediate approach, using replicated trials on farms with some participation of the farmer, is another way to get both real world conditions for the trial and enough control to get meaningful data. There is yet much to be done in designing and implementing meaningful research that relates work on the stations to real problems on the farm.

There is concern in the research community that the job is not finished until results have reached the farm and there is some evaluation of what has happened as a result of the introduction of new technology. In order to measure impact of a change in technology, it is necessary to know current production levels or income before the introduction of something new. If there is a careful survey of farming practices and problems in the current production systems, as outlined in the section on problem identification and priorities for research, this will help to establish a baseline for a given region, crop, or production system. If only conventional wisdom is used to determine what problems should have priority in research, there is less chance that this baseline will be available. Again, the farmer can participate in the process of identifying current production levels and problems. This is the best way to involve the primary producer in the total effort, and will help to ensure that research is correctly oriented to real problems in the zone.

The setting of baseline levels for production also ensures that there will be something against which to measure the impact of the new technology. An important part of this process is to correctly identify the objectives of the farmer. The commonly accepted goals of maximum production and net profit that are used in most developed countries may not apply to farmers with limited resources. Among the criteria that a small farmer may use to evaluate success in the system are production of a range of crop and animal products for consumption by the family, stability of production and income, distribution of this production and income through the year, minimal costs of inputs to adopt the new technology, and minimal risk associated with a conversion from traditional practices to the new methods (Francis, 1981b). These factors are difficult to measure, and are outside the competence and appreciation of most scientists trained in agriculture, especially agronomy and animal science.

The creation of interdisciplinary teams for this type of assessment is becoming more commonplace. The inclusion of rural sociologists and agricultural economists on teams that formerly were made up entirely of agronomists is a welcome step in the right direction. With the correct training and orientation in the field, this type of biological/social science team can make considerable progress identifying the cultural and social aspects of a farming system as well as the biological constraints to production. Some of these factors can be quantified, and some have to be evaluated on a more qualitative basis. The input of social sciences to agricultural development as part of this complex evaluation of impact is important in measuring the effects of new technology, and in the design of research systems for solving further constraints to production (see Chap. 12).

The view for the future looks promising. On-farm research in an untapped area with limited prior work. As information systems and agricultural computer networks become more commonplace, it will become possible to have a network of on-farm experiments augmented by experiment station research. In addition more thought needs to be given to the trade-offs between replicating results over years vs. locations. Simulations based on environmental parameters tempered with records of time-series data can provide insight into this problem.

Another area of great potential is the application of multiobjective programming techniques to multiple cropping research methodology. A hierarchy of research objectives, such as economic, technical, and social goals, can be formulated along with associated resource requirements (growing degree days, energy, rainfall limitations). A "best" strategy can be identified and limiting conditions under which the strategy will remain best can be established. Researchers would then have a systematic way of controlling the decision-making process.

REFERENCES

Brakke, J. P., Francis, C. A., Saeed, M., Youngquist, W. C., and Nelson, L. A. 1983. Efficiency of testing maize in different environments and cultural systems, *Agron. Abstr.*, 1983, p. 42.

Byerlee, D., and Collinson, M. 1980. *Planning Technologies Appropriate to Farmers: Concepts and Procedures,* CIMMYT Economics Program, Mexico.

Davis, J. H. C., Amezquita, M. C., and Muñoz, J. E. 1981. Border effects and optimum plot sizes for climbing beans (*Phaseolus vulgaris*) and maize in association and monoculture, *Exp. Agric.* 17:127–135.

Dear, K. B. G., and Mead, R. 1983. The use of bivariate analysis techniques for the presentation, analysis and interpretation of data, *Statistics in Intercropping,* Tech. Rep. 1, Dep. Applied Statistics, University of Reading, Reading, U.K.

Francis, C. A. 1967. *Downey Mildew Disease of Maize in the Philippines,* M.S. thesis, Cornell University, Ithaca, New York.

———. 1981a. Development of plant genotypes for multiple cropping systems, in: *Plant Breeding Symposium II,* Chap. 7, (K. J. Frey, ed.), Iowa State University Press, Ames, Iowa.

———. 1981b. Rationality of farming systems practiced by small farmers, in: *KSU Symposium on Small Farms in a Changing World—Prospects for the Eightees,* Kansas State University, Manhattan, Kansas.

———. 1983. Research methodology for intercropping experiments, *Agron. Abstr.,* 1983, p. 44.

———. 1985. Variety development for multiple cropping systems, in: *CRC Critical Reviews in Plant Science,* CRC Press, Boca Raton, Florida, 3:133–168.

Francis, C. A., Prager, M., and Tejada, G. 1982a. Density interactions in tropical intercropping. I. Maize (*Zea mays* L.) and climbing beans (*Phaseolus vulgaris* L.), *Field Crops Res.* 5:163–176.

———. 1982b. Density interactions in tropical intercropping. II. Maize (*Zea mays* L.) and bush beans (*Phaseolus vulgaris* L.), *Field Crops Res.* 5:253–264.

Garrity, D. P., Harwood, R. R., Zandstra, H. G., and Price, E. C. 1979. *Determining Superior Cropping Patterns for Small Farms in a Dryland Rice Environment: Test of a Methodology,* International Rice Research Institute (IRRI), Los Baños, Philippines.

Gilbert, E. H., Norman, D. W., and Winch, F. E. 1980. Farming systems research: a critical appraisal, *MSU Rural Devel. Papers* 6.

Gomez, A. A., and Gomez, K. A. 1983. *Multiple Cropping in the Humid Tropics of Asia,* IDRC, Ottawa, Canada, IDRC-176e.

Gutierrez, U., Infante, M., and Pinchinat, A. 1975. *Situación del cultivo de frijol en America Latina,* in: Boletín Informe ES-19, Centro International de Agricultura Tropical (CIAT), Cali, Colombia.

Harwood, R. R. 1979. *Small Farm Development: Understanding and Improving Farming Systems in the Humid Tropics,* Westview Press, Boulder, Colorado.

Hildebrand, P. E. 1979. Generating technology for traditional farmers—the Guatemalan experience, *Symposium on Socio-Economic Constraints to Crop Protection,* IX Intl. Congress of Plant Protection, Washington, D.C.

Huxley, P. A., and Maingu, Z. 1978. Use of a systematic spacing design as an aid to the study of intercropping: Some general considerations, *Exp. Agric.* 14:49–56.

Mead, R., and Riley, J. 1981. A review of statistical ideas relevant to intercropping research, *J. Roy. Stat. Soc. A* 144:462–509.

Mead, R., and Stern, R. D. 1980. Designing experiments for intercropping research, *Exp. Agric.* 16:329–341.

Norman, D. W., Modiakgotla, E., Seibert, J., and Tjirongo, M. 1983. Helping the limited resource farmer through the farming systems approach to research, *Culture Agric.* 19:1–8.

Parkhurst, A. M., and Francis, C. A. 1984a. Time-honored experimental designs and complex cropping systems, in: *Proc. 144th Ann. Meeting Amer. Stat. Assoc.*, Philadelphia, Pennsylvania, p. 81.

———. 1984b. Research methodology and analysis for complex cropping systems, in: *Proc. Int. Biometrics Conf.*, Tokyo, pp. 100–109.

Perrin, R. K., Winkleman, D. L. Moscardi, E. R., and Anderson, J. R. 1979. *From Agronomic Data to Farmer Recommendations: an Economics Training Manual,* CIMMYT Information Bull. 27, Mexico City, Mexico.

Saeed, M., Francis, C. A., and Rajewski, J. F. 1984. Maturity effects on genotype x environment interactions in grain sorghum, *Agron. J.* 76:55–58.

Willey, R. 1979a. Intercropping—its importance and research needs. Part 1. Competition and yield advantages, *Field Crop Abstr.* 32:1–10.

———. 1979b. Intercropping—its importance and research needs. Part 2. Agronomy and research approaches, *Field Crop Abstr.* 32:73–85.

Chapter 14

Statistical Methods for Multiple Cropping

Roger Mead

The design and analysis of intercropping experiments is a complex and important topic that has been substantially neglected in the development of intercropping research programs. My experience over the last 6 years with a wide range of such programs is that a large proportion of intercropping experiments are designed and/or analyzed without any direct involvement of a statistician. It is also true that few statisticians have sought to become involved. The result has been poorly designed experiments and inadequate analysis of data. My intention in this chapter is to provide guidelines to help overcome this situation, but it must be emphasized that the efficient design and analysis of experiments will be achieved only when statisticians are directly involved in the planning and interpretation of all individual experiments and of research programs.

The first point to emphasize is that there is no single, appropriate form of analysis. Indeed in any intercropping experiment it is extremely important to consider and use several different forms of analysis. This is discussed in some detail by Mead and Stern (1980, 1981). Intercropping is different from monocropping in that two, or more, crop yields are measured, and it is further different in that not all experimental plots have both crops grown. The data structure is therefore complex and this alone would prompt consideration of different forms of analysis.

A major difficulty in the analysis of intercrop experimental yields is ensuring that measurements are compared only if they are genuinely comparable. Obviously maize (*Zea mays* L.) and bean (*Phaseolus vulgaris* L.) yields cannot be directly compared. Treatments that each produce a maize yield and a bean yield may be compared in terms of maize yields, or of bean yields, or of an appropriately defined combination of the two yields. However if treatments involve different sets of crops, for example, (1) maize, (2) maize and bean, and (3) maize and cassava (*Manihot esculenta*), then such treatments may be compared only in terms of maize yields or of a "value" variable such as financial return.

In the four sections of this chapter I shall discuss four forms of analysis. It will often be sensible to analyze each yield variable separately considering the effects of applied treatments and possibly, in addition, the effect of the presence or absence of the second crop. The principles of analysis for a single variable should be well known, but experience shows that they are often unknown, misunderstood, or simply ignored. The main principles as they relate to intercropping experiments are presented in the first section, and the principles of experimental design, which are equally ignored, misunderstood, or unknown, are presented next. The presentation is necessarily brief, and for a more extended presentation of the principles the reader is referred to Mead and Curnow (1983).

Because plots on which two crops are grown produce two yields and because it is likely that the two yields will be interrelated, it is extremely desirable that the two sets of crop yields be considered together in a bivariate analysis. A bivariate analysis is designed to examine simultaneously the patterns of variation for the two crop yields. A joint pattern cannot be detected by considering only the two separate analyses. Examination of only the separate analyses may also lead to interpreting an underlying effect twice, or not at all, because of the joint variation of the two yields. Bivariate analysis is discussed in the next section and should be a standard component of analysis of intercropping data.

The third form of analysis involves the use of indices of combined yield. There are always dangers in the use of indices because they reduce the information presented. There is sometimes an assumption that an index provides a sufficient summary of the data and this may be a dangerous illusion. Nevertheless the indices discussed in the third section can provide a valuable insight into a complex situation by identifying a single dimension in which comparisons of overall performance may be made.

The last section is on the measurement of the stability of intercropping. Much of what has been written previously on stability measures is statistically unsound and a new approach to stability is described and illustrated.

A general review, from a statistician's viewpoint, of the problems of design and analysis of intercropping experiments is given by Mead and Riley (1981). This contains more technical detail than is given here, and it also includes a very wide-ranging discussion at the Royal Statistical Society with contributions from agronomists, economists and statisticians.

GENERAL PRINCIPLES OF STATISTICAL ANALYSIS AND DESIGN

Analysis

Analysis of Variance The initial stage for most analyses of experimental data is the analysis of variance for a single variate, or measurement. The analysis of variance has two purposes. The first is to provide, from the error mean square, an estimate of the background variance between the experimental units. This variance estimate is essential for any further analysis and interpretation. It defines the precision of information about any mean yields for different experimental treatments. One major requirement often neglected is that the error mean square must be based on variation between the experimental units to which treatments are applied. If treatments are applied to plots 10×3 m, then the variance estimate used for comparing treatments must be that which measures the variation between whole plots. Measurements on subplots or on individual plants are of *no value* for making comparisons between treatments applied to whole plots.

The second purpose of the analysis of variance is to identify the patterns of variability within the set of experimental observations. The pattern is assessed through the division of the total sum of squares (SS) into component sums of squares and the interpretation of the relative sizes of the component mean squares. To illustrate the simple analysis of variance, and for illustration of other techniques, later in this chapter, I shall use data from a maize/cowpea (*Vigna unguiculata*) intercropping experiment conducted by Dr. Ezumah at IITA, Nigeria. The experimental treatments consisted of three maize varieties, two cowpea varieties, and four nitrogen levels (0, 40, 80, 120 kg/ha) arranged in three randomized blocks of 24 plots each. The data for cowpea and maize yields are given in Table 14.1. The analysis of variance and tables of mean yields for the cowpea yields are shown in Table 14.2. The analysis of variance shows that there is very substantial variation in cowpea yield for the different maize varieties; there is also a clearly significant (5 percent) interaction between cowpea variety and nitrogen level and a nearly significant variation between mean yields for different nitrogen levels. The tables of means for cowpea yield that should be presented are therefore for (1) maize varieties and (2) cowpea variety $\times$ nitrogen levels, with the mean yields for nitrogen levels as a margin to the table. The analysis of variance implies strongly that no other means should be presented.

The interpretation indicated by the analysis and mean yields is as follows. Yield of cowpea is substantially determined by the maize variety grown with the cowpea. Higher cowpea yields are obtained when maize variety 1 is grown. For cowpea variety *B*, cowpea yield is reduced as increasing amounts of nitrogen are applied (presumably because of correspondingly improved maize yield). Yields for cowpea variety *A* are not affected in this manner.

Assumptions in the Analysis of Variance The interpretation of an analysis of variance and of the subsequent comparisons of treatment means depends critically on the correctness of three assumptions made in the course of the

Table 14.1 Cowpea and Maize Yields in Intercrop Trial at IITA, Nigeria

		Yield (kg/ha)[a]								
		1			2			3		
Cowpea variety	Nitrogen level	I	II	III	I	II	III	I	II	III
		Cowpea								
A	N_0	259	645	470	523	540	380	585	455	484
A	N_1	614	470	753	408	321	448	427	305	387
A	N_2	355	570	435	311	457	435	361	586	208
A	N_3	609	837	671	459	483	447	416	357	324
B	N_0	601	707	879	403	308	715	590	490	676
B	N_1	627	470	657	351	469	602	527	321	447
B	N_2	608	590	765	425	262	612	259	263	526
B	N_3	369	499	506	272	421	280	304	295	357
		Maize								
A	N_0	2121	2675	3162	2254	3628	4069	2395	2975	4576
A	N_1	3055	3262	3749	3989	3989	4429	4429	4135	4429
A	N_2	3922	3955	4095	4642	4135	4642	5589	4429	5156
A	N_3	4129	4129	4022	3975	4789	4282	5990	5336	5663
B	N_0	2535	2535	2288	4209	3989	2321	2901	4429	3482
B	N_1	2675	3402	3122	4789	4936	3342	3555	4936	4135
B	N_2	3855	3815	3535	5083	4496	3702	6023	5296	4069
B	N_3	3815	4202	3749	5656	5516	5223	5516	5083	5369

[a]Yields grouped by maize variety (1, 2, 3) and planting block (I, II, III).
Source: Data from Dr. Ezumah, IITA, unpublished.

analysis. If the assumptions are not valid, the conclusions drawn may also be invalid and, therefore, misleading. Evidence available from the analyses of intercropping experiments suggests that failure of the assumptions is at least not less frequent than in monoculture experiments. It is therefore vital that the experimenter deliberately consider the assumptions before completing the analysis. The three assumptions are:

1. That the variability of results does not vary between treatments
2. That treatment differences are consistent over blocks
3. That observations for any particular treatment for units within a single block would be approximately normally distributed

There is an element of subjectivity about the assessment of these assumptions. For a more extensive discussion the reader is referred to Chap. 7 of Mead and Curnow (1983). In brief, the experimenter should ask:

1. Does it seem reasonable, and do the data appear to confirm, that the ranges of values for each treatment are broadly similar and that there

Table 14.2 Analysis of Variance and Tables of Means for Cowpea Data in Intercrop Trial at IITA, Nigeria

	Analysis of variance			
Source	SS	df	MS	*F*
Blocks	73,000	2	36,500	2.8
Maize varieties (M)	409,400	2	204,700	15.7[a]
Cowpea varieties (C)	6,000	1	6,000	0.5
Nitrogen (N)	113,100	3	37,700	2.9
M × C	9,900	2	4,950	0.4
M × N	67,600	6	11,267	0.9
C × N	172,400	3	57,433	4.4[b]
M × C × N	135,400	6	22,567	1.7
Error	599,300	46	13,000	

Table of means (cowpea yield (kg/ha))

	Nitrogen level					
Cowpea variety	0	40	80	120	Maize variety	Mean
A	482	459	413	511	1	582
B	597	497	479	367	2	430
Mean	539	478	446	439	3	415
SE of difference for *N* means = 50					SE of difference	43
SE of difference for combinations = 71						

Note: SS = sums of squares; df = degrees of freedom; MS = mean square; F = F value for testing; SE = standard error.
[a]Significant at 0.1% level.
[b]Significant at 5% level.
Source: Data from Dr. Ezumah, IITA, unpublished.

is no trend for treatments giving generally higher yields to display a correspondingly greater range? In biological material it is more reasonable to suppose that treatments with a high mean yield also have a rather higher variance of yield, and so an experimenter should be prepared to recognize this occurrence and to use a transformation of yield before analysis.

2 Are treatment differences similar in the "good" blocks and in the "bad" blocks? Again if the pattern of bigger differences in better blocks, which might reasonably be expected, is found, then a transformation of yield is necessary.

3 Do I believe that an approximately normal distribution is a sensible assumption?

For the data in Table 14.1 a visual inspection reveals no reason to doubt the assumptions. The only peculiarity of the data is the repetition of some values in the set of maize yields, but since no obvious explanation could be found the data were used for analysis and interpretation as shown in Table 14.2.

Comparisons of Treatment Means Many sets of experimental results are wasted through an inadequate analysis of the results. In many cases this results from the use of multiple comparison tests of which the most prevalent, and therefore the one that causes most damage, is Duncan's multiple range test. The reason that multiple comparison tests lead to a failure to interpret experimental data properly is that such tests ignore the structure of experimental treatments and hence fail to provide answers to the questions that prompted the choice of experimental treatments.

Two particular situations in which multiple range tests should *never* be used are for factorial treatment structures and if the treatments are a sequence of quantitative levels. In the former the results should be interpreted through examination of main effects and interactions. In the second the use of regression to describe the pattern of response to varying the level of the quantitative factor should be obligatory. Thus, for the cowpea yield example, the effect of nitrogen on yield for cowpea variety B can best be summarized by the regression equation

$$\text{Yield} = 591 - 1.77\,N$$

where yield and N are both measured in kg/ha. The predicted yields for the four nitrogen levels (0, 40, 80, 120 kg/ha) are 591, 520, 449, and 379, which obviously agree very closely with the observed means.

Examples of the failure of experimenters to interpret their data properly occur regularly in all agricultural research journals wherever multiple comparison methods are widely used. Examples of misuse and discussion of alternative forms of analysis are given by Bryan-Jones and Finney (1983), Morse and Thompson (1981), and many other authors. The only sensible rule to adopt when analyzing experimental data is *never use multiple range tests or other multiple comparison methods*.

Presentation of Results The prime consideration in presenting experimental results should be to provide the reader with all necessary information for a proper interpretation of results, without unnecessary detail. This principle leads to some particular advice:

1. Tables of mean yields should always be accompanied by standard errors for differences between mean yields and the degrees of freedom for those standard errors.
2. When multiple levels of analysis are used, as for split plot designs then *all* the different standard errors must be given.
3. When results are presented in graphic form the data should always be shown (plotting mean yields). A graph showing only a fitted line or curve deprives the reader of the opportunity to assess the reasonableness of the fitted model.
4. Standard errors are much more effective with tables of means than with graphs where standard errors are represented by bars.

5 All standard errors or other measures of precision should be defined unambiguously. The statement below a set of means "standard error = 11.3" is ambiguous because it does not specify if it is for a mean or a difference of means or, even, for a single value rather than a mean.

Design

Proper Use of Replication, Blocking, and Randomization Good experimental design is at least as important in intercropping experiments as in monocropping experiments. The choice and structuring of experimental plots are extremely important and the third basic principle of design requires a proper system of allocation of treatments to units.

First, the experimental plot must be chosen so that sufficient plot replication is possible and so that the yields from a plot may be considered properly representative. Intercropping tends to need rather larger plots than monocropping both because of the need to have sufficient plants of both crops and because of the need for large guard areas. If the guard areas are a very substantial proportion of the experimental area, as may easily happen when spacing treatments are involved, then a systematic arrangement of treatments may be appropriate (see Mead and Stern, 1981, for further discussion). Note that when treatments are arranged systematically, the analysis of variance is not the appropriate initial form of analysis; rather an analysis of the response to spacing should be based on fitting response models for each systematic set of plots, and then analyzing the set of fitted response models.

Because the experimental plots for intercropping experiments are often large there is a danger of allowing blocks of plots, for a randomized block design, to become large also. Large blocks are a contradiction of the blocking principle. Intercropping experiments are often performed on land only recently prepared for experimental research and may therefore be expected to be heterogeneous. Both these considerations mean that very considerable care must be taken in the identification and use of blocks. Certainly standard randomized block designs with long thin blocks of 8 to 16 plots are almost certainly inefficient. Two positive suggestions can be made. First, the experimenter should examine the set of available plots and should judge, in the field, which plots might genuinely be expected to behave very similarly and which would be different. Second, much greater use should be made of designs other than randomized complete blocks. Blocks should be restricted to eight or fewer plots; confounded designs or more general incomplete block designs should be encouraged. The availability of computer programs for analysis means that the classical designs are unnecessarily restrictive and a much more flexible set of designs using a properly defined, and possibly unequally sized, set of blocks should be used.

Finally the importance of random allocation of treatments to blocks in all blocks cannot be overemphasised. Randomization provides the justification for the analysis of variance because it ensures that the data are genuinely a random sample. There is no excuse for copying designs from textbooks without randomizing treatment allocation within each block for each experiment.

Factorial Structures The use of factorial treatment structure with many factors in each experiment is essential if intercropping research programs are to be efficient. Compared with monocropping research programs, there are more factors to be considered in intercropping research, basically because of the two crops (discussed in Chap. 10). Thus whereas in monocropping one might consider varying:

1 Genotype
2 Sowing data
3 Several nutrient factors
4 Crop arrangement
5 Crop density
6 Irrigation regimes

For intercropping research the factors that have to be considered include:

1 Two genotype factors
2 Two sowing dates
3 Two crop densities
4 Two crop arrangements
5 The relative arrangement (intimacy) of the two crops
6 Several nutrient factors
7 Irrigation regimes

Instead of, say, 8 factors with 28 two-factor interactions we have to consider 13 factors with 78 two-factor interactions. Only by the use of factorial experiments with 3 or more factors can we hope to assess the effects of these factors. (Chapter 13 describes methods of setting priorities on these factors.)

Since the early days of monocropping research, the knowledge of the advantages of factorial structure, particularly in the early stages of research programs, has become well established. The major contribution of statistics to efficient use of experimental resources has been the demonstration that it is much more efficient to ask several questions in a single experiment through the use of factorial structure than by the previous philosophy of controlling all factors except one in each experiment. It would be appalling if experimenters were to ignore this knowledge and continue as before.

There are two further aspects of design connected with factorial treatment structure. Many factorial experiments are arranged in a split plot design. In some cases this is essential because of the nature of the treatments, and in such cases of practical necessity the split plot design is useful. However, in general split plot designs are inefficient and should be avoided. The inefficiency results from the splitting of information into two levels at one of which (the main plots) the precision of comparisons is usually very poor with few degrees of freedom. It is sometimes suggested that split plot designs are appropriate when interactions are of more interest than one set of main effects. This is misleading. The loss of information on the main effect is much greater than the gain on the interaction

SS and the gain is largely illusory because results are presented as tables of means for which the main effect is required in addition to the interaction.

In contrast to the popular but inefficient split plot design, confounded designs provide the solution to the conflicting requirements of many factors and small blocks. The construction of designs using small blocks from which all the main effects and two-factor interactions of interest can be estimated is extremely simple. Mead (1984) gives a detailed explanation of the scope for and ease of construction and analysis of confounded designs.

Resources A useful approach to assessing whether an experiment is efficient is to consider the allocation of resources. Most agronomically sensible experiments will usually contain between 30 and 80 experimental plots. The concept of the analysis of variance allows us to consider an equation for the resources for the experiment expressed in terms of the degrees of freedom (df) for the different components:

$$\text{Total df} = \text{blocks df} + \text{treatments df} + \text{error df}$$

The total df is the total number of plots minus 1. The blocks df will be one less than the number of blocks which will typically be between one-quarter and one-twelfth of the number of plots. The error df must be sufficient to provide a reliable estimate of the plot variance within blocks; a reasonable minimum is about 12 df for error. However, for efficiency it is wasteful to have many more df than this minimum. Twenty degrees of freedom should be regarded as the maximum to be permitted in a resource-efficient experiment. The remaining component of the df equation is the treatments df which is the number of treatment effect comparisons that it is intended to estimate from the experiment. It follows that the number of treatment df should be

$$\text{Total df} - \text{blocks df} - \text{error df}$$

and any design that does not include sufficient treatment df is failing to use efficiently the resources available for the experiment.

BIVARIATE ANALYSIS

What is a Bivariate Analysis?

A bivariate analysis is a joint analysis of the pairs of yields for two crops intercropped on a set of experimental plots. The philosophy is that because two yields are measured for each plot, and the yields will be interrelated, they should be analyzed together. The interrelationship is important since it implies that conclusions drawn independently from two separate analyses of the two sets of yields may be misleading. There are two major causes of interdependence of yield of two crops grown on the same plot. If the competition between the two crops is intense, then it might be expected that on those plots where crop *A* performs unusually well, crop *B* will perform unusually badly and vice versa. This would lead to a negative background correlation between the two crop yields, quite apart from any pattern of joint variation caused by the applied treatments. Failure to take this negative correlation into account could lead to high standard errors of means for each crop analyzed separately, which could mask real differences between treatments.

Alternatively it may be that on apparently identical plots, the two crops respond similarly to small differences between plots producing a positive background correlation. Again looking at separate analyses for the two crops distorts the assessment of the pattern of variation.

To see how consideration of this underlying pattern of joint random variation is essential to an interpretation of differences in treatment mean yields some hypothetical data are shown in Fig. 14.1. Individual plot yields are shown for

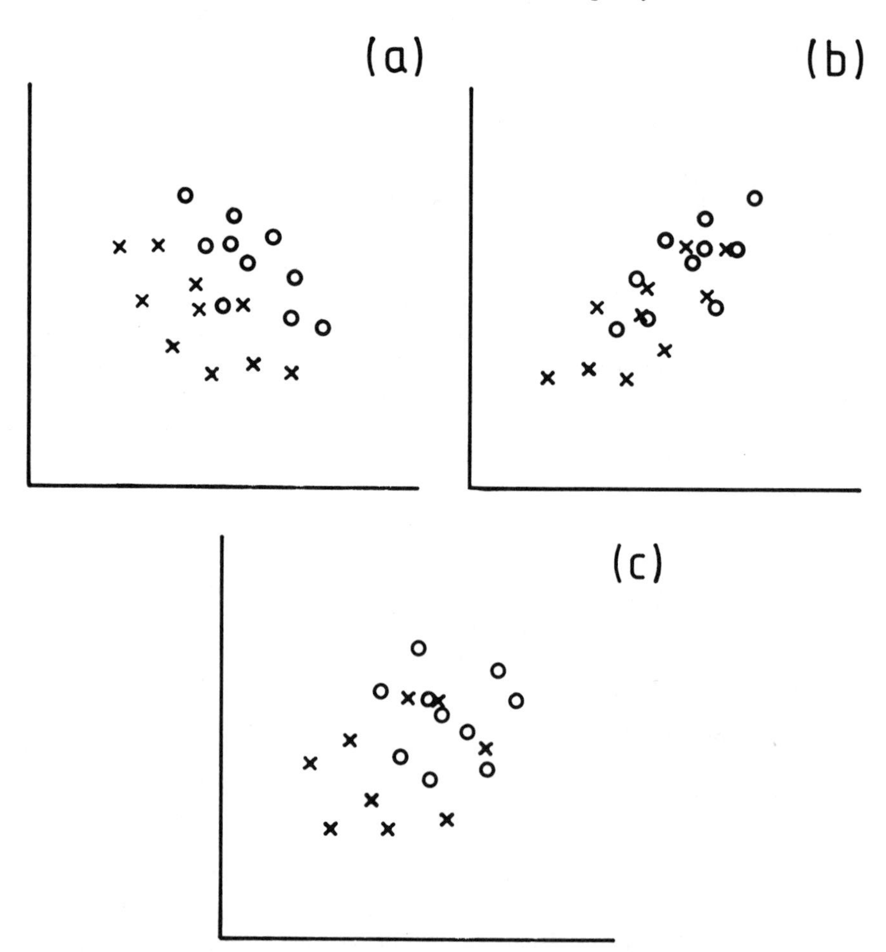

Figure 14.1 Different correlation patterns for yields with the same values of the individual crop yields: (a) negative correlation, (b) positive correlation, (c) no correlation. The two axes are for the yields of the two crops. Two intercrop systems give yields represented by × and o.

two intercrop systems (X and O), the mean crop yields for the two systems being identical for three situations. In Fig. 14.1a the pattern of background variation corresponds to a strongly competitive situation (negative correlation), whereas for Fig. 14.1b there is a positive correlation of yields over the replicate plots for each treatment. In Fig. 14.1c there is no correlation between the two crop yields. In all three cases the comparisons in terms of each crop yield separately would show no strong evidence of a difference between the two systems. However the joint consideration of the pair of yields against the background variation shows that the difference between the systems is clearly established in Fig. 14.1a, that Fig. 14.1b suggests strongly that the apparent effect is attributable to random variation, and that in Fig. 14.1c the separation of the two systems is rather more clear than could be established by an analysis for either crop considered alone.

The Form of Bivariate Analysis The calculations for a bivariate analysis are formally identical with those required for covariance analysis. The difference is that, whereas in covariance analysis there is a major variable and a secondary variable whose purpose is to improve the precision of comparisons of mean values of the major variable, in a bivariate analysis the two variables are treated symmetrically. Bivariate analysis of variance consists of an analysis of variance for X_1, analysis of variance for X_2, and a third analysis (of covariance) for the products of X_1 and X_2. Computationally this third analysis of sums of products is most easily achieved by performing three analyses of variance for X_1, X_2, and $Z = X_1 + X_2$. The covariance terms are then calculated by subtracting corresponding SS for X_1 and for X_2 from that for Z and dividing by 2. The bivariate analysis including the intermediate analysis of variance for Z are given in Table 14.3 for the maize/cowpea experiment discussed earlier.

The bivariate analysis of variance, like the analysis of variance, provides a structure for interpretation. In addition to the sums of squares and products

Table 14.3 Bivariate Analysis of Variance for Maize/Cowpea Yield Data (0.001 kg/ha) in Intercrop Trial

Source	df	Maize SS (X_1)	Cowpea SS (X_2)	SS for ($X_1 + X_2$)	Sum of products	F	Correlation
Blocks	2	0.29	0.0730	0.247	−0.058	1.75	−0.40
M variety	2	17.52	0.4094	12.665	−2.632	11.90	−0.98
C variety	1	0.03	0.0060	0.062	0.013	0.44	1.00
Nitrogen	3	28.50	0.1131	25.081	−1.766	10.59	−0.98
M × C	2	1.11	0.0099	0.922	−0.099	0.82	−0.95
M × N	6	1.25	0.0676	0.920	−0.199	0.64	0.93
C × N	3	0.24	0.1724	0.152	−0.130	2.40	−0.64
M × C × N	6	1.28	0.1354	1.349	−0.033	1.40	−0.08
Error	46	15.90	0.5993	13.671	−1.414		−0.46
(MS)		(0.346)	(0.0130)		(−0.031)		
Total	71	66.13	1.5861	55.080	−6.318		

Note: See Table 14.1.

for each component of the design, the table includes an error mean square line which provides a basis for assessing the importance of the various component sums of squares and products. The general interpretation of this analysis is quite clear and is essentially similar to the pattern of analysis of cowpea yield. There are large differences attributable to the different maize varieties and to the variation of nitrogen level; there is also a suggestion that there may be an interaction between cowpea variety and nitrogen level.

Diagrammatic Presentation We have argued earlier that interpreting the patterns of variation in maize and cowpea yields without allowing for the background pattern of random variation can be misleading. The primary advantage of the bivariate analysis is that it leads to a simple form of graphic presentation of the mean yields for the pair of crops making an appropriate allowance for the background correlation pattern. The graphic presentation uses skew axes for the two yields instead of the usual perpendicular axes. If the yields are plotted on skew axes with the angle between the axes determined by the error correlation, and if, in addition, the scales of the two axes are appropriately chosen, then the resulting plot, such as Fig. 14.2, has the standard error for comparing two mean yield pairs equal in all directions. The results in Fig. 14.2 are for the three maize varieties from the example, and the size of the standard error of a difference between two mean pairs is shown by the radius of the circle.

Construction of the skew axes diagram is based on the original papers of Pearce and Gilliver (1978, 1979) and detailed instructions for construction are given by Dear and Mead (1983, 1984). The form of the diagram given in Fig. 14.2 treats the two crops symmetrically, in contrast to the original suggestion of Pearce and Gilliver, in which one yield axis is vertical and the other is diagonally above or below the horizontal axis, depending on the sign of the error

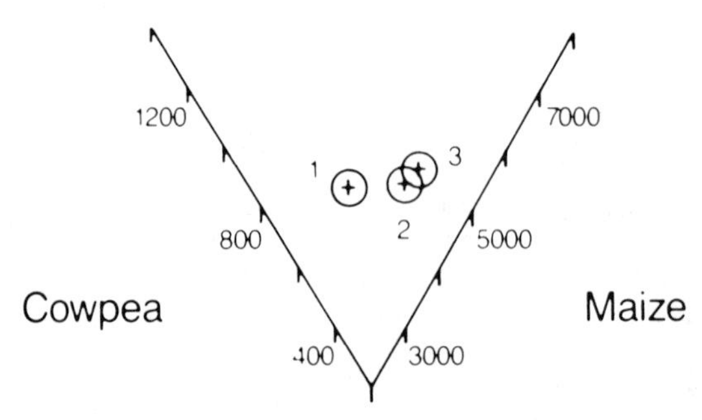

Figure 14.2 Bivariate plot of pairs of mean yields for three maize varieties (1, 2, 3). Maize and cowpea yields are in kilograms per hectare. (Data from Table 14.1.)

correlation. A summary of the method for construction of the symmetric diagram is as follows:

If the error mean squares for the two crops are $V_1(= 0.346$ in the example) and $V_2(= 0.0130)$, and the covariance is $V_{12}(= -0.0310)$, then the angle between the axes θ is defined by

$$\cos\theta = \frac{V_{12}}{(V_1V_2)^{1/2}}$$

If the range of values for the two yields X_1 and X_2 are (x_{10}, x_{11}) and (x_{20}, x_{21}) respectively, then we define two new variables y_1 and y_2,

$$y_1 = \frac{x_1}{\sqrt{V_1}} = K_1x_1$$

and

$$y_2 = \frac{x_2 - V_{12}x_1/V_1}{(V_2 - V_{12}^2/V_1)^{1/2}} = k_2\left(x_2 - \frac{V_{12}x_1}{V_1}\right)$$

and ranges

$$y_{10} = k_1x_{10}$$

$$y_{11} = k_1x_{11}$$

$$y_{20} = k_2\left(x_{20} - \frac{V_{12}x_{10}}{V_1}\right)$$

$$y_{21} = k_2\left(x_{21} - \frac{V_{12}x_{11}}{V_1}\right)$$

Plot the four pairs of y values $O(y_{10}, y_{20})$, $A(y_{11}, y_{20})$, $B(y_{10}, y_{21})$ and $C(y_{11}, y_{21})$ on standard rectangular axes, using the *same scale* for y_1 and for y_2. The x_1 axis is constructed by joining the points O and A, the x_2 axis by joining O and B. The x_1 scale is defined by $O(x_1 = x_{10})$ and $A(x_1 = x_{11})$; the x_2 scale is defined by $O(x_2 = x_{20})$ and $B(x_2 = x_{21})$. Further points on both axes may be marked using a ruler and the two defining points. The rotation of the x_1 and x_2 axes to achieve symmetry can be performed subjectively or by simple trigonometry. Individual points for pairs of mean yields may be plotted by first measuring x_1 along the x_1 axis, and x_2 parallel to the x_2 axis. More details of the diagram construction are given by Dear and Mead (1983, 1984).

The interpretation of the diagrams is extremely straightforward. The results in Fig. 14.2 show that the differences among the three maize varieties are important for both maize and cowpea yields, with the difference between varieties

2 and 3 clearly less than between either variety and variety 1. There is a clear consistency through the sequence of varieties 1 to 2 to 3, with the increase in maize yield being directly reflected in a decrease in cowpea yield. The three points fall nearly on a line illustrating the strong relation between the two crop yields over the three varieties. (Note also that the correlation for maize varieties, shown in Table 3, is -0.98). Remember that random correlation between the two yields has been allowed for by the skewness of the axes and the displayed pattern is additional to the background correlation pattern.

The results for nitrogen main effects and the interaction of cowpea variety with nitrogen are shown in Figs. 14.3 and 14.4. The four nitrogen levels produce four pairs of mean yields in an almost straight line. The dominant effect is on the yield of maize which increases consistently with increasing nitrogen. In addition there is a clear pattern of compensation between the two crop yields with cowpea yield decreasing as maize yield increases. The pattern of yields for the cowpea variety/nitrogen interaction emphasizes the two effects of yield increase for one crop and compensation between crops. For variety *A* the effect of increasing nitrogen is simply an increase of maize yield, the "line" of the nitrogen level means being almost exactly parallel to the maize yield axis. In contrast for variety B the dominant effect is the change in the balance of maize/cowpea yields with the maize yield increasing consistently with increasing nitrogen and the cowpea yield showing a corresponding decline.

Significance Testing There are two forms of test that are useful in bivariate analysis, and these correspond to the *t* and *F* tests used in the analysis of a single variate. We have already mentioned in the discussion of the skew axes plot that the standard error of a difference is the same in all directions in these diagrams. Because of the scaling of axes which is part of the construction of the diagram the standard error per observation is 1 (measured in the units of y_1 and y_2). The

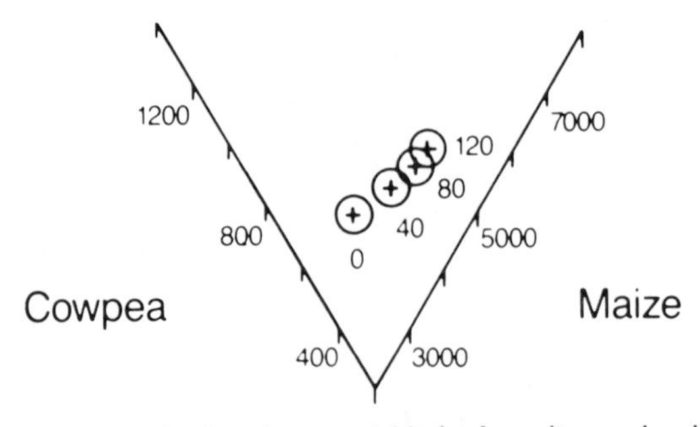

Figure 14.3 Bivariate plot of pairs of mean yields for four nitrogen levels (0, 40, 80, and 120 kg/ha). Maize and cowpea yields are in kilograms per hectare. (Data from Table 14.1.)

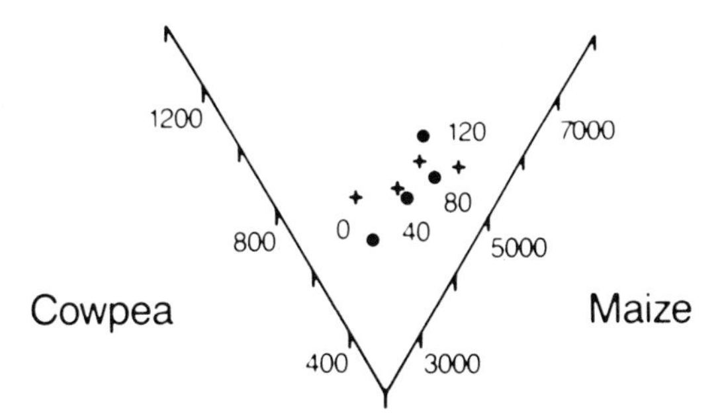

Figure 14.4 Bivariate plot of pairs of mean yields for two cowpea varieties: *A* (•) and *B* (+) for four nitrogen levels (0, 40, 80, and 120 kg/ha). Maize and cowpea yields are in kilograms per hectare. (Data from Table 14.1.)

standard error of a mean of n observations is therefore $1/\sqrt{n}$ and the standard error of a difference between two points is $\sqrt{2/n}$.

Confidence regions for individual treatment means can be constructed as circles with radius $\sqrt{2F/n}$, where F is the appropriate percentage point of the F distribution on 2 and e degrees of freedom (e is the error degrees of freedom).

The analogue of the F test in a univariate analysis of variance is also an F test. The basic concept on which the test is based is the determinant constructed from the two sums of squares and the sum of products. Suppose that the error SSP are E_1, E_2, and E_{12}, then the determinant is

$$E_1 \times E_2 - E_{12}^2$$

and it reflects both the sizes of E_1 and E_2 and the strength of the linear relationship between x_1 and x_2. To assess the treatment variation for a treatment SSP with values T_1, T_2, and T_{12} we calculate a statistic, L, which compares the determinant of treatment plus error with that for error

$$L = \frac{(T_1 + E_1)(T_2 + E_2) - (T_{12} + E_{12})^2}{E_1E_2 - E_{12}^2}$$

The test of significance then involves comparing

$$F = (\sqrt{L} - 1)\,\frac{e}{t}$$

with the F distribution on $2t$ and $2(e - 1)$ degrees of freedom, where t is the degrees of freedom for the treatment component. For the maize variety effect

$$T_1 = 17.52 \qquad T_2 = 0.4094 \qquad T_{12} = -2.634$$
$$E_1 = 15.90 \qquad E_2 = 0.5993 \qquad E_{12} = -1.414$$

$$L = \frac{(33.42)(1.0087) - (-4.046)^2}{(15.90)(0.5993) - (-1.414)^2} = 2.303$$
$$F = 11.90 \text{ on 4 and 90 df and is very clearly significant}$$

The results of the various bivariate F tests are seen in Table 14.3 for the maize/cowpea data. Perhaps even more than univariate F ratios, such F statistics should be interpreted only as general indications of overall changes in variation pattern. It is still true, as in univariate F tests that the overall F test may not reach significance while particular treatment comparisons are significant. In addition, because the bivariate F test summarizes variation over all treatment levels for both variables, the scope for particular comparisons to be significant is greater while the overall statistic is not.

Nonetheless the analysis produces a fairly clear picture, supplemented by the plots in Figs. 14.2 through 14.4. The effects of maize varieties and nitrogen levels are large (much greater than the 0.1 percent level) and the pattern of response of the two crops is shown in Figs. 14.2 and 14.3. There is a clear suggestion that the pattern of nitrogen is modified by an interaction with cowpea varieties (interaction significant at 5 percent on F on 6 and 90 df).

Assumptions and Some Problems of Bivariate Analysis For a bivariate analysis of variance, all three of the assumptions discussed earlier are still required, but in addition the correlation between the two yields should be consistent over treatments. Equivalently we require that the joint pattern of variation of the two yields, which could be observed only if a sample of pairs of yields were available for a single treatment, should be the same, even though the mean yields of both crops may be changed by different treatments. Examples of results that comply with the assumptions were shown earlier in Fig. 14.1.

Even more than in the case for the three assumptions in a univariate analysis of variance, it is difficult to test, in a formal manner, the assumption of constant correlation. This is basically because to demonstrate similarity of relationships requires a large number of observations, and it is rare to have the 10 or more replications that would be required for any useful test.

Methods of splitting the error sums of squares and products into separate components, which can be compared using an F test similar to that employed in the previous section, have been described by Dear and Mead (1984), and a wide range of data sets have been examined by them. The broad conclusion from their analysis of 40 experiments is that there is no evidence to suggest that the assumption of uniform correlation is unreasonable. Some other unpublished research has suggested that there may be situations where this reassuring finding is not supported, and it is clearly necessary to examine the assumption wherever this is possible. In contrast with their findings on the homogeneity of correlation,

Dear and Mead found that the assumptions of homogeneity of variance for the separate crop yields were often unsatisfied for the data they examined and suggested that logarithmic or other transformations would often be necessary for the bivariate (or univariate) analysis.

One problem with the use of the bivariate analysis, which is not yet solved, occurs when there are several levels of error variation as in a split plot design. The estimated correlations for the different levels of error will be different and may be substantially different. This would lead to different degrees of skewness of the axes for different levels of treatment comparison. While this difficulty provides a further reason for avoiding split plot designs, it does also provide a statistical problem for research. Clearly one might hope to have a common correlation pattern with possibly differing variances for the two error levels, and to develop appropriate statistical models and methods of estimation for this situation.

INDICES

Why Indices?

Indices are used in many forms of data summary to combine information. If we wish to compare systems of cropping involving two or more measurements then we must either accept that the systems are not comparable because they differ in several aspects, or convert yields onto a single scale. Consider, for example, mean yields for four treatments in the maize/cowpea experiment discussed earlier.

Treatment	Maize	Cowpea
1. $M_1C_2N_0$	2365	458
2. $M_2C_2N_3$	5323	319
3. M_3N_3	4722	—
4. C_2N_3	—	1490

Treatments 1 and 2 involve both crops; the other two treatments are for sole maize and sole cowpea. In any assessment of the overall performance of treatments 1 and 2 we have to balance the lower maize yield and higher cowpea yield of 1 against the higher maize yield and lower cowpea yield of 2. It is simply not possible to say in any absolute terms that 2 is better than 1 because such a judgment would imply some limit on the relative values of maize and cowpea.

The same problem occurs if we compare intercrop treatments with sole crops. The comparison of 2 with 3 is clear in that 2 must be preferred to 3 since it provides more yield for both crops. However even in this case the question of how much better 2 is than 3 does not have a simple answer. The comparison of 2 with 4 is, once again, a question of balancing gains against losses.

Those arguments lead naturally to the conclusion that only comparisons involving simultaneous consideration of both crop yields are valid in that they retain all the available information. However, practical considerations require that conclusions are sometimes based on a single value for each treatment. The gain from being able to make quantitative comparisons of treatments simply on a single scale will often outweigh the loss of information. Further, different indices lose different components of information and it must therefore be expected that if several indices are used, contradictory conclusions will result. Experimenters, and even statisticians, sometimes express surprise that two different indices appear to lead to different comparative conclusions; such a reaction merely means that they have not realized that they are expecting the impossible.

Simple Value Indices Simple indices involve the allocation of a value to each crop and the subsequent calculation of a total value from the sum of the values for the separate crops. If the values of two crops are assessed as K_1 and K_2, then the total value of an intercrop treatment producing mean yields Y_1 and Y_2 is

$$V = K_1Y_1 + K_2Y_2$$

The most frequently used value index is that of financial return. Other value indices include protein and dry matter. The main criticism made specifically of financial indices is that prices fluctuate and hence the ratio of K_1 to K_2 may vary considerably. A partial answer to this criticism is to employ several price ratios. Thus the results for the four treatments discussed earlier in this section might be presented for five price ratios as follows:

Price Ratio for Maize/Cowpea

Treatment	1:1	1:2	1:3	1:4	1:5
1	3111	3569	4027	4485	4943
2	5642	5961	6280	6599	6918
3	4722	4722	4722	4722	4722
4	1490	2890	4470	5960	7450

While some comparison patterns, such as (2 vs. 1) or (2 vs. 3), remain consistent for this range of price ratios others, such as (1 vs. 3) or (2 vs. 4), do not.

One other form of single measurement comparison which is exactly equivalent to the financial value index is the crop equivalent. In calculating a crop equivalent, yield of one crop is "converted" into yield equivalent of the other crop by using the ratio of prices of the two crops. The exact equivalence of crop equivalent yield to financial index is immediately obvious algebraically but may be perceived clearly also by considering the four treatments for a 1 : 3 price ratio. This ratio implies that a unit yield of cowpea is worth 3 units of maize. We can therefore calculate yields as maize equivalents or cowpea equivalents as follows:

Treatment	Maize equivalent	Cowpea equivalent
1	2653 + 3(458) = 4027	458 + 2653/3 = 1342
2	5323 + 3(319) = 6280	319 + 5323/3 = 2093
3	4722 = 4722	4722/3 = 1574
4	3(1490) = 4470	1490 = 1490

The relative comparisons are identical for the two equivalents.

Biological Indices of Advantage or Dominance The most important index of biological advantage is the relative yield total (RYT) introduced by de Wit and van den Bergh (1965) or land equivalent ratio (LER) reviewed by Willey (1979). The index is based on relating the yield of each crop in an intercrop treatment mixture to the yield of that crop grown as a sole crop. If the two crop yields in the intercrop mixture are M_A, M_B, and the yields of the crops grown as sole crops are S_A, S_B, then the combined index is

$$L = \frac{M_A}{S_A} + \frac{M_B}{S_B} = L_A + L_B$$

The interpretation embodied in LER is that L represents the land required for sole crops to produce the yields achieved in the intercropping mixture. A value of L greater than 1 indicates an overall biological advantage of intercropping. The two components of the total index, L_A and L_B represent the efficiency of yield production of each crop when grown in a mixture, relative to sole crop performance. For the maize/cowpea yields treatment 2 may be assessed relative to treatments 3 and 4 to give an LER

$$L = \frac{5323}{4722} + \frac{319}{1490} = 1.13 + 0.21 = 1.34$$

Other indices have been proposed as measures of biological performance. There are two different objectives for which such indices have been proposed. The first is the assessment of the benefit, or overall advantage, of intercropping, or mixing. The second is the assessment of the relative performance of the two crops, the concept of dominance or competitiveness. It is important not to confuse these two objectives, which should be quite separate conceptually.

The RYT or LER is the main index of advantage currently used. The other index which has been used is the relative crowding coefficient (de Wit, 1960), which can be defined in terms of the LER components as

$$\frac{L_A}{1 - L_A} \times \frac{L_B}{1 - L_B}$$

The two main indices of dominance are the aggressivity coefficient, introduced by McGilchrist and Trenbath (1971) defined essentially as

$$L_A - L_B$$

and the competition ratio proposed by Willey and Rao (1980) and defined essentially as

$$\frac{L_A}{L_B}$$

The full definition of each index as originally given involves proportions of the two crops in the mixture. However, for applications in intercropping, this masks the underlying concepts involved in the ideas of advantage or dominance. Each of these four indices is based clearly on the LER components L_A and L_B. [Indeed since there are only four simple arithmetical operations ($+$, $-$, $\times$, $\div$) it could be argued that the set of possible indices is now complete!] Crucially, however, the components L_A and L_B are ratios, and the value of a ratio is determined as much by the divisor as by the number divided. Hence the interpretation of L_A and L_B, and therefore of any index based on L_A and L_B, depends on the choice of divisor.

This question of interpretation is extremely important, and becomes even more important when comparison of LERs is considered in the next section. For the LER to be interpreted as the efficiency of land use the sole crop yields, S_A and S_B must represent some well-defined, achievable, optimal yields. It is therefore necessary that the choice of sole crop yield used in the calculation of the LER be clearly defined and justified as appropriate to the objective that the LER is intended to achieve. To illustrate this argument consider the yields for several intercrop and sole crop treatments in the maize/cowpea experiment. The mean yields for two maize varieties, two cowpea varieties and two nitrogen levels are shown in Table 14.4.

If we consider a particular intercrop combination, for example M_1 C_1 N_0, we could assess the biological advantage of intercropping as

$$L = \frac{2653}{2568} + \frac{458}{1036} = 1.03 + 0.44 = 1.47$$

This is simply interpretable as the benefit in the situation where the only varieties available are M_1 and C_1 and no nitrogen is available. It also implies that the sole crop yields of 2568 and 1036 could not be improved by modifying the spatial arrangement or the management of the sole crop since we are assessing the intercrop performance in relation to the land required to produce the same yields by sole cropping. No one would deliberately use an inefficient method of sole cropping to try to match the intercropping performance. Suppose the combination

Table 14.4 A Subset of Yields from the Maize/Cowpea Experiment

	Intercrop yields		Sole-crop yields			
Treatment	**Maize**	**Cowpea**	**M_1**	**M_3**	**C_1**	**C_2**
$M_1C_1N_0$	2653	458				
$M_3C_1N_0$	3315	508	2568	3555	1036	787
$M_1C_2N_0$	2453	731				
$M_3C_2N_0$	3604	585				
$M_1C_1N_3$	4093	706				
$M_3C_1N_3$	5663	366	3651	4722	1795	1490
$M_1C_2N_3$	3922	458				
$M_3C_2N_3$	5323	320				

Note: Data from intercrop trial (Table 14.1).

$M_1C_2N_0$ is now considered. Since the sole crop yield for C_1 is better than that for C_2, the advantage of intercropping might be argued to be overestimated if we compare $M_1C_2N_0$ with M_1 and C_2 for which the LER would be

$$L = \frac{2453}{2568} + \frac{731}{787} = 0.96 + 0.93 = 1.89$$

If we measure $M_1\ C_2\ N_0$ against M_1 and C_1 we obtain an LER value

$$L = \frac{2453}{2568} + \frac{731}{1036} = 0.96 + 0.71 = 1.67$$

We could go further and argue that if M_3 is available as an alternative to M_1 then we should compare $M_1\ C_2\ N_0$ with the best available varieties, M_3 and C_1, which could be used as a sole cropping alternative. We would then have

$$L = \frac{2453}{3555} + \frac{731}{1036} = 0.69 + 0.71 = 1.40$$

This last L value represents the most stringent assessment of advantage of the intercropping combination $M_1C_2N_0$ and alternative forms of L could all be criticized as presenting an illusory benefit of intercropping as compared with sole cropping.

What about using sole crop yields for N_3 rather than for N_0? Here the argument becomes more complicated. It may well be that in the farming situation for which the conclusions drawn are to be relevant, there is no real possibility of using extra nitrogen as required in N_3. The advantage of 1.40 would then be assessed in the most stringent manner possible for the practical situation considered.

The purpose of this example is not to define rules for calculating LER measures of advantage but to demonstrate that the choice of divisors for the LER is a matter requiring careful thought. The divisor in LER calculations cannot be assumed to be obvious, and discussions about LER values when the choice of divisor is not clearly defined should be treated with suspicion.

One distinction that might usefully be made is between the LER or RYT as a measure of biological sufficiency of a particular combination without any implications of agronomic benefit and the use of the LER to assess the greater efficiency of the use of land resources. The former concept developed naturally from competition studies and is a strictly nonagronomic idea. The latter is an agronomically justifiable concept, but, as explained in the example, it is also an inherently more complex measure. Perhaps we should use RYT for the non-practical biological concept and LER for the agronomic concept!

Comparison and Analysis of LER Values The assessment of advantage of a single intercrop combination requires careful thought. When it is desired to compare different intercrop treatments using LER values, the need to calculate the LER to produce meaningful comparisons is accentuated. There are now two problems.

The first is the choice of divisor, and I believe that comparisons of LER values are valid in their practical interpretation only if the divisors are constant for all the values to be compared. If different divisors are used for different intercrop treatments then the quantities being compared may be considered as

$$L_1 = \frac{M_{A1}}{S_{A1}} + \frac{M_{B1}}{S_{B1}}$$

and

$$L_2 = \frac{M_{A2}}{S_{A2}} + \frac{M_{B2}}{S_{B2}}$$

The interpretation of any difference between L_1 and L_2 cannot be assumed to be the advantage of intercropping treatment 1 compared with intercropping treatment 2, since the difference could equally well be caused by differences between sole cropping treatments S_{B1} and S_{B2} or between S_{A1} and S_{A2}.

Although LER values using different divisors are often compared, the concept that is being used as the basis for comparison is the vague one of efficiency which is not interpretable in any practically measurable form of yield difference between different intercropping treatments. We should recognize that such comparisons are of a theoretical nature only and are not practically useful.

The form of the LER which is the sum of two ratios of yield measurements has prompted concern about the possibility of using analysis of variance methods for LER values. More generally the question of the precision and predictability of LER values has been felt by some to be a problem.

The comparison of LER values within an analysis of variance is, I believe, usually valid provided that a single set of divisors is used over the entire set of intercropping plot values. Some statistical investigations of the distributional properties of LERs were made by Oyejola and Mead (1981) and Oyejola (1983). They considered various methods of choice of divisors including the use of different divisors for observations in different blocks. Allowing divisors to vary between blocks provided no advantage in precision or in the normal distributional assumptions: variation of divisors between treatments was clearly disadvantageous. The recommendation arising from these studies is therefore that analysis of LER values is generally appropriate, provided that constant divisors are used, and with the usual caveat that the assumptions for the analysis of variance for any data should always be checked by examination of the data before, during and after the analysis.

The question of precision of LERs and, by implication, their predictability, is an unnecessarily confusing one. If LERs are being compared within experiments the standard errors of comparison of mean LERs are appropriate for comparing the effects of different treatments. Experiments are inherently about comparisons of the treatments included rather than about predictions of performance of a single treatment. The precision of a single LER value must take into account the variability of the divisors used in calculating the LER value. However a more appropriate question concerns the variation to be expected over changing environments and this must be assessed by observation over changing environments. No single experiment can provide direct information about the variability of results over conditions outside the scope of the experiment. This, of course, does not imply that single experiments have no value since we may reasonably expect that the precision of estimation of treatment differences will be informative for the prediction of the differential effects of treatments.

Extensions of LER In the last section it was mentioned that there were two problems in making comparisons of LER values for different intercropping treatments. The second problem is that the concept of the LER as a measure of advantage of intercropping assumes that the relative yields of the two crops are those that are required. The calculation of the land required to achieve, with sole crops, the crop yields obtained from intercropping makes this assumed ideal of the actual intercropping yields clear. However with two (or more) intercropping treatments the relative yield performance $L_A : L_B$ will inevitably vary and hence the comparison of LER values for two different treatments can be argued to require that two different assumptions about the ideal proportion $L_A : L_B$ shall be simultaneously true.

This difficulty led to the proposed "effective LER" of Mead and Willey (1980) which allows modification of the LER to provide the assessment of advantage of each intercropping treatment at any required ratio $\lambda = L_A/(L_A + L_B)$. The principle is that to modify the achieved proportions of yield from the two crops we consider a "dilution" of intercropping by sole cropping. The achieved proportion of crop A could be increased by using the intercropping treatment on part of the land and sole crop A on the remainder, the land proportions being chosen so as to achieve the required yield proportions. Details of the calculations are given in Mead and Willey (1980). It is important if the use of a modification of the LER is proposed that the reason for using the effective LER is clearly understood. It is not primarily a form of practical adjustment but arises from the philosophical basis of the LER.

It may be that in using the LER as a basis for comparison of different treatments the emphasis is not on the biological advantage of intercropping but on the combination of yields onto a single scale, in terms of yield potential. In this view the LER becomes another form of value index, the two values being the reciprocals of the sole crop yields. When a range of price ratio indices is used, it is almost invariably found that the ratio of the LER values is well in the center of the price ratio range. The principle of the argument for using an effective LER is no longer essential but there may still be advantages in making practical comparisons or treatments in terms of performance at a particular value of λ. There are, however, other possible ways of modifying the LER as a value index, not arising from the philosophy of the LER, and the most important of these is the calculation of combined yield performance to achieve a required level of crop yield A. Arguments for, and details of, this alternative modified LER are given by Reddy and Chetty (1984) and Oyejola (1983).

Implications for Design The particular implications to be considered here concern the use of sole crop plots. If the arguments about the choice of divisors are followed then it will not be necessary to include many sole crop treatments within the designed experiment. The investigation of the agronomy of monocropping has been extensive and in most intercropping experiments there should rarely be any need for an experimental investigation of the optimal form of monocropping. Therefore, there should often be no need for more than a single, sole cropping treatment for each crop.

The reduction in the number of sole crop plots in intercropping experiments would be of great benefit because it would enable a greater part of the resources for an experiment on intercropping to be used for investigating intercropping. Many intercropping experiments which I have seen have used between one-third and one-half of the plots for sole crops. To some extent this reflects a propensity for continuing to ask whether intercropping has an advantage, when this is widely established, instead of asking the practically more important question of how to grow a crop mixture.

It is possible to take the reduction of sole crop treatments further. The analysis in this chapter and the previous chapter do not require sole crop treatments within the experiment to be treated like other treatments. For the bivariate analysis no sole crop information is essential though sole crop information does provide a standard against which to compare the pairs of yields. For the analysis and interpretation of LERs, estimates of mean yields for the two sole crops are needed as divisors. However there is no need for the sole crops to be randomized and grown on plots within the main experiment. Sufficient information for the calculation and interpretation of LERs can be obtained from sole crop areas alongside the experimental area. This will tend to improve the precision of the experiment by reducing block sizes and also simplifies the pattern of plot size.

STABILITY AND RISK

What is Stability?

One of the first questions that I encountered as a statistician advising on research in intercropping was, "How shall we measure the stability of intercropping compared with monocropping?" It seems to be widely believed that intercropping provides a more stable system of agriculture than does monocropping. However the concept of stability is variously and often inadequately defined, and many of the attempts to express stability in quantitative terms have been statistically unconvincing. In this section I review methods of assessment of stability and relative risk for different systems. I then outline a new approach based on bivariate distributions and described in a recent paper by Mead et al. (1984) in which more details are given. Data to illustrate the new method are taken from a series of trials performed by the All-India Coordinated Sorghum Improvement Project.

Stability Measured in Terms of Variance The use of measures of simple variation to characterize stability relies on the intuitive idea of "stable" as unchanging. Hence systems showing less variation of yield are considered more stable. Instead of using absolute variation, measured by the sample variance or standard deviation, most authors prefer the coefficient of variation, so that comparisons are in terms of relative variation. Statistically the description of relative variability, which does seem much more appropriate than absolute variability for most biological variation, is much better achieved through the use of a logarithmic scale. The standard deviation of the sample of log-transformed yields is the best simple statistic of stability based on variation.

However all stability measures based on variation are unsatisfactory because they ignore information about the structure of the data samples. It is obviously desirable, when intercropping and monocropping systems are being compared, that data from the same set of locations/years should be used for both systems. Otherwise the data are not really comparable. However since the various location/year combinations provide pairs of observations (intercrop, monocrop) any analysis should relate to this structure of data. Simple measures of variance do not recognize the structure of data and are potentially inefficient or misleading. Consider the data shown below for two systems at six sites.

	Site A	Site B	Site C	Site D	Site E	Site F
System 1	20	24	21	5	26	23
System 2	19	22	19	5	24	21

System 1 gives higher yields than Section 2 at all sites except Site D where the yields are equal. The advantage of system 1 over system 2 is consistent over the other five sites and, considering the results in pairs, it is clear that system 1 is not less stable than system 2. However, if the pairing of yields is ignored both variance and CV are larger for system 1 than for system 2.

Stability Measured in Terms of Environment Dependence The idea of assessing stability through the dependence of yield on environment stems from the discussion of stability for genotype/environment interactions in monocropping. The underlying concept is that a stable system will react less to changes in the environment than will an unstable system. If a suitable form of relationship can be developed between yield of a system and an environmental index, then systems for which the slope of the relationship is smaller may be described as more stable. The choice between using absolute or relative variation, discussed in the previous subsection is again relevant. The most suitable expression of stability through environment dependence is probably to use a linear regression of log yield on a suitable index of environment and to compare the slopes of regressions calculated for different systems.

The major problem with using regressions of yield on environmental index to provide an assessment of stability is the choice of environmental index. It is necessary to find an index that represents the variation in environment that is relevant to the variation in yield. If an attempt is made to construct an index, then inevitably there are many possible components to select from, and the selection and testing of the proposed index requires lengthy and complex data analysis.

An alternative approach has been to adapt the method developed by Finlay and Wilkinson (1963) in which the mean yield of all genotypes is defined to be the index for each environment. This method does have some general logic problems, and when applied to a comparison of two or three systems it becomes quite inappropriate. Essentially if the stability of intercrop yield X is to be compared with that for the corresponding sole crop yields Y and Z, then the Finlay-Wilkinson method involves comparing three regressions:

$$X \text{ on } \frac{X + Y + Z}{3}$$

$$Y \text{ on } \frac{X + Y + Z}{3}$$

$$Z \text{ on } \frac{X + Y + Z}{3}$$

These regressions are clearly highly interdependent and comparison of the slopes using the usual standard errors, and therefore assuming the regressions are independent is clearly statistically invalid.

Stability Measured in Terms of Risk The risk concept is a practical one. The subsistence farmer wants to reduce the chance of a poor yield. Hence the stability of different systems should be assessed in terms of the probability of poor yields. However, "poor" is not uniquely definable and so a range of possible poor yield levels must be considered so that the risk of falling below such a yield can be estimated for each system for each poor level. Any investigation of risk probabilities requires large amounts of data from a range of locations and years. There has been considerable discussion over the differences and similarities of variation over locations and over years. It is clear that the possible difference in pattern of variation between years and between locations must be recognized, but from a practical point of view sufficient information is likely to be available only if both forms of variation can be considered together. Thus we need to check that the two forms of variation are broadly similar before using all the data to provide estimates of risk probabilities.

As with the use of simple variance measures the structure of the data should be considered in the assessment of stability in terms of risks. Thus it is necessary to recognize that in comparing an intercropping system with a monocropping system we have to consider the data as bivariate. For each site/year combination we have yields for the two systems and we must consider an appropriate model for the joint variation of the two yields.

Model-Based Approach to Relative Risk Assessment

Assumptions of the Method The philosophy of the model-based approach to risk assessment assumes that:

1. Risk probabilities are the most suitable form of expression of stability
2. The pattern of the bivariate distribution of yields for the two systems can be described by a simple model
3. Yields for each system can be expressed on a single scale (in practice this almost inevitably means a financial scale and introduces all the problems of combining two yields into an index of overall value, as discussed earlier)
4. The variation between years and between sites is essentially similar so that the whole data set can be treated as one

Data and Initial Analysis Data for a sorghum (*Sorghum bicolor*)/pigeon pea (*Cajanus cajan*) intercrop and sorghum sole crop are available from 51 site/year combinations, deriving from 12 sites and 7 years; the experiments were performed within the research project of the All-India Coordinated Sorghum Improvement Project under the control of Dr. S. P. Singh (unpublished data). Yields for the two crops in the intercropping system are combined using a relative price ratio of 1.8 for pigeon pea/sorghum, which is the relevant average ratio for India over the period of the experiments. The bivariate plot of intercrop return and sorghum sole crop return is shown in Fig. 14.5. It is clear that there

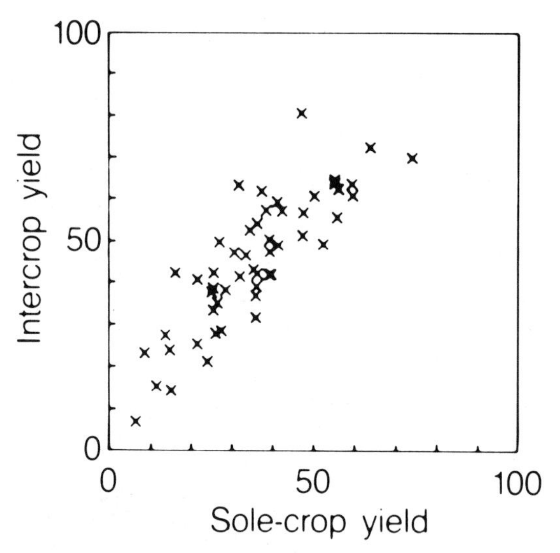

Figure 14.5 Bivariate plot of intercrop return against sole crop sorghum return for 51 experiments. (Data from Dr. S.P. Singh, All-India Coordinate Sorghum Improvement Project, unpublished.)

is a strong, and slightly curved, relationship between the two yield returns, and fairly clear also that intercropping return is generally higher than sole cropping return.

To examine whether the pattern of joint variation is consistent over temporal, spatial, and temporal/spatial variation a preliminary bivariate analysis of variance is calculated and this is shown in Table 14.5. The location and year effects are not orthogonal since only a subset of the possible combinations are available, and therefore the sums of squares are affected by the order of fitting; both orders are therefore used. In addition to direct examination of the analysis of variance, which reveals that site-to-site variation is more substantial than year-to-year variation, regressions of one system yield return on the other were calculated for each line of the analysis of variance and the regressions for different lines were compared. The comparisons showed negligible indications of inconsistency between the regressions, and it was concluded that the patterns of variation between sites, years, and site/year combinations were broadly similar. The data set can therefore be treated as a single sample.

Empirical Relative Risk Treating the data as a sample from a single bivariate population distribution, we could consider, for any level of return d, the proportions of the sample for which intercropping or monocropping returns fell below d. The ratio of these proportions calculated for intercropping and monocropping for any value of d is an estimate of the relative risk of the two systems. If we plot:

$$\text{prob(intercrop return} < d)$$

Table 14.5 Bivariate Analysis of Variance for Sorghum/Pigeon Pea Intercropping and Monocropping Experiments

Source of variation	df	Intercropping yield SS	Sum of cross products	Sorghum sole crop yield SS
Sites (ignoring years)	10	4,685	4,249	4,613
Years (eliminating sites)	6	792	469	736
Years (ignoring sites	6	1,133	686	1,219
Sites (eliminating years)	10	4,344	4,032	4,130
Years and sites	16	5,477	4,718	5,349
Residual	34	5,392	5,145	7,353
Total	50	10,869	9,863	12,702

Source: Data from Dr. S. P. Singh, All-India Coordinated Sorghum Project, unpublished.

against

$$\text{prob(sole crop sorghum return} < d)$$

over the complete set of possible d values, then we have a visual representation of relative risk, and this is shown in Fig. 14.6. Because the relative risk is

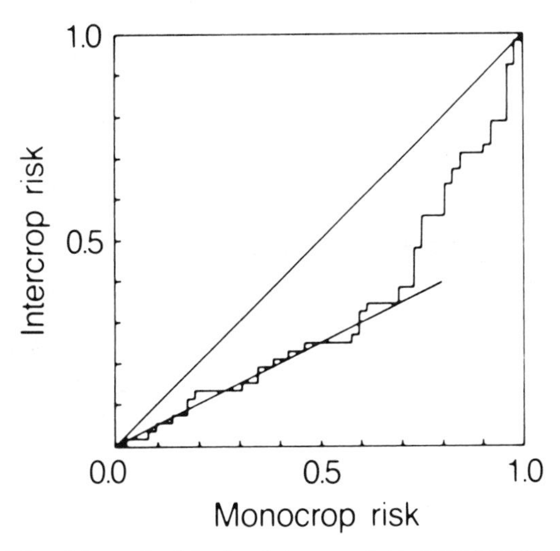

Figure 14.6 Relative risk graph: risk of an intercrop yield return less than d plotted against the risk of a sorghum sole crop yield return less than the same value, d.

calculated empirically from the data, the curve is irregular. Nevertheless it is clear from the figure that for risks up to 50 percent for the sorghum sole crop system there is an approximate linear relationship with a slope 0.5. Such a line is drawn in Fig. 14.6.

A Bivariate Distribution Model Although the empirical relative risk curve is informative, it relies on using a large sample. If a suitable bivariate distribution could be found which fitted the data, then a theoretical relative risk curve could be estimated from the fitted model and smaller samples could then be analyzed by the method.

Examination of Fig. 14.5 shows clearly that there is a strong positive relationship between the two returns and there is also a clear suggestion that the relationship is not linear. In such situations it is often sensible to consider instead of the two original variables, X and Y, the sum and difference

$$S = X + Y \qquad D = X - Y$$

Examination of the distribution of S suggests that a normal distribution would fit the data well and the normal distribution with mean 80.3 and variance 853 provides an excellent fit. The distribution of D depends on the value of S (the pattern of data in Fig. 14.5 is banana shaped) and the fitted model for D is

$$\text{Mean} = 11.00 - \frac{(\text{sum} - 80.9)^2}{365}$$

$$\log(\text{Variance}) = 4.3 - \frac{(\text{sum} - 79.8)^2}{2322}$$

A contour plot of this model is shown in Fig. 14.7 for comparison with the data plot in Fig. 14.5. Visually, the fit seems adequate and similar agreement is obtained for four other data sets in Mead et al. (1984), though the distributional pattern of S and D do vary considerably.

Relative Risk Estimation from the Fitted Model The immediate benefit of the fitted model is that the risk probability for any level of minimum required return can be calculated, producing a smooth risk curve. The actual calculations are numerically complicated but not difficult with modern computers and details are again given in Mead et al. (1984). The fitted risk curves for each system are shown in Fig. 14.8 and the relative risk curve for the intercropping and sole crop system is shown in Fig. 14.9.

It can be seen that for low risks, the relative risk curve is almost straight. In other words, the ratio of the risks for the two systems is almost constant. The ratio is about 0.5 confirming our earlier assessment from the empirical relative

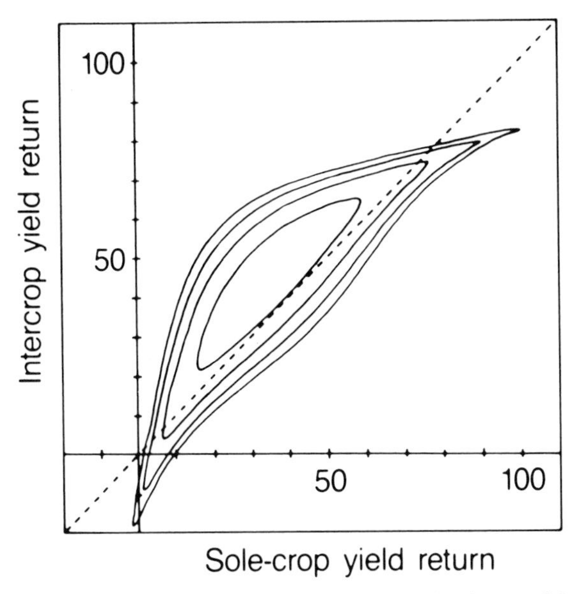

Figure 14.7 Contour plot of the fitted bivariate model.

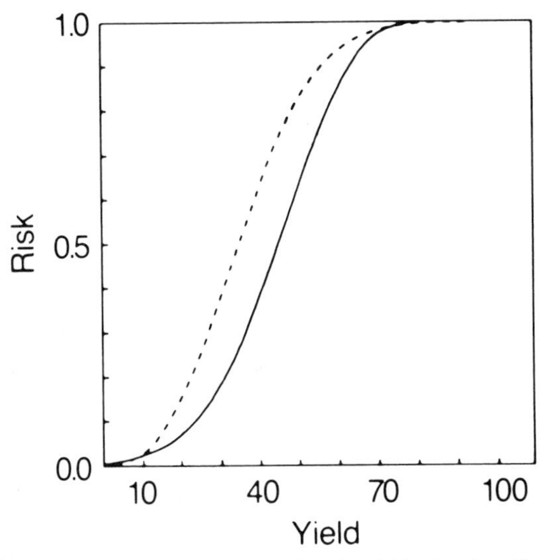

Figure 14.8 Fitted risk curves showing probability of a yield return less than the yield plotted against that yield for sorghum sole crop (broken line) and intercrop (solid line).

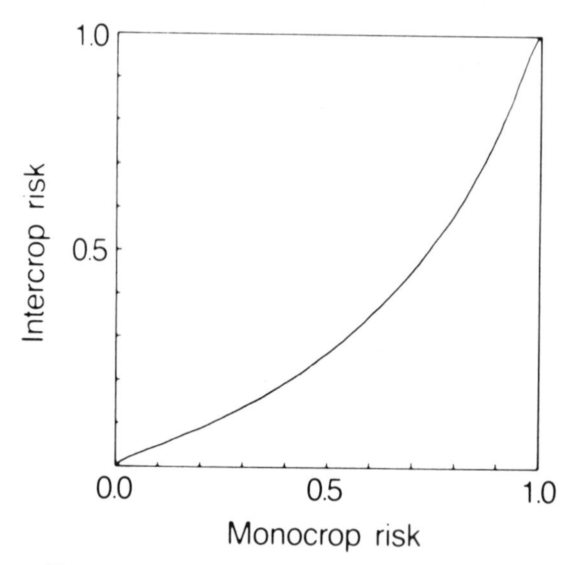

Figure 14.9 Relative risk graph for the fitted model.

risk diagram that the risk for any low level of minimum income is roughly twice as high for sole crop sorghum as for the intercrop of sorghum and pigeon pea.

CONCLUSIONS

The inclusion of a new method in a chapter whose purpose is mainly to review the state of the statistical art needs some justification. There are two related justifications. One is that the methods currently in use are not satisfactory and a model-based approach which overcomes the statistical objections is much needed and should be made available and used. The second is the belief that stability is an important aspect of intercropping studies and that it is important to be able to measure stability in a comparative manner.

The general principle of fitting models to structured data is important. The general approach of fitting the model in two stages to the sum and the difference is likely to be appropriate for most data samples. The appropriate form of distribution for the sum, S, may not always be a normal distribution. The distribution of S may be skew in either direction and an appropriate skewed distribution may be appropriate or possibly a transformation of S to produce approximate normality.

The conditional distribution of D, the difference, given a particular value of S, will probably behave in a fashion that can be modeled by simple regression models for the mean value of D and for the logarithm of the variance of D. The choice of models should not be too difficult and may be much aided by plotting D against S (essentially rotating Fig. 14.5).

I hope that this model-based approach to stability will be tested on various sets of data, and I would be interested to hear of the results or to be involved in the use of the method for new data.

ACKNOWLEDGMENTS

I am greatly indepted to Dr. H. C. Ezumah of the International Institute of Tropical Agriculture, Ibadan, Nigeria, and Dr. S. P. Singh of the All-India Coordinated Research Programme, Hyderabad, India, for the data for the first three, and the fourth sections, respectively.

REFERENCES

Bryan-Jones, J., and Finney, D. J. 1983. On an error in "Instructions to Authors," *Hort. Sci.* 18:279–282.

Dear, K. B. G., and Mead R. 1983. The use of bivariate analysis techniques for the presentation, analysis and interpretation of data, in: *Statistics in Intercropping,* Tech. Rep. 1, Dep. Applied Statistics, University of Reading, Reading U.K.

———. 1984. Testing assumptions, and other topics in bivariate analysis, in: *Statistics in Intercropping,* Tech. Rep. 2, Dep. Applied Statistics, University of Reading, Reading, U.K.

de Wit, C. T. 1960. On competition, *Versl. Landbouwk. Onderzook* 66:(8):1–82.

de Wit, C. T. and Van den Bergh, J. P. 1965. Competition among herbage plants, *Neth. J. Agric. Sci.* 13:212–221.

Finlay, K. W., and Wilkinson, G. N. 1963. The analysis of adaptation in plant breeding programmes, *Aust. J. Agric. Res.* 14:742–754.

McGilchrist, C. A., and Trenbath, B. R. 1971. A revised analysis of plant competition experiments, *Biometrics* 27:659–671.

Mead, R. 1984. Confounded experiments are simple, efficient and misunderstood, *Exp. Agric.* 20:185–201.

Mead, R., and Curnow, R. N. 1983. *Statistical Methods in Agriculture and Experimental Biology,* Chapman and Hall, London, U.K.

Mead, R., and Riley, J. 1981. A review of statistical ideas relevant to intercropping research (with discussion), *J. Royal Stat. Soc.* 144:462–509.

Mead, R., Riley, J., Dear, K. B. G., and Singh, S. P. 1984. Stability comparison of intercropping and monocropping systems, in: *Proc. XIIIth Int. Biometric Conf.,* Tokyo, Japan.

Mead, R., and Stern, R. D. 1980. Designing experiments for intercropping research, *Exp. Agric.* 16:329–341.

———. 1981. Statistical considerations in experiments to investigate intercropping, in: *Proc. Int. Workshop on Intercropping, 1979,* (R. W. Willey, ed.) International Crops Research Institute for the Semi-Arid Tropics (ICRISAT), pp. 263–276.

Mead R., and Willey R. W. 1980. The concept of a land equivalent ratio and advantages in yields from intercropping, *Exp. Agric.* 16:217–228.

Morse, P. M., and Thompson, B. K. 1981. Presentation of experimental results, *Can. J. Plant Sci.* 61:799–802.

Oyejola, B. A. 1983. Some statistical considerations in the use of the land equivalent ratio to assess yield advantages in intercropping, Ph.D. thesis, University of Reading, Reading, U.K.

Oyejola, B. A., and Mead, R. 1981. Statistical assessment of different ways of calculating land equivalent ratios (LER), *Exp. Agric.* 18:125–138.

Pearce, S. C., and Gilliver, B. 1978. The statistical analysis of data from intercropping experiments, *J. Agric. Sci.* 91:625–632.

———. 1979. Graphical assessment of intercropping methods, *J. Agric. Sci.* 93:51–58.

Reddy, M. N., and Chetty, C. K. R. 1984. Stable land equivalent ratio for assessing yield advantage from intercropping, *Exp. Agric.* 20:171–177.

Willey, R. W. 1979. Intercropping—its importance and research needs. Parts I and II, *Field Crop Abstr.* 32:1–10, 73–85.

Willey, R. W. and Rao, M. R. 1980. A competitive ratio for quantifying competition between intercrops, *Exp. Agric.* 16:117–125.

Chapter 15

Future Perspectives of Multiple Cropping

Charles A. Francis

There is an increasing recognition by scientists and agricultural administrators of the current and potential future importance of multiple cropping systems. This importance was discussed in the introductory chapter. In a conference inauguration, the Minister of Agriculture of Tanzania, Hon. John S. Malacela (1982) observed that,

> The fact that mixed cropping is the way of life for the subsistence farmer in the tropics underscores the reasons for studying it in some detail. Theoretically, there are a variety of reasons why farmers have adopted this practice. Insurance against the vagaries of weather, diseases, and pests is a major reason. By planting more than one crop in the same field, the farmer is also maximizing moisture, maintaining soil fertility, and minimizing soil erosion, which are some of the serious drawbacks of monocultural farming. . . . Intercropping . . . offers unlimited opportunity to increase the productivity of arable land, and research efforts should be mounted to tackle problems that limit the efficiency of the practice.

Even though multiple cropping systems have been considered important by farmers in many parts of the world for centuries, the interest of the research community has accelerated only in the past two decades. In the literature search cited in Table 1.1, there were only 187 published articles up to 1960, but 359 papers were published from 1961 to 1970. In the decade of the 1970s there were 1440 published reports in the survey, and this trend continues. The majority of these research reports describe agronomic systems, interactions among crop species, crop interactions with the environment, and the effects of cultural practices on multiple crop performance. It is this wealth of recent information and its interpretation that causes many in the agricultural research and development community to seriously examine the future of multiple cropping systems.

The future global importance of multiple cropping is difficult to predict. Current trends indicate an expansion in double cropping and other sequential systems in many areas, both in the developing world and in developed countries. The potentials of short-cycle rice (*Oryza sativa*) and other cereal varieties have made possible two or three crops per year in areas with ample resources, such as the intensive, high-input rice and catch crop patterns practiced in the Central Visayas of the Philippines. Yet the rapid expansion of mechanization and increase in farm size has resulted in a reduction of intercropping or other intensive multispecies practices in the more climatically and edaphically favored parts of developing countries. It is unlikely that intensive cereal/grain legume intercrops will expand in the near future except where there is a clear complementarity between component species.

There is ample evidence that some forms of intensive sequential or relay cropping systems will continue to expand:

- Winter wheat (*Triticum sativum*)/soybean (*Glycine max*) double cropping in southeast United States
- Overseeding legume cover into growing maize (*Zea mays*), wheat, and soybeans
- Strip cropping of maize/soybeans or sorghum (*Sorghum bicolor*)/soybeans
- Double and triple cropping of high-value vegetable crops
- Intensive use of relay and sequential systems in China and Southeast Asia
- Use of multistoried perennial and perennial/annual crop mixtures in Southeast Asia
- Alley cropping in West Africa and other intensive terrace systems, such as those in Burundi

Thus, a number of intensive cropping systems are being practiced by both low-resource and high-technology farmers, both in the developed world and in developing countries.

This importance warrants continued study of multiple cropping potentials. The evidence is that intensive cropping systems are not merely a vestige of the historical roots of crop culture, as eloquently described by Plucknett and Smith in Chap. 2, but would appear to be increasing in importance in much of the world in specific situations where there is an economic, biological, environmental, or social advantage to this type of crop culture.

In each of these production situations, it is important to examine current and potential future cropping systems. Some methods for setting specific research priorities are outlined in Chap. 13, but in the broader picture a look at the total system is essential. The research emphasis placed on either intensive intercropping or relay cropping, as well as that focused on intensive sequential systems, depends on learning (1) why farmers practice these systems today, (2) how likely is the practice or expanded use of these systems in the future, (3) what constraints are faced by the farmer in improving food production and increasing profits in these systems, and (4) how likely can research and development efforts solve these constraints (J. Sanders, Purdue University, personal communication). There is a need to examine these questions from the points of view of several disciplines.

The future potential of multiple cropping to help meet burgeoning world food demand needs to be put into perspective by looking at the biological potentials of these systems and the ecological and environmental consequences of their use, as described in Chaps. 3, 4, and 5. Of equal importance are the economic and social impacts of multiple cropping systems, discussed in Chaps. 11 and 12. Multiple cropping systems represent a response by farmers to production resource scarcity, as well as an attempt to maintain the lowest possible costs of production and most stable, least-risk strategies to produce food and income. To some degree, the expansion of multiple cropping systems or intensification of their use will depend on how research drives the technology or information base in this area, and how national political decisions encourage or discourage the small-farm sector.

The farther one projects into the future, the less confidence can be placed on the analysis. The previous chapters describe in detail the advantages of multiple cropping: more efficient and complete use of resources such as solar energy, nutrients, and water; reduced risk with increased diversity in crops and income sources; and greater biological stability in the cropping systems and the environment. Finite supplies of fossil fuels obligate us to carefully consider alternatives, and these intensive systems appear to provide such biological efficiencies as reliance on nitrogen fixation, reduced pesticide use through genetic resistance and integrated pest management, and potential biological production compensation by multiple species in a system. Thus, there are clear indications that in some areas these intensive systems offer promise as innovative approaches to crop resource use that can be sustained and are appropriate for adequate long-term food production. Multiple cropping systems provide a potential for the majority of farmers who operate in a low-resource situation. Research in multiple cropping systems and the analysis of their applications are both timely and important in the overall strategy to meet world food needs.

BIOLOGICAL POTENTIALS OF MULTIPLE CROPPING

Research has shown that multiple cropping has several advantages over monoculture under a range of circumstances. According to Swindale (1981), intercropping

> Appears to make better use of the natural resources of sunlight, land and water. It may have some beneficial effects on pest and disease problems, . . . and there is an advantage of mixing a legume with a nonlegume to save on the use of nitrogenous fertilizers.

Results from current research on crop rotations (Chap. 8), publications on pest incidence (Chap. 9), and detailed physiology studies (Chap. 4) are beginning to unravel some of the intricacies of these sytsems and explain how they function. Yet many of the biological efficiencies of the systems have evolved over centuries as farmers, through a directed trial-and-error process, brought together new combinations of plant species and cultural practices to meet their needs (Fig. 15.1).

Plucknett and Smith (Chap. 2) have outlined some of the known and probable origins of multiple species systems. It is unlikely that most farmers understood why, in a biological sense, their systems produced more food at a lower risk. Yet they undoubtedly did understand the benefits of diversity and intensive culture of several crops together. This led to the development of a diversity of species combinations and practices for producing them. By combining and integrating the qualitative indigenous knowledge about cropping systems and the quantitative analytic techniques used in scientific research, a clearer picture of the biology of multiple cropping systems is possible. An understanding of the soil/plant/water environment, the potentials of newer crop genotypes, and improved management (Rao, Chap. 6) makes possible the development of systems

Figure 15.1 Intercropped mixture of *Xanthosoma* sp., cassava (*Manihot esculenta*), pineapple (*Anana* sp.), and *Bixa orellana,* near Lázaro Cárdenas, Teapa, Tabasco, Mexico. (Photo by Dr. Stephen Gliessman.)

with increased resource use efficiency, agronomic sustainability, and thus greater food production security and its concomitant rewards (Greenland, 1975). Long-term viability also depends on taking into account the environmental consequences of the intensive resource use that comes from an exploitation of these biological potentials. What are some of these biological potentials?

The intensification of use of land and growth resources in the time dimension is one obvious potential for multiple cropping. Although extended maturity of individual crops can take advantage of certain entire temperature-favorable or moisture-favorable growing seasons, especially in the temperate regions, there are many more options with shorter-cycle component crops in multiple cropping systems. The soybean/winter wheat double cropping or relay cropping in the temperate United States is a successful combination of a summer annual with a winter annual, which makes efficient use of both potential growing seasons while maintaining vegetative cover on the land for most of the year (Langdale et al., 1984; Lewis and Phillips, 1976). Two- and three-crop rice patterns in the Philippines and China, as well as rice/grain legume and rice/rice/grain legume systems are examples of the same type of intensification in the tropics and subtropics where moisture rather than temperature may be most limiting (Gomez and Gomez, 1983). A combination of summer grain crop (maize or soybean) with an overseeded relay legume species which grows in fall and again in spring for nitrogen production is another example of the concentrated use of time in agronomic systems (Hofstetter, 1984). Intensive cropping patterns use the growth resources and land area through the maximum possible period of the year, and thus result in as much dry matter and food as potentially can be produced by the farmer. Intercropping and relay cropping are examples of systems that make the most efficient use of potential total growing season.

The intensification of land and resource use in the space dimension is another important aspect of multiple cropping, especially with intercropping and relay systems. Enhanced efficiency of incident light use is possible with two or more species that occupy the same land area during a significant part of the growing season and have a different pattern of foliage display (see references and discussion in Trenbath, Chap. 4). Different rooting patterns can explore a greater total soil volume with roots of different depths. These combinations of crops can result in tapping of deeper strata and bringing nutrients to the surface for succeeding crop cycles (Harwood, 1984). When there are differences in the timing or intensity of uptake of critical growth elements (both nutrients and water) from the soil, then some combination of species could make greater total use of available growth factors than would be possible with a single species in monoculture (Barker and Francis, Chap. 8). The several chapters on agronomic practices (Chaps. 6 through 8) give examples of these potentials in cereal grains, legumes, and root crops. These biological potentials can be realized when interspecific competition for growth resources is less than intraspecific competition in the same environment, and the result is greater resource use and overyielding by the multiple cropping system (Andrews, 1972).

The intensification of resource use in both time and space is one of the ultimate biological potentials of multiple cropping systems. Different maturities and rooting patterns of grossly different species provide an agronomic mix of plants which is far different from a monoculture in the same place, and the potential yields may be far greater. A relay or intercropping system of maize and sorghum in Central America or of sorghum and pigeon pea (*Cajanus cajan*) in northern Nigeria illustrate the potentials of crops which have different temporal and spatial exploitation of growth factors (Rao, Chap. 6). These systems can make use of an entire rainy season in a manner more efficient than any known monoculture. Their potentials could be compared to a double crop of one species or of two short-cycle species that could be grown in this type of rainfall regime to determine which alternative is most advantageous to the farmer. Combinations of annual and perennial species such as alley cropping maize, sorghum, or grain legumes between strips of *Leucaena* or other woody legume illustrate both spatial and temporal complementarity.

Potentially important for the limited resource farmer is the accumulative use of nutrients that are internally produced or available and that cycle through systems on the farm. Nitrogen fixed by a woody legume species or by an annual leguminous grain or green manure crop can provide a partial or complete alternative to expensive and/or unavailable chemical fertilizer. Accumulated organic matter from intensive systems can provide additional nitrogen. If carefully managed, this can be a more sustainable system, and one which might be called "regenerative" in its improvement of the soil fertility resource for future cropping seasons (Harwood, 1983).

Another set of biological variables in multiple cropping involves the complexity of insect, plant pathogen, and weed interactions discussed by Altieri and Liebman (Chap. 9). Potentials of multiple species systems, whether intensified in time or space or both, appear to contribute to integrated pest management. Competition with weeds from an increased density of crops or from a combination of species occupying two or more niches in the cropping environment can effectively reduce weed germination and growth, and thus increase crop productivity without expensive herbicide inputs. The potential of multispecies systems, as well as the upland/lowland rotations or wet season/dry season alternating culture in some parts of the tropics can contribute to weed control with a cropping pattern that is counter-cyclical to that of the weed species (Francis and Harwood, 1985). The weeds that predominate with one crop in the sequence may be quite different species from those in a succeeding crop, and thus their reproductive cycle may be broken, resulting in a cultural control of weeds.

Insect patterns may also differ in multiple species cropping systems compared to monoculture. Although the same insects may be present, their distribution and reproduction appear to be altered by the mixture of species. This generally results in less damage to crops (Altieri and Liebman, Chap. 9). There are a few cases reported where increased insect damage occurred in multiple cropping, but these reports are minimal. A number of factors may explain why insects are inhibited by a multispecies system. There may be more natural enemies—predators and parasites—in the association of crops, and these would provide a biological control of damaging insect species. The diverse species present in a multiple-crop situation may promote and maintain a wider range of these natural enemies due to the greater and more diverse habitats provided. There may be less opportunity for an insect to land on a crop of choice, when more than one crop is present in the field. "Visual and chemical stimuli from both host and nonhost plants affect both the rate of colonization of herbivores and their behavior" (Altieri and Liebman, Chap. 9). Again, these methods of suppressing insect populations and reducing their damage can contribute to a nonchemical control that makes use of the biological structuring of the system and internal resources on the farm.

Diversity of crop species in an intercropping pattern gives a dispersion of potential host plants for pathogens that is similar to natural ecosystems. The differences in occurrence and damage from plant pathogens are not as striking nor as consistent as the differences with insects (Altieri and Liebman, Chap. 9). Yet there is evidence that mixtures of crops buffer against losses to plant diseases; there may be reduced spore dissemination, delayed infection of plants, or modification of the microenvironment in a way that reduces pathogen development and economic loss. With a few diseases, this modification in microenvironment may promote greater pathogen spread and damage, but there are fewer of these cases listed in the review (Chap. 9). A multiline variety of wheat or other cereal is a parallel breeding solution to disease problems, and this is one approach that could be used to diversify a monoculture and promote this type of "resistance" in the crop (Jensen, 1952).

Another biological reality is the complex set of physiological reactions of plants when grown in mixtures, as explored by Trenbath (Chap. 4) and reviewed by Willey (1979a, 1979b). When greater total density is used in a mixture of species, there is a greater total demand for resources, although this may be spread differently over time or space. This may result in stress on one or more components of the mixture or sequence; the stress may be different from that found in monoculture. One example is the reduced light intensity on a lower story crop, such as bean or soybean grown under maize, while the maize suffers only from an increased competition for moisture or nutrients. Careful examination of an intercrop system and the physical organization of the components could lead to ideas about how to better arrange or sequence two or more crop species. More precise measurement of their yields, total dry matter production, and yield components could confirm these hypotheses and lead to design of more productive multiple cropping systems.

Breeding crop varieties or hybrids for complex systems presents some different challenges than breeding for monoculture (Francis, 1985b; Smith and Francis, Chap. 10). There is a wide range of traits that will be needed in new varieties for either system. These traits include general adaptation to temperature and rainfall patterns in a region, resistance to prevalent insect and disease problems, and seed color and quality characteristics that are acceptable to the producer and the marketplace. Yet other traits may be more important to success in an intercrop: competitive ability, tolerance to stress under lower light levels or less available moisture, ability to climb the maize in the case of indeterminate beans, and efficient use of available nutrients.

There also may be differences in the economic and resource bases of the farmers who predominate in the use of these systems, and the crop quality factors may be more important to the subsistence farmer. Productivity may be important, but profitability also must be considered by most farmers. Some production constraints which have been identified cannot be solved easily through agronomic manipulation of the cropping system. When plant breeding holds promise to solve these constraints, an important focus must be on breeding objectives, and how these vary among systems. Although a number of the same traits may be important for certain species in several systems, there will no doubt be a difference in the relative importance of these traits. Since progress in a breeding program is inversely proportional to the number of traits for which selection is practiced, it is important to concentrate on those traits that are most critical for success of the crop variety in the system of interest. The most critical final evaluation in a breeding program is testing new varieties in the system or systems of interest. These need to approximate as closely as possible the systems into which the varieties will be introduced. A logical final step in the process is testing by farmers under their own farm conditions.

The methodology for deciding on research priorities and for analyzing and interpreting results has been examined by Parkhurst and Francis (Chap. 13) and by Mead (Chap. 14). There are some unique differences and complexities introduced into cropping systems by including more than one species. The number of potential interactions is increased vastly by increasing the number of species in a system, and this makes the choice of research problems even more critical than in monoculture research. There is some debate about the optimum types or combinations of experiments that should be utilized in the field to efficiently explore the questions of component technology in multiple cropping systems. At one end of the spectrum is the complex factorial design with all factors included in at least two levels. This gives the maximum number of comparisons and the most comprehensive evaluation of interactions in the systems. At the other extreme are simple experiments which study only one or two factors at a time, with many of these experiments carried out each season to get an idea of how to manipulate the many potential changes in a cropping system. This gives less information on interactions, although there is greater experimental control over each trial and less potential loss if a few treatments are lost from the trials. Careful reading of the two cited chapters will reveal advantages and disadvantages of each of these approaches, and the best recommendation is probably to work with some reasonable combination of factors in each trial, to explore the most important interactions, and then to move on to another series of experiments in subsequent seasons. There is a growing interest in designs and treatments that provide illustrated results as response surfaces rather than as comparison of discrete treatments (Barker et al., 1985). It is critical to combine the research on the experiment station with testing under farm conditions to ensure that information and recommendations are relevant to the farmer. This is important with multiple cropping systems, since the limited-resource conditions of many farmers who practice multiple cropping systems may dictate a need for realistic low-input recommendations. This focus is quite different than the emphasis at many experiment stations on high-input technology.

These factors in the biological environment—crops, weeds, insects, and pathogens—can be manipulated through agronomic practices and genetic change. The several chapters that deal with agronomic and breeding approaches, plus those on methodology for research, provide a survey of the importance of the biological potentials of multiple cropping systems and how to study them. To complicate the application of this information, the economic and social factors involved in limited-resource agriculture, often practiced by small farmers, must be considered during the research and development process. No biological potential can be successfully exploited if the practices or varieties are not accepted by the farmer. These aspects are explored in a later section.

ECOLOGICAL AND ENVIRONMENTAL ASPECTS OF ALTERNATIVE CROPPING SYSTEMS

There are some obvious parallels between multiple species cropping systems and naturally occurring plant communities. These often include (from Francis, 1985b):

1. Genetic diversity in plant species
2. Resulting diversity in the insect and pathogen populations that are associated with crops
3. Nutrient cycles that are relatively closed, with much of the nutrient requirement of succeeding crops supplied by a previous crop or cover crop residue (in low-input systems)
4. Vegetative cover over the land through much of the year
5. High total use of available light and water through the year because of the presence of growing crops
6. Low risk of complete loss of crops in a given season or year because of the different ecological niches they occupy and the different patterns of demand for growth factors
7. High level of production stability (compared to monoculture) as a result of compensation by other components of the system when one component fails

These characteristics of multiple cropping systems, especially those with two or more species together in the field at the same time, make them desirable for the limited-resource farmer. The comparative biological advantage of multiple cropping systems under a high-input situation is less dramatic, although the double cropping (winter wheat/soybean or rice/rice), ratoon cropping [sugarcane (*Saccharum officinarum*), sorghum, rice], and relay cropping [maize/sesame (*Sesamum indicum*) or maize/soybean] of a number of commercial crops all illustrate the benefits of these systems when resources are not limiting.

One of the widely debated theories in multiple cropping and ecology circles is that greater genetic diversity leads to greater productivity. Hart (Chap. 3) presents evidence on both sides of the question, and gives examples of where this is not true. Systems with high productivity, such as intensive rice or sugar cane production, are the result of high nutrient subsidy and low genetic diversity. Subsidizing a system with external resources—fertilizers, pesticides, irrigation water—can bring high levels of productivity through dominance of the production environment, but these systems are sustainable only at a high external cost. The diversity and cropping intensity of multiple cropping systems, especially an intercrop pattern such as maize/bean, can bring moderate to high levels of productivity through manipulation and exploitation of the resources internal to the farm, and this can be sustained at a lower cost (Francis and Harwood, 1985). In this way, the stability of production or sustainability over time in an intensive multiple species cropping system is similar to a natural ecosystem.

The concept of energy and nutrient cycling is important in cropping systems that are designed for sustainability over a long period. In multiple cropping systems, and especially in low-input systems with reliance primarily on internal resource cycling, the increased production of crop residues and their use by subsequent crops are important factors in promoting stability of production. Systems that are gaining in favor, such as minimum or conservation tillage ("ecofallow" production systems), conserve and rely on residues for reducing runoff losses of nutrients and moisture and for supplying organic matter and nutrients to crops in the next year. The cycling of carbon, nitrogen, phosphorus, potassium, and other elements in an agricultural system is promoted by those cropping patterns that include a range of species, especially when the crops are dissimilar in rooting patterns and growth cycle.

The so-called "phosphorus pumping" activity of some deep-rooted species can bring this critical nutrient up to the annual crop root zone and keep it cycling in the strata where needed. The preservation of a high proportion of nutrients in living and decaying organic matter also ensures that these nutrients will be available for subsequent crops rather than leaching down through the profile and being lost from the immediate crop environment (Fig. 15.2). This is how tropical rain forests and other natural ecosystems maintain fertility and plant growth (Nye and Greenland, 1960).

Hart (Chap. 3), Harwood (1983), and Rodale (1983, 1985) further suggest that multiple cropping systems could apply what is known about natural ecosystems, plant communities with their associated microorganisms, and plant populations to design useful alternative cropping systems that would meet the objectives of the farmer and be dependent on internal resources. It is critical to understand how individual components of technology that are proposed to improve a farmer's cropping system fit into the overall farm system and the ecology of the region. If this concept is taken into account in the evaluation of new technology, there is a greater chance that the eventual choice of new practices will be more ecologically sound and be more sustainable over time. Francis and Kauffman (1985) present a series of resource-efficient technologies that rely on internal resources and can apply both to large and to small farms. These include

Figure 15.2 Multistoried tree crop agroecosystem with coffee, cacao, and *Erythrina* shade trees; note sapling of forest tree *Cedrela mexicana* in the right foreground, near Cárdenas, Tabasco, Mexico. (Photo by Dr. Stephen Gliessman.)

fertility regeneration through multiple cropping and nutrient cycling, pest protection through species diversity in crops and genetic resistance, integrated pest management, minimizing tillage and coexistence with a low level of weed population, mixed cropping of annuals and perennials, and integration of animals into the cropping/farming system. These factors all contribute to development of systems that are more sustainable, in part through a mimic of natural ecosystems in each area.

ECONOMIC AND SOCIAL IMPACT OF MULTIPLE CROPPING

Discussion of economic and social factors must necessarily focus on two distinct applications of the concepts of multiple cropping. First is the intensive cash grain double cropping or relay cropping of commercial crops in the developed world or in favored areas of developing countries. Where these crops are grown with high technology, adequate fertility and other inputs, and with mechanization of planting, cultivation, and harvest, there is little to distinguish the systems from commercial monoculture, at least in the economic and social sense. Cropping decisions are made on costs of production and market factors, and the commercial nature of these systems leads to decision making that is similar in most ways to decisions for monoculture systems.

In the small-farm situations where most intercropping is practiced by farmers with limited resources, the economic and social situation is much different. The complexity of these farming systems is due to the many climatological, biological, economic, and social factors that interact in the total small farm environment (Francis, 1985a). In addition to the complexity of these factors and their interactions, there is a wide range of objectives of the farm family, including production of food and income, minimizing risk and providing stability of production, and sustaining both food and income through as much of the year as possible. This must all be accomplished with a minimal land resource and with limited or no capital on many farms. Multiple cropping is one of the strategies that farmers use to meet these challenges.

Lynam et al. (Chap. 11) suggest that degree of market integration of the farmer is important, and that there are few truly subsistence farms; most farmers in the developing world have some commercial and some subsistence objectives, and thus the consumption and production decisions are interrelated. They also describe the uncertain nature of agriculture, and that with limited resources to invest in the production process and an uncontrolled environment, the farmer becomes more concerned with security and maintaining subsistence needs. Intercropping is seen by the farmer as a strategy to achieve both biological and economic, as well as nutritional, diversity which has a better chance of sustaining the family through a wide range of variable and usually uncontrolled cropping seasons.

Profitability of the cropping system depends on the biological success of the component crops and on relative prices, costs of production, and the ways crops complement each other over time. Net profit thus is site, time, and input-level specific. Multiple crop systems are not always more profitable, but they do appear to be more stable over time and give a higher probability of providing the farmer with a specified level of net income (Francis and Sanders, 1978; Rao and Willey, 1980). Thus biological diversity appears to provide greater economic stability. The economic diversification that results from multiple species plantings can also be achieved by planting a diverse series of crops in monoculture on the same farm. These two strategies for providing greater stability need to be recognized and compared.

Structural changes in the rural economy, such as prices of inputs, availability of a market, and relative prices of commodities can influence the decisions of the farmer on what crops to plant and in what proportions (Lynam et al., Chap. 11). Greater access to markets, wider availability of inputs such as fertilizers, pesticides, and irrigation, plus credit or other government incentives to use these inputs, can influence the farmer toward higher-technology approaches to food production. This shift may result in a higher proportion of export crops and less basic food production, a dependence on external resources that must be purchased and transported to the farm, a reliance on government participation and infrastructure, and timely payment and favorable world market price for commodities (Francis and Harwood, 1985). In spite of the apparent advantages of specialization and the relative comparative advantage for producing one specific well-adapted crop, many farmers are currently short of food and income because they have deemphasized food and subsistence crops.

Also important are the sociocultural factors involved in the farmer's decisions on type of cropping system to employ (Bradfield, Chap. 12). There are strong psychological factors involved in the adoption of new technology. The behavior of farmers is rational, but the extension agent and researcher must understand the total economic, cultural, and nutritional environment within which decisions are made.

There are institutional factors that influence decisions as well. High government prices, and thus incentives to produce cotton for export, may be viewed as a viable solution for income generation and purchase of food by small farmers. On the other hand, farmers may have had a previous experience where markets have disappeared, or prices have gone down drastically, or government agencies have not delivered the needed inputs for production on a timely schedule. The farmer may be rational in rejecting the apparent short-term incentives to produce this new export crop based on past experience that led to shortage of income and food as a result of factors beyond the family's control. Many economists expect that subsistence food production will become less important as agricultural development moves ahead and markets become better developed and more stable.

Other factors in the governmental and political area may influence decisions to adopt or not to adopt a new technology. Bradfield (Chap. 12) presents the example of a recommendation to introduce certain soil conservation, cropping pattern, or other technology practices which would enhance the long-term fertility status of the soil. Perhaps not known to the extensionist, the land ownership patterns, tenancy conditions, lack of credit to buy inputs, lack of assured markets for the products, or lack of other infrastructure make the recommendation impossible or very difficult to adopt. This could easily cancel its obvious biological and physical environmental advantages in the region. Although many new technologies in developing countries were thought to be scale neutral in application, they are still out of reach for the majority of farmers with limited resources. The development of more productive intercropping techniques may in fact be the scale-specific type of technology that will be well suited to the resource base and the multiple and complex needs of the small farmer.

FUTURE RESEARCH IN MULTIPLE CROPPING

Each of the preceding chapters has presented a section on future research in multiple cropping from the perspective of an expert on that topic and discipline. In this chapter an overview of research is presented from the perspective of total cropping and farming systems. The needs for research in broader aspects of multiple cropping that transcend disciplines and require a team approach to research and extension would appear to be a valuable route to understanding complex systems, and to working with farmers to improve them (Gilbert et al., 1980).

In the areas of ecology and environment, there are a number of research directions that could be pursued to develop more productive cropping systems in the context of sustainability of food supply and regeneration of the production environment (Chap. 3 and 5). Continued study of natural ecosystems and how plants interact in those systems can lead to basic information on crop competition, complementation, and interdependence. This information can be applied to cropping systems in an attempt to make them more sustainable and more ecologically sound than currently recommended monoculture or available multiple species systems. The processes of soil fertility maintainance, nutrient cycling, and pest suppression are especially active in natural ecosystems, and information from this research could lead to useful clues about how to design cropping system alternatives. The active and growing field of agroecology is providing improved methodology for this research.

Another fertile area for study is the identification and characterization of successful traditional and sustainable cropping systems still used by small farmers (Gliessman et al., 1981; Chap. 5). This study could lead to an understanding of the vital elements of those systems that promote their productivity and success, and provide an information base to inform other farmers about successful and proven practices. When useful practices and systems are characterized and their elements understood, this information can be combined with the practices that have come from technical research into new or modified cropping practices that have a high probability of success. This combining of traditional knowledge with scientific approaches to gaining insight on their productivity and stability is a new and promising avenue to pursue. Much of the current success of monoculture technologies in the maize belt of the United States and other productive zones has come from the innovation of farmers combined with the potentials of modern technology.

A growing body of information on agroforestry will become a useful resource to the agronomist and research administrator interested in multiple cropping systems. This integration of the long-term sustainability and contributions of forest species to food and income, and the close integration of annual crop plants with perennials holds great promise for improvement of farming systems. In many areas where clear cutting of economic species has occurred, there is a potential to regenerate a part of this forest resource concurrently with development of food crop growing potentials. There are hillside areas that are prone to erosion and nutrient and water loss where some combination of permanent or semipermanent trees with shorter-term economic crops would be highly desirable. This field is receiving increasing interest and support in the research community, and the International Center for Research in Agroforestry (ICRAF) in Kenya is leading efforts in this research.

Agronomists and plant physiologists agree on the need for more research on the components of cropping systems and how the components interact in the use of growth resources. There may be additional data that can be collected from existing experiments that will make them more useful in gaining an understanding of how systems work. With yield trials of intercrops, the study of components of yield can often reveal some insight on the timing of competition for growth resources (Carter et al., 1983). This can lead to new combinations or physical/spatial organizations of crops that will reduce competition at a crucial stage, or help the crop mixture compensate in some way for a reduction in one component crop. These studies of the detailed yield components and biomass of an intercrop can also give the agronomist insight on system design and the breeder direction in setting priorities in a selection and testing program, and help researchers focus on the most important factors in the evaluation of new alternative systems and varieties.

There is a serious need to investigate systems under conditions of low levels of production inputs and stress on the crops. Many of the most food-deficient areas of the world are in the arid and semiarid regions, and there has been less research carried out in these areas than in more favorable zones. Efficient use of low levels of resources may be more easily accomplished using a mixture of species, and this could lead to new lower-risk strategies and cropping systems for the farmer.

Studies of weed interactions with crops are important to an understanding of resource use and competition for scarce moisture in multiple cropping systems. Research now in progress shows that low densities of weeds are not necessarily harmful to yields, and may in fact contribute to organic matter and water and nutrient retention in the soil (Rodale Research Center, unpublished). This has not been studied adequately in multiple cropping systems. Root interactions also are poorly understood in comparison to the wealth of information on competition for light. This is a promising and little-understood area of competition, and an improved appreciation of rooting systems and uptake patterns would lead to design of genotypes and systems that can take best advantage of scarce growth resources.

Improvement of crop genotypes specifically for multiple cropping systems sometimes will be necessary and sometimes not. There is no doubt that the current understanding of intercropping potentials is constrained by the lack of availability of varieties that have been developed for specific intercrop systems. The improved genetic component has come directly from monoculture breeding programs, and thus there is no reason why it should be adapted to the different types of interspecific competition that are unique to intercropping. Although some methodology has been proposed, there have been few programs implementing these procedures in the field. The agronomist will be saddled with this limitation for some time into the future. Current studies on genotype by cropping system interaction do provide some insight on which species are more likely to demonstrate specific adaptation to multiple species cropping systems (Francis, 1985b).

A more generic question revolves around the applicability of other available component technology from studies in monoculture to the complexities of competition and resource use in an intercrop. A systematic study is under way (Francis, unpublished data) to statistically test a series of recommended technologies and their interactions with cropping systems. This is analogous to the genotype by cropping system interaction evaluations. This study will lead to some guidelines on which types of results are most applicable to new cropping systems, such as those with multiple species, and which technologies need to be developed and tested specifically for these intensive systems.

Several of the authors of these chapters cite the importance of on-farm research (OFR) and evaluation of new technologies by researchers working together with farmers. The farming systems approach is mentioned frequently as the methodology which offers the most promise in this activity. There are many forms of farming systems research (FSR), and almost as many interpretations as there are practitioners of the trade. Yet a generalized experience is emerging from this activity and it is one that will be useful for multiple cropping research. When the search for successful traditional technology is combined with evaluation of current constraints to production, and when the farmer is directly involved with the choice of alternatives and the field testing, there is a high probability that the right questions will be answered and the technologies will be appropriate and adopted. The feedback activities that are emphasized in this process are also critical to its long-term success. A functional and rewarding association is thus created between the people involved in research, extension, and on-farm application of results.

There is a continuing need to develop more efficient evaluation tools for on-farm research and testing, and the elaboration of minimum data sets that will give the relevant information without unnecessary, though interesting, details on systems and families who are participants. The emerging methods for assessing research priorities will be useful for multiple cropping systems, as well as for other research areas. The development and articulation of new or revised statistical techniques are powerful contributions to the researcher's set of tools to effectively evaluate multiple cropping systems. In addition to the review by Mead (Chap. 14), there is a book in preparation by Dr. W. Federer at Cornell

University which will be another major contribution to this field (personal communication).

Finally, the use of simulation modeling has been suggested as a useful tool for multiple cropping research. When adequate data sets are available, or when enough is known about crop species interactions and responses to major agronomic practices to make informed assumptions about response curves, then modeling becomes a potentially valuable tool for the researcher. This does not replace the field trial. Models can be used to simulate a wide range of new production alternatives, using different combinations of inputs or techniques which may better exploit resources internal to the farm. Given the complexity and number of factors in multiple cropping systems, simulation could be more important here than in monoculture research. The models can take into account long-term rainfall and temperature data, and can project the relative success of a near-infinite number of combinations of practices, genotypes, and combinations. From these, the most promising can be chosen for testing in the field. As the basic data set grows, so does the power and predictability of the simulation exercise. This could be an efficient way to approach the complexity of site-specific recommendations for small farmers with different levels of limited resources.

Successful application of multiple cropping computer simulation models depends on accurate selection of evaluation criteria. Much more needs to be done than to just measure or predict yield, biomass, or net income. Food production from the system, the distribution of food or income through the year, the labor requirements for each alternative system, and the risk that would be assumed, to name a few criteria, need to be included. These can be established through discussions with farmers during the FSR approach, and can be modified through subsequent runs with the computer. The farmer can be given a list of potential consequences of a given set of practices. For example, if this new variety is planted on this date with these other two species, the effect on yield, income, food supply, risk of no food, and long-term fertility implications could be detailed, for this and alternative practices, as well as compared with the current variety and practices. The modeling approach is not well tested, but is just in the conceptual stage for multiple cropping. Given the power of the microcomputer, there is no reason why this type of analysis could not be run, given proper software and instructions on how to modify it, by people working at any experiment station or regional extension office. This is one tool that may soon be available to the researcher, extensionist, and farmer.

FUTURE PROJECTIONS FOR MULTIPLE CROPPING

In projecting the future importance of multiple cropping systems in different regions of the world, it is necessary to look at past trends and try to anticipate any modifications to those trends in the future. There is no question about the influence of mechanization on rural population and labor required in agriculture. This has promoted greater labor use efficiency and expansion of monoculture.

Farm work is difficult, and advances in mechanization have made food production a more enjoyable and profitable way of life for those farmers with the land resources and capital to develop them. High-technology, commercial exploitation has moved successfully to some areas of the developing world, and the impact on food production through intensive use of technology has been significant. This has also reduced the use of intercropping systems that may have been prevalent in those areas.

There are multiple cropping systems that are well suited to a high-technology approach to agriculture, and their use will certainly expand as food needs increase. Double and triple cropping, ratoon cropping of additional species, such as sorghum and rice in certain conditions, and overseeding of legumes into growing cereal grains and grain legumes will become more prevalent as the technology is developed for a wider range of climatic and cropping conditions. These are "high-technology" applications of multiple species systems, and they are likely to be adopted by progressive farmers in most countries where the alternatives are demonstrated to be biologically successful and adapted to production resources as well as profitable.

An informed prediction of the future of multiple cropping on low-input farms with limited-resource farmers is more difficult. The authors of these chapters are consistent in their prediction that multiple cropping systems will continue to be important as new technology becomes available and the advantages of intensive systems become better understood. Natural ecosystems have been successful as a result of centuries of evolution to fit specific conditions. Cropping systems have followed this same path, though the evolution was directed to a different set of objectives by the cultivator and family. Multiple cropping systems combine a number of attributes of natural systems, while taking advantage of the resources available to the farm family, to produce food and income according to some of the criteria listed above. This has not happened by accident, but rather has been the result of a concerted attempt by farmers to continuously improve their cropping systems to fit the climatic and resource constraints. A number of external factors have entered the farmer's environment (e.g., government programs to promote export crops and discourage production of basic food crops and commodities for local sale and consumption). This and other political decisions that impinge on the farmer must be examined as a part of any strategy for the rural sector in a developing country.

Multiple cropping systems continue to predominate on many farms with limited resources in the world. The research community is becoming increasingly interested in the potentials offered by this type of cropping pattern, as every possible alternative is examined as a part of the solution to the challenge of world food production and income for rural families. Those in the research and extension organizations of both developed and developing countries need to carefully examine the potentials of these traditional systems to contribute food, provide income, and minimize risk of failure for farmers with limited resources. These farmers and the systems they have preserved may be improved by modern science to provide a significant part of the future world food supply.

REFERENCES

Andrews, D. J. 1972. Intercropping with sorghum in Nigeria, *Exp. Agric.* 8:139–150.

Barker, T. C., Francis, C. A., and Crause, G. F. 1985. Resource-efficient experimental designs for on-farm research, in: *Proc. Farming Systems Symp.,* Kansas State University, Manhattan, Kansas, 16 pp.

Carter, D. C., Francis, C. A., Pavlish, L. A., Heinrich, G. M., and Matthews, R. V. 1983. Sorghum and soybean density interactions in one intercrop pattern, *Agron. Abstr.*, 1983, p. 43.

Francis, C. A. 1985a. Rationality of new technology for small farmers in the tropics, *Agric. Human Values.* Vol. 2 (2):55–60, Univ. Florida, Gainesville.

———. 1985b. Variety development for multiple cropping systems, *Crit. Rev. Plant Sci.* CRC Press, Boca Raton, Florida 3:133–168.

Francis, C. A., and Harwood, R. R. 1985. *Enough food,* Rodale Press and Regenerative Agriculture Association, Emmaus, Pennsylvania, 24 pp.

Francis, C. A., and Kauffman, C. S. 1985. Regenerative agriculture: An environmentally sound approach to sustained food production, Global Meeting on Environment and Development, Environment Liaison Center, 4–8 February 1985, Nairobi, Kenya, Document E3/E (reprinted as *Global Concept Paper* 10, Rodale Research Center, Kutztown, Pennsylvania.

Francis, C. A., and Sanders, J. H. 1978. Economic analysis of bean and maize systems: Monoculture versus associated cropping, *Field Crops Res.* 1:319–335.

Gilbert, E. H., Norman, D. W., and Winch, F. E. 1980. Farming systems research: a critical appraisal, *MSU Rural Devel. Paper 6,* Michigan State Univ., E. Lansing, Michigan.

Gliessman, S. R., Garcia, E., and Amador, M. A. 1981. The ecological basis for application of traditional agricultural technology in the management of tropical agro-ecosystems, *Agro-ecosystems* 7:173–185.

Gomez, A. A., and Gomez, K. A., 1983. Multiple cropping in the humid tropics of Asia, IDRC, Ottawa, Ontario, Canada, IDRC–176e.

Greenland, D. J. 1975. Bringing the revolution to the shifting cultivators, *Science* 190:841–844.

Harwood, R. R. 1983. International overview of regenerative agriculture, in: *Proc. Resource-Efficient Farming Methods for Tanzania,* Rodale Press, Emmaus, Pennsylvania, pp. 24–35.

———. 1984. Organic farming at the Rodale Research Center, in: *Organic Farming: Current Technology and Its Role in a Sustainable Agriculture,* Amer. Soc. Agron. Spec. Publ. 46, pp. 1–17.

Hofstetter, R. 1984. Overseeding research results, 1982–1984, Agronomy Dep., Rodale Research Center, Kutztown, Pennsylvania, RRC/AG-84/29, 30 pp.

Jensen, N. F. 1952. Intra-varietal diversification in oat breeding, *Agron. J.* 44:30–34.

Langdale, G. W., Hargrove, W. L., and Gibbens, J. 1984. Residue management in double crop conservation tillage systems, *Agron. J.* 76:689–694.

Lewis, W. M., and Phillips, J. A. 1976. Double cropping in the Eastern United States, in: *Multiple Cropping* (R. I. Papendick, P. A. Sanchez, and G. B. Triplett, eds.), Amer. Soc. Agron. Spec. Publ. 27, pp. 41–50.

Malacela, J. S. 1982. Opening address, in: *Intercropping: Proc. Second Symp. Intercropping in Semi-Arid Areas,* 4–7 August 1980, Morogoro, Tanzania. pp. 12–14.

Nye, P. H., and Greenland, D. J. 1960. *The Soil Under Shifting Cultivation,* Commonwealth Bureau of Soils, Harpenden, Bucks, U.K. Tech. Commun. No. 51, 156 pp.

Rao, M. R., and Willey, R. W. 1980. Evaluation of yield stability in intercropping studies on sorghum and pigeon pea, *Exp. Agric.* 16:105–116.

Rodale, R. 1983. Breaking new ground: The search for a sustainable agriculture. *The Futurist,* February, 1983, 17 (1):15–20. World Future Society, Bethesda, Maryland.

———. 1985. Pioneer enterprises in regeneration zones: The role of small business in regional recovery, *Whole Earth Review,* July 1985, pp. 34–38.

Swindale, L. D. 1981. Foreword, in: *Proc. Int. Workshop on Intercropping,* 10–13 January 1979, Hyderabad, India, pp. vii–viii.

Willey, R. W. 1979a. Intercropping—its importance and research needs. Part 1. Competition and yield advantages, *Field Crop Abstr.* 32:1–10.

———. 1979b. Intercropping—its importance and research needs. Part 2. Agronomy and research approaches, *Field Crop Abstr.* 32:73–85.

Author Index

Subject Index